HSPA+
EVOLUTION TO RELEASE 12

HSPA+
EVOLUTION TO RELEASE 12
PERFORMANCE AND OPTIMIZATION

Edited by

Harri Holma
Nokia Siemens Networks, Finland

Antti Toskala
Nokia Siemens Networks, Finland

Pablo Tapia
T-Mobile, USA

This edition first published 2014
© 2014 John Wiley & Sons, Ltd

Registered office
John Wiley & Sons Ltd, The Atrium, Southern Gate, Chichester, West Sussex, PO19 8SQ, United Kingdom

For details of our global editorial offices, for customer services and for information about how to apply for
permission to reuse the copyright material in this book please see our website at www.wiley.com.

Library of Congress Cataloging-in-Publication Data

HSPA+ Evolution to release 12 : performance and optimization / edited by Harri Holma, Antti Toskala, Pablo Tapia.
    pages cm
  Includes index.
  ISBN 978-1-118-50321-8 (hardback)
 1. Packet switching (Data transmission)   2. Network performance (Telecommunication)   3. Radio–Packet
transmission.   I. Holma, Harri, 1970–   II. Toskala, Antti.   III. Tapia, Pablo.   IV. Title: HSPA plus Evolution to
release 12.   V. Title: High speed packet access plus Evolution to release 12.
  TK5105.3.H728 2014
  621.382′16–dc23

                                                                                        2014011399

A catalogue record for this book is available from the British Library.

ISBN: 9781118503218

Set in 10/12pt Times by Aptara Inc., New Delhi, India

1   2014

To Kiira and Eevi
—Harri Holma

To Lotta-Maria, Maija-Kerttu and Olli-Ville
—Antti Toskala

To Lucia
—Pablo Tapia

Contents

6   Voice Evolution   **103**
Harri Holma and Karri Ranta-aho

7   Heterogeneous Networks   **117**
Harri Holma and Fernando Sanchez Moya

13 Smartphone Performance   293
Pablo Tapia, Michael Thelander, Timo Halonen, Jeff Smith, and Mika Aalto

14 Multimode Multiband Terminal Design Challenges   337
Jean-Marc Lemenager, Luigi Di Capua, Victor Wilkerson, Mikaël Guenais,
Thierry Meslet, and Laurent Noël

Foreword

Our industry has undergone massive change in the last few years and it feels like only yesterday that T-Mobile USA, and the industry as a whole, was offering voice centric services and struggling to decide what to do about "data." Once the smartphone revolution started and consumption of new services literally exploded, wireless operators had to quickly overcome many new challenges. This new era of growth in the wireless industry continues to create great opportunity coupled with many new challenges for the wireless operators.

In my role of Chief Technical Officer at T-Mobile USA, I have enjoyed leading our company through a profound technology transformation over a very short period of time. We have rapidly evolved our network from a focus on voice and text services, to providing support to a customer base with over 70% smartphones and a volume of carried data that is growing over 100% year on year. At the time of writing, we offer one of the best wireless data experiences in the USA which is one of the fastest growing wireless data markets in the world. We provide these services through a combination of both HSPA+, and more recently, LTE technologies.

Many wireless operators today have to quickly address how they evolve their network. And with consumer demand for data services skyrocketing, the decision and choice of technology path are critical. Some operators may be tempted to cease investments in their legacy HSPA networks and jump straight to LTE. In our case, we bet heavily on the HSPA+ technology first, and this has proved instrumental to our success and subsequent rollout of LTE. As you will discover throughout this book, there are many common elements and similarities between HSPA+ and LTE, and the investment of both time and money into HSPA will more than pay off when upgrading to LTE. It certainly did for us.

Furthermore, industry forecasts indicate that by the end of the decade HSPA+ will replace GSM as the main reference global wireless technology and HSPA+ will be used extensively to support global voice and data roaming. This broad-based growth of HSPA+ will see continued development of existing economies of scale for both infrastructure and devices.

This book provides a great combination of theoretical principles, device design, and practical aspects derived from field experience, which will help not only tune and grow your HSPA+ network, but also LTE when the time comes. The book has been written by 26 experts from multiple companies around the world, including network infrastructure and chip set vendors, mobile operators, and consultancy companies.

As worldwide renowned experts, Harri and Antti bring a wealth of knowledge to explain all the details of the technology. They have helped create and develop the UMTS and HSPA technologies through their work at NSN, and more recently pushed the boundaries of HSPA+ from Release 8 onward.

Pablo has been a key player to our HSPA+ success at T-Mobile, working at many levels to drive this technology forward: from business case analysis and standardization, to leading technology trials through design and optimization activities. His practical experience provides a compelling perspective on this technology and is a great complement to the theoretical aspects explored in this book.

I hope you will find this book as enjoyable as I have, and trust that it will help advance your understanding of the great potential of this key and developing technology.

Neville Ray
CTO
T-Mobile USA

Preface

HSPA turned out to be a revolutionary technology, through making high-speed wide-area data connections possible. HSPA is by far the most global mobile broadband technology and is deployed by over 500 operators. Data volumes today are substantially higher than voice volumes and mobile networks have turned from being voice dominated to data dominated. The fast increase in data traffic and customer expectation for higher data rates require further evolution of HSPA technology. This book explains HSPA evolution, also called HSPA+. The book is structured as follows. Chapter 1 presents an introduction. Chapter 2 describes the basic HSDPA and HSUPA solution. Chapter 3 presents multicarrier and multiantenna evolution for higher efficiency and higher data rates. Chapter 4 explains continuous packet connectivity and high speed common channels. Multiflow functionality is described in Chapter 5, voice evolution in Chapter 6, and heterogeneous networks in Chapter 7. Advanced receiver algorithms are discussed in Chapter 8. ITU performance requirements for IMT-Advanced and HSPA+ simulation results are compared in Chapter 9. Chapters 10 to 13 present the practical network deployment and optimization: HSPA+ field measurements in Chapter 10, network planning in Chapter 11, network optimization in Chapter 12, and smartphone optimization in Chapter 13. Terminal design aspects are presented in Chapter 14. The inter-working between LTE and HSPA is discussed in Chapter 15 and finally the outlook for further HSPA evolution in Chapter 16. The content of the book is summarized here in Figure P.1.

Acknowledgments

The editors would like to acknowledge the hard work of the contributors from Nokia, from T-Mobile USA, from Videotron Canada, from Teliasonera, from Renesas Mobile and from Signals Research Group: Mika Aalto, Luigi Dicapua, Ryszard Dokuczal, Mikael Guenais, Timo Halonen, Matthias Hesse, Thomas Höhne, Maciej Januszewski, Jean-Marc Lemenager, Thierry Meslet, Laurent Noel, Brian Olsen, Hisashi Onozawa, Hannu Raassina, Karri Ranta-aho, Jussi Reunanen, Fernando Sanchez Moya, Alexander Sayenko, Jeff Smith, Mike Thelander, Jeroen Wigard, Victor Wilkerson, and Carl Williams.

We also would like to thank the following colleagues for their valuable comments: Erkka Ala-Tauriala, Amar Algungdi, Amaanat Ali, Vincent Belaiche, Grant Castle, Costel Dragomir, Karol Drazynski, Magdalena Duniewicz, Mika Forssell, Amitava Ghosh, Jukka Hongisto, Jie Hui, Shane Jordan, Mika Laasonen, M. Franck Laigle, Brandon Le, Henrik Liljeström, Mark McDiarmid, Peter Merz, Randy Meyerson, Harinder Nehra, Jouni Parviainen, Krystian Pawlak, Marco Principato, Declan Quinn, Claudio Rosa, Marcin Rybakowski, David

A. Sánchez-Hernández, Shubhankar Saha, Yannick Sauvat, Mikko Simanainen, Dario Tonesi, Mika Vuori, Dan Wellington, Changbo Wen, and Taylor Wolfe.

The editors appreciate the fast and smooth editing process provided by Wiley and especially by Sandra Grayson, Liz Wingett, and Mark Hammond.

We are grateful to our families, as well as the families of all the authors, for their patience during the late night writing and weekend editing sessions.

The editors and authors welcome any comments and suggestions for improvements or changes that could be implemented in forthcoming editions of this book. The feedback is welcome to the editors' email addresses harri.holma@nokia.com, antti.toskala@nokia.com, and pablo.tapia.m@gmail.com.

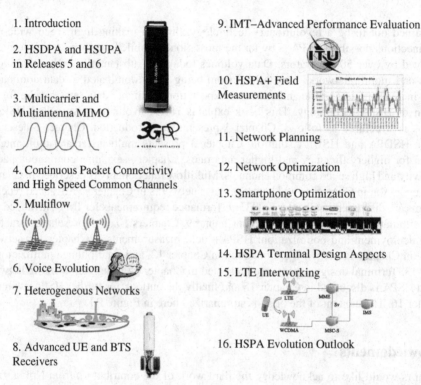

1. Introduction

2. HSDPA and HSUPA in Releases 5 and 6

3. Multicarrier and Multiantenna MIMO

4. Continuous Packet Connectivity and High Speed Common Channels

5. Multiflow

6. Voice Evolution

7. Heterogeneous Networks

8. Advanced UE and BTS Receivers

9. IMT–Advanced Performance Evaluation

10. HSPA+ Field Measurements

11. Network Planning

12. Network Optimization

13. Smartphone Optimization

14. HSPA Terminal Design Aspects

15. LTE Interworking

16. HSPA Evolution Outlook

Figure P.1   Contents of the book.

Abbreviations

3GPP	Third Generation Partnership Project
ACK	Acknowledgment
ACL	Antenna Center Line
ACLR	Adjacent Channel Leakage Ratio
ACP	Automatic Cell Planning
ADC	Analog Digital Conversion
AICH	Acquisition Indicator Channel
AL	Absorption Losses
ALCAP	Access Link Control Application Part
AM	Acknowledged Mode
AMR	Adaptive Multirate
ANDSF	Access Network Discovery and Selection Function
ANQP	Access Network Query Protocol
ANR	Automatic Neighbor Relations
APE	Application Engine
APNS	Apple Push Notification Service
APT	Average Power Tracking
aSRVCC	Alerting SRVCC
AWS	Advanced Wireless Services
BBIC	Baseband Integrated Circuit
BCH	Broadcast Channel
BH	Busy Hour
BiCMOS	Bipolar CMOS
BLER	Block Error Rate
BOM	Bill of Material
BPF	Band Pass Filter
BSC	Base Station Controller
BT	Bluetooth
BTS	Base Station
CA	Carrier Aggregation
CAPEX	Capital Expenses
CCCH	Common Control Channel
CDMA	Code Division Multiple Access
CIO	Cell Individual Offset

CL	Closed Loop
CL-BFTD	Closed Loop Beamforming Transmit Diversity
CM	Configuration Management
CMOS	Complementary Metal Oxide Semiconductor
CoMP	Cooperative Multipoint
CPC	Continuous Packet Connectivity
C-PICH	Common Pilot Channel
CPU	Central Processing Unit
CQI	Channel Quality Information
CRC	Cyclic Redundancy Check
CS	Circuit-Switched
CSFB	CS Fallback
CSG	Closed Subscriber Group
CTIA	Cellular Telecommunications and Internet Association
DAC	Digital Analog Conversion
DAS	Distributed Antenna System
DASH	Dynamic Adaptive Streaming over HTTP
DC	Direct Current
DC	Dual-Carrier
DCA	Direct Conversion Architecture
DCH	Dedicated Channel
DC-HSDPA	Dual-Cell HSDPA
DC-HSPA	Dual-Cell HSPA
DDR	Double Data Rate
DF	Dual Frequency
DFCA	Dynamic Frequency and Channel Allocation
DL	Downlink
DM	Device Management
DMIPS	Dhrystone Mega Instructions Per Second
DPCCH	Dedicated Physical Control Channel
DRAM	Dynamic Random Access Memory
DRX	Discontinuous Reception
DS	Deep Sleep
DSL	Digital Subscriber Line
DSP	Digital Signal Processing
DTX	Discontinuous Transmission
E-AGCH	Enhanced Absolute Grant Channel
EAI	Extended Acquisition Indicator
Ec/No	Energy per Chip over Interference and Noise
E-DCH	Enhanced Dedicated Channel
EDGE	Enhanced Data rates for GSM Evolution
E-DPCH	Enhanced Dedicated Physical Channel
eF-DPCH	Enhanced Fractional Dedicated Physical Channel
EGPRS	Enhanced GPRS
E-HICH	E-DCH Hybrid ARQ Indicator Channel
eICIC	Enhanced Inter-Cell Interference Cancellation

EIRP	Equivalent Isotropical Radiated Power
EMI	Electro-Magnetic Interference
EPA	Extended Pedestrian A
EPC	Evolved Packet Core
EPS	Evolved Packet System
E-RGCH	Enhanced Relative Grant Channel
eSRVCC	Enhanced SRVCC
ET	Envelope Tracking
EU	European Union
E-UTRA	Enhanced Universal Terrestrial Radio Access
EVM	Error Vector Magnitude
FACH	Forward Access Channel
FBI	Feedback Information
FDD	Frequency Division Duplex
F-DPCH	Fractional Dedicated Physical Channel
FE	Front End
FE-FACH	Further Enhanced Forward Access Channel
feICIC	Further Enhanced Inter-Cell Interference Coordination
FEM	Front End Module
FGA	Fast Gain Acquisition
FIR	Finite Impulse Response
FM	Frequency Modulation
FOM	Figure of Merit
FS	Free Space
FTP	File Transfer Protocol
GAS	Generic Advertisement Service
GCM	Google Cloud Messaging
GGSN	Gateway GPRS Support Node
GoS	Grade of Service
GPEH	General Performance Event Handling
GPRS	General Packet Radio Service
GPS	Global Positioning System
GPU	Graphical Processing Unit
GS	Gain Switching
GSM	Global System for Mobile Communications
GSMA	GSM Association
GTP	GPRS Tunneling Protocol
HARQ	Hybrid Automatic Repeat-reQuest
HD	High Definition
HDMI	High Definition Multimedia Interface
HDR	High Dynamic Range
HEPA	High Efficiency PA
HLS	HTTP Live Streaming
HO	Handover
HPF	High Pass Filter
H-RNTI	HS-DSCH Radio Network Temporary Identifier

HSDPA	High Speed Downlink Packet Access
HS-DPCCH	High Speed Downlink Physical Control Channel
HS-DSCH	High Speed Downlink Shared Channel
HS-FACH	High Speed Forward Access Channel
HSPA	High Speed Packet Access
HS-RACH	High Speed Random Access Channel
HS-SCCH	High Speed Shared Control Channel
HSUPA	High Speed Uplink Packet Access
HTTP	Hypertext Transfer Protocol
IC	Integrated Circuit
IEEE	Institute of Electrical and Electronics Engineers
IIP	Input Intercept Point
IIR	Infinite Impulse Response
ILPC	Inner Loop Power Control
IMEISV	International Mobile Station Equipment Identity and Software Version
IMSI	International Mobile Subscriber Identity
IMT	International Mobile Telephony
IO	Input–Output
IP	Intellectual Property
IP	Internet Protocol
IQ	In-phase/quadrature
IRAT	Inter Radio Access Technology
ISD	Inter-Site Distance
ISMP	Inter-System Mobility Policy
ISRP	Inter-System Routing Policy
ITU-R	International Telegraphic Union Radiocommunications sector
JIT	Just In Time
KPI	Key Performance Indicator
LA	Location Area
LAC	Location Area Code
LAU	Location Area Update
LCD	Liquid Crystal Display
LDO	Low Drop Out
LNA	Low Noise Amplifier
LO	Local Oscillator
LS	Light Sleep
LTE	Long Term Evolution
MAC	Medium Access Control
MAPL	Maximum Allowable Pathloss
MBR	Maximum Bitrate
MCL	Minimum Coupling Loss
MCS	Modulation and Coding Scheme
MDT	Minimization of Drive Tests
MGW	Media Gateway
MIMO	Multiple Input Multiple Output
MIPI	Mobile Industry Processor Interface

ML	Mismatch Loss
MLD	Maximum Likelihood Detection
MME	Mobility Management Entity
MMMB	Multimode Multiband
MMSE	Minimum Mean Square Error
MO	Mobile Originated
MOS	Mean Opinion Score
MP	Multiprocessing
MPNS	Microsoft Push Notification Service
MSC	Mobile Switching Center
MSC-S	MSC-Server
MSS	Microsoft's Smooth Streaming
MSS	MSC-Server
MT	Mobile Terminated
NA	North America
NACK	Negative Acknowledgement
NAIC	Network Assisted Interference Cancellation
NAS	Non-Access Stratum
NB	Narrowband
NBAP	NodeB Application Part
NGMN	Next Generation Mobile Network
OEM	Original Equipment Manufacturer
OL	Open Loop
OLPC	Open Loop Power Control
OMA	Open Mobile Alliance
OPEX	Operational Expenses
OS	Operating System
OSC	Orthogonal Subchannelization
OTA	Over The Air
PA	Power Amplifier
PAE	Power Added Efficiency
PAM	Power Amplifier Module
PAPR	Peak to Average Power Ratio
PCB	Printed Circuit Board
PCH	Paging Channel
PDCP	Packet Data Convergence Protocol
PDP	Packet Data Protocol
PDU	Payload Data Unit
PIC	Parallel Interference Cancellation (PIC)
PICH	Paging Indicator Channel
PLL	Phase Locked Loop
PLMN	Public Land Mobile Network
PM	Performance Management
PRACH	Physical Random Access Channel
PS	Packet Switched
P-SCH	Primary Synchronization Channel

PSD	Power Spectral Density
QAM	Quadrature Amplitude Modulation
QBE	Quadband EGPRS
QoE	Quality of Experience
QoS	Quality of Service
QPSK	Quadrature Phase Shift Keying
QXDM	Qualcomm eXtensible Diagnostic Monitor
R99	Release 99
RAB	Radio Access Bearer
RAC	Routing Area Code
RACH	Random Access Channel
RAN	Radio Access Network
RANAP	Radio Access Network Application Part
RAT	Radio Access Technology
RB	Radio Bearer
RET	Remote Electrical Tilt
RF	Radio Frequency
RFFE	RF Front End
RFIC	RF Integrated Circuit
RLC	Radio Link Control
RNC	Radio Network Controller
ROHC	Robust Header Compression
RoT	Rice over Thermal
RRC	Radio Resource Control
RRM	Radio Resource Management
RRSS	Receiver Radio Signal Strength
RSCP	Received Signal Code Power
RSRP	Reference Signal Received Power
rSRVCC	Reverse SRVCC
RSSI	Received Signal Strength Indicator
RTP	Real Time Protocol
RTT	Round Trip Time
RUM	Real User Measurements
RX	Receive
SAW	Surface Acoustic Wave
S-CCPCH	Secondary Common Control Physical Channel
SCRI	Signaling Connection Release Indication
SD	Secure Digital
SD	Sphere Decoding
S-DPCCH	Secondary Dedicated Physical Control Channel
S-E-DPCCH	Secondary Dedicated Physical Control Channel for E-DCH
S-E-DPDCH	Secondary Dedicated Physical Data Channel for E-DCH
SF	Single Frequency
SF	Spreading Factor
SFN	System Frame Number
SGSN	Serving GPRS Support Node

SHO	Soft Handover
SI	Status Indication
SIB	System Information Block
SIC	Successive Interference Canceller
SIM	Subscriber Identity Module
SINR	Signal to Interference and Noise Ratio
SIR	Signal to Interference Ratio
SMSB	Single Mode Single Band
SNR	Signal to Noise Ratio
SoC	System on Chip
SON	Self-Organizing Network
SRVCC	Single Radio Voice Call Continuity
S-SCH	Secondary Synchronization Channel
SSID	Service Set Identifier
STTD	Space Time Transmit Diversity
SW	Software
TA	Tracking Area
TAC	Tracking Area Code
TAU	Tracking Area Update
TDD	Time Division Duplex
TD-SCDMA	Time Division Synchronous Code Division Multiple Access
TFCI	Transport Format Control Indicator
TIS	Total Isotropic Sensitivity
TM	Transparent Mode
TRP	Total Radiated Power
TRX	Transceiver
TTI	Transmission Time Interval
TVM	Traffic Volume Measurement
TX	Transmit
UARFCN	UTRAN Absolute Radio Frequency Channel Number
UDP	User Datagram Protocol
UE	User Equipment
UHF	Ultrahigh Frequency
UI	User Interaction
UL	Uplink
UM	Unacknowledged Mode
UMI	UTRAN Mobility Information
UMTS	Universal Mobile Telecommunications System
URL	Uniform Resource Locator
USB	Universal Serial Bus
U-SIM	UMTS SIM
UTRAN	Universal Terrestrial Radio Access Network
VAM	Virtual Antenna Mapping
VCO	Voltage Controlled Oscillator
VIO	Input Offset Voltage
VoIP	Voice over IP

VoLTE	Voice over LTE
vSRVCC	Video SRVCC
VST	Video Start Time
VSWR	Voltage Standing Wave Ratio
WB	Wideband
WCDMA	Wideband Code Division Multiple Access
Wi-Fi	Wireless Fidelity
WiMAX	Worldwide Interoperability for Microwave Access
WLAN	Wireless Local Area Network
WPA	Wi-Fi Protected Access
WW	World Wide
XPOL	Cross-Polarized
ZIF	Zero Insertion Force

1

Introduction

Harri Holma

1.1 Introduction

GSM allowed voice to go wireless with more than 4.5 billion subscribers globally. HSPA allowed data to go wireless with 1.5 billion subscribers globally. The number of mobile broadband subscribers is shown in Figure 1.1. At the same time the amount of data consumed by each subscriber has increased rapidly leading to a fast increase in the mobile data traffic: the traffic growth has been 100% per year in many markets. More than 90% of bits in mobile networks are caused by data connections and less than 10% by voice calls. The annual growth of mobile data traffic is expected to be 50–100% in many markets over the next few years. Mobile networks have turned from voice networks into data networks. Mobile operators need to enhance network capabilities to carry more data traffic with better performance. Smartphone users expect higher data rates, more extensive coverage, better voice quality, and longer battery life.

Most of the current mobile data traffic is carried by HSPA networks. HSPA+ is expected to be the dominant mobile broadband technology for many years to come due to attractive data rates and high system efficiency combined with low cost devices and simple upgrade on top of WCDMA and HSPA networks. This book presents HSPA evolution solutions to enhance the network performance and capacity. 3GPP specifications, optimizations, field performance, and terminals aspects are considered.

1.2 HSPA Global Deployments

More than 550 operators have deployed the HSPA network in more than 200 countries by 2014. All WCDMA networks have been upgraded to support HSPA and many networks also support HSPA+ with 21 Mbps and 42 Mbps. HSPA technology has become the main mobile broadband solution globally. Long Term Evolution (LTE) will be the mainstream solution in

HSPA+ Evolution to Release 12: Performance and Optimization, First Edition.
Edited by Harri Holma, Antti Toskala, and Pablo Tapia.
© 2014 John Wiley & Sons, Ltd. Published 2014 by John Wiley & Sons, Ltd.

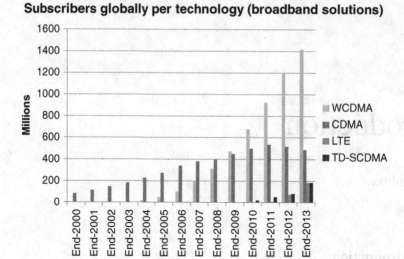

Figure 1.1 Number of subscribers with mobile broadband technologies

the long term for GSM and HSPA operators and also for CDMA and WiMAX operators. HSPA traffic will continue to grow for many years and HSPA networks will remain in parallel to LTE for a long time. The technology evolution is shown in Figure 1.2.

HSPA has been deployed on five bands globally, see Figure 1.3. The most widely used frequency band is 2100 MHz 3GPP Band 1 which is used in many countries in Europe, Asia, the Middle East, and Africa. The other high-band options are Band 4 at 2100/1700 MHz and Band 2 at 1900 MHz. In many cases operators have deployed HSPA on two bands: high band for capacity and low band for coverage. The two low-band options are 900 MHz and 850 MHz. The low-band rollouts have turned out to be successful in improving the network coverage, which gives better customer performance and better network quality. Low bands have traditionally been used by GSM. Using the same bands for HSPA is called refarming. GSM and HSPA can co-exist smoothly on the same band.

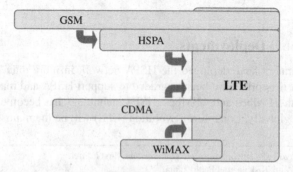

Figure 1.2 Radio technology evolution

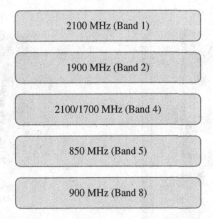

Figure 1.3   HSPA main frequency bands

1.3   Mobile Devices

The very first 3G devices ten years ago suffered from high power consumption, high cost, poor quality, and lack of applications. Battery power consumption during a voice call has dropped from more than 400 mA to below 100 mA in the latest devices, 3G smartphone prices have dropped below 50 EUR, and a huge number of applications can be downloaded to the devices. The attractive price points for 3G smartphones are enabled by the low-cost chip sets available from multiple vendors. The major improvement in power consumption shows the importance of RF and baseband technology evolution. The large HSPA market size brings high volume production and enables low cost devices. The attractive price points are needed to make the devices available to billions of subscribers. HSPA devices are available for many operating systems including Android, iOS, and Windows Phones. Those smartphone platforms offer hundreds of thousands of applications for the users. Not only has the data capability been improved but also the voice quality has been enhanced with the introduction of High Definition (HD) voice based on Adaptive Multirate Wideband (AMR-WB) codec.

Since the number of global HSPA frequency variants is just five, it is possible to make a global HSPA device that can be used in any HSPA network worldwide. That makes the logistics simpler. The LTE frequency options are more complex and, therefore, multiple variants of the same devices are required for the global market.

An example low end 3G device is shown in Figure 1.4: HSPA Release 7 capability with 7.2 Mbps data rate, dual band support, 2.4″ screen, 90 g weight, and up to 500 hours of standby time. Such devices can act as the first access to the Internet for millions of people.

1.4   Traffic Growth

The new smartphones and their applications are increasing traffic volumes rapidly in mobile networks. The combined data volume growth of a few major operators over a two-year period is shown in Figure 1.5: the growth rate has been 100% per year. The fast data growth sets challenges for network capacity upgrades. Higher spectral efficiency, more spectrum, and

Figure 1.4 Low end HSPA phone – Nokia 208. Source: Nokia. Reproduced by permission of Nokia.

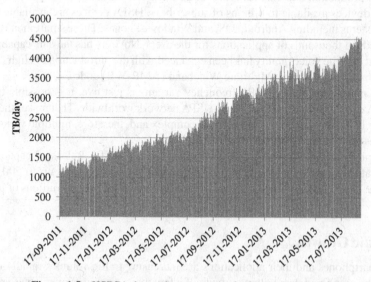

Figure 1.5 HSDPA data volume growth of a few major operators

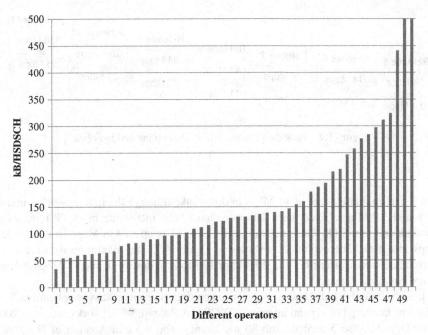

Figure 1.6 Average data volume for HS-DSCH channel allocation

more base stations will be needed. If the traffic were to grow by 100% per year for a 10-year period, the total traffic growth would be 1000 times higher.

Not only the data volume but also the signaling volumes are increasing in the network. Smartphone applications typically create relatively small packet sizes that are transmitted frequently, causing continuous channel allocations and releases. Figure 1.6 shows the average data volume for each High Speed Downlink Shared Channel (HS-DSCH) allocation. Each bar represents a different mobile operator. Those operators with a low data volume per allocation (below 100 kB) have high penetration of smartphones while those operators with a large data volume per allocation (above 200 kB) have relatively higher penetration of USB modems and laptop connectivity. If the average data volume per allocation is low, then more allocations and more related signaling will be required. If we assume 50–100 kB per allocation and the data volume 1 GB/sub/month, this corresponds to 330–660 allocations per subscriber per day. Assuming that the busiest hour takes approximately 6% of the daily allocations, we have then on average 20–40 allocations per busy hour. That means the network allocates resources on average every 2 minutes for every smartphone subscriber during the busy hour. Such frequent channel allocations require high signaling capacity. The high signaling also causes more uplink interference because of the control channel overhead. Therefore, solutions for uplink interference control solutions are needed in the smartphone dominated networks.

1.5 HSPA Technology Evolution

The first version of HSDPA was defined in 3GPP Release 5 during 2002, the backwards compatibility started in 2004, and the first HSDPA network was launched in 2005. Release 5

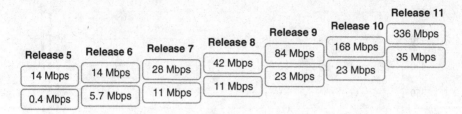

Figure 1.7   Peak data rate evolution in downlink and in uplink

allowed a maximum data rate of 14 Mbps in downlink, although the first networks supported only 1.8 and 3.6 Mbps. Since then, data rates have been increasing in 3GPP releases both in downlink and in uplink. The peak rate in Release 11 is up to 336 Mbps in downlink and 35 Mbps in uplink. The evolution is shown in Figure 1.7. The maximum data rates in the commercial networks currently (mid 2014) are 42 Mbps in downlink and 11 Mbps in uplink, which are based on Release 8.

The latency evolution helps end user performance since many interactive applications benefit from the low latency. The typical latency in WCDMA Release 99 networks was 150–200 ms while HSDPA Release 5 enabled sub-80 ms latency, and the combination of HSDPA and HSUPA in Release 6 gave even sub-30 ms latency. Latency here refers to round-trip time, which is the two-way latency through the mobile and the core network. The HSPA latency improvements are shown in Figure 1.8. Latency consists of the transmit delay caused by the air interface frame sizes and by the processing delays in UE and by the delays in the network elements and transport network. The channel allocation delay has also improved considerably from typically 1 second in the first WCDMA networks to below 0.4 seconds in the latest HSPA networks.

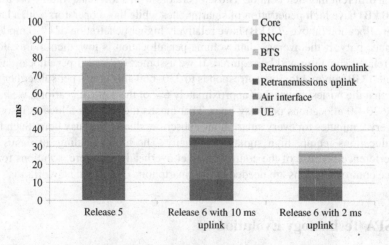

Figure 1.8   Round-trip time evolution

1.6 HSPA Optimization Areas

A number of further optimization steps are required in HSPA to support even more users and higher capacity. The optimization areas include, for example, installation of additional carriers, minimization of signaling load, optimization of terminal power consumption, low band refarming at 850 and 900 MHz, control of uplink interference, and introduction of small cells in heterogeneous networks. Many of these topics do not call for major press releases or advanced terminals features but are quite important in improving end user satisfaction while the traffic is increasing.

The most straightforward solution for adding capacity is to add more carriers to the existing sites. This solution is cost efficient because existing sites, antennas, and backhaul solutions can be reused. The load balancing algorithms can distribute the users equally between the frequencies.

The limiting factor in network capacity can also be the signaling traffic and uplink interference. These issues may be found in smartphone dominated networks where the packet sizes are small and frequent signaling is required to allocate and release channels. The signaling traffic can increase requirements for RNC dimensioning and can also increase uplink interference due to the continuous transmissions of the uplink control channel. Uplink interference management is important, especially in mass arenas like sports stadiums. A number of efficient solutions are available today to minimize uplink interference.

The frequent channel allocations and connection setups also increase the power consumption in the terminal modem section. Those solutions minimizing the transmitted interference in the network tend to also give benefits in terms of power consumption: when the terminal shuts down its transmitter, the power consumption is minimized and the interference is also minimized.

Low-band HSPA refarming improves network coverage and quality. The challenge is to support GSM traffic and HSPA traffic on the same band with good quality. There are attractive solutions available today to carry GSM traffic in fewer spectra and to squeeze down HSPA spectrum requirements.

Busy areas in mobile networks may have such high traffic that it is not possible to provide the required capacity by adding several carriers at the macro site. One optimization step is to split the congested sector into two, which means essentially the introduction of a six-sector solution to the macro site. Another solution is the rollout of small-cell solutions in the form of a micro or pico base station. The small cells can share the frequency with the macro cell in co-channel deployment, which requires solutions to minimize the interference between the different cell layers.

1.7 Summary

HSPA technology has allowed data connections to go wireless all over the world. HSPA subscribers have already exceeded 1 billion and traffic volumes are growing rapidly. This book presents the HSPA technology evolution as defined in 3GPP but also illustrates practical field performance and discusses many of the network optimization and terminal implementation topics.

2

HSDPA and HSUPA in Release 5 and 6

Antti Toskala

2.1 Introduction

This chapter presents the Release 5 based High Speed Downlink Packet Access (HSDPA) and Release 6 based High Speed Uplink Packet Access (HSUPA). This chapter looks first at the 3GPP activity in creating the standard and then presents the key technology components introduced in HSDPA and HSUPA that enabled the breakthrough of mobile data. Basically, as of today, all networks have introduced as a minimum HSDPA support, and nearly all also HSUPA. Respectively, all new chip sets in the marketplace and all new devices support HSDPA and HSUPA as the basic feature set. This chapter also addresses the network architecture evolution from Release 99 to Release 7, including relevant core network developments.

2.2 3GPP Standardization of HSDPA and HSUPA

3GPP started working on better packet data capabilities immediately after the first version of the 3G standard was ready, which in theory offered 2 Mbps. The practical experience was, however, that Release 99 was not too well-suited for more than 384 kbps data connections, and even those were not provided with the highest possible efficiency. One could have configured a user to receive 2 Mbps in the Dedicated Channel (DCH), but in such a case the whole downlink cell capacity would have been reserved to a single user until the channel data rate downgraded with RRC reconfiguration. The first studies started in 2000 [1], with the first specification in 2002 for HSDPA and then 2004 for HSUPA, as shown in Figure 2.1.

While the Release 99 standard in theory allows up to 2 Mbps, the devices in the field were implemented only up to 384 kbps. With HSDPA in Release 5 the specifications have been followed such that even the highest rates have been implemented in the marketplace, reaching

HSPA+ Evolution to Release 12: Performance and Optimization, First Edition.
Edited by Harri Holma, Antti Toskala, and Pablo Tapia.
© 2014 John Wiley & Sons, Ltd. Published 2014 by John Wiley & Sons, Ltd.

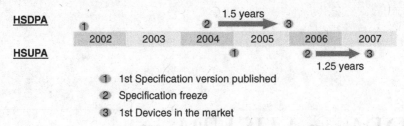

Figure 2.1   3GPP HSDPA and HSUPA specification timeline

Figure 2.1   3GPP HSDPA and HSUPA specification timeline

up to 14 Mbps, and beyond this in later Releases with different enhancements, as covered in the later chapters of this book. The first HSDPA networks were opened at the end of 2005 when the first data products appeared in the marketplace, some 18 months after the 3GPP specification freeze.

With HSUPA, commercial implementations appeared in the marketplace in mid-2007, some 15 months after the specification freeze of early 2006, as shown in Figure 2.1. With smart phones it took generally a bit longer for the technology to be integrated compared to a USB dongle-style product.

The Release 6 uplink with HSUPA including up to 5.8 Mbps capability has also been widely implemented, while the uplink data rate has also been enhanced beyond the Release 6 capability in later 3GPP Releases.

2.3   HSDPA Technology Key Characteristics

The Release 99 WCDMA was basically very much circuit switch oriented in terms of resource reservation, which restricted the practical use of very high data rates while amply providing the CS services (voice and video) that did not require a too high bit rate [2]. The Release 5 HSDPA included several fundamental improvements to the WCDMA standard, with the key items being:

- BTS based scheduling;
- physical layer retransmissions;
- higher order modulation;
- link adaptation;
- dynamic code resource utilization.

BTS-based scheduling moved the scheduling of the common channels from the RNC to the BTS, closer to the radio with actual knowledge of the channel and interference conditions and resource availability. While the scheduling for HSDPA was moved to the BTS, lots of other intelligence remained in the RNC, and further HSDPA functionalities were in most cases additional functionalities, not replacing RNC operation. The Release 99 functional split for the radio resource management is shown in Figure 2.2.

From the protocol stack point of view, the new functionality is added to the MAC layer, with the new MAC-hs added in the BTS. This is illustrated in Figure 2.3, with the remaining MAC-d in the RNC responsible for possible MAC layer service multiplexing.

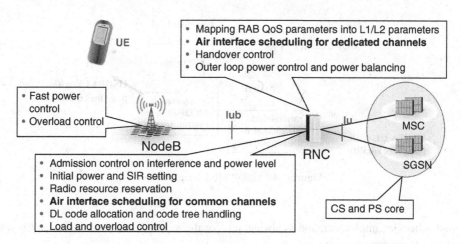

Figure 2.2 Release 99 radio resource management functional split

BTS-based link adaptation allowed reaction to the momentary downlink link quality with all the information of the radio link situation available in the BTS via the new physical layer Channel Quality Information (CQI) feedback. When a user comes closer to the BTS, the received power level is better than needed. This is due to the limited downlink power control dynamic range, since signals received by the users cannot have too large a power difference. This means that a user close to the BTS has a symbol power level tens of dBs greater than necessary. This excessive symbol power level is utilized with the link adaptation. For a user in a good link quality situation, higher order modulation and less channel coding can be used. Respectively, for a user with lower link quality, more robust modulation and more channel encoding redundancy can be used for increased protection. This is illustrated in Figure 2.4.

The link adaptation thus ensures successful transmission even at the cell edge. When data first arrives at the buffer in the base station, the base station scheduler will, however, first prefer to check that the UE has good link condition. Normally, the link condition changes all the time, even for a stationary user. By placing the transmissions in such time instants, based in the CQI, the resource usage is minimized and the system capacity is maximized. The time window considered in the base station varies depending on the service requirements. The

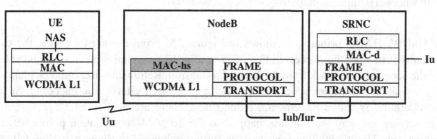

Figure 2.3 HSDPA user plane protocol stack

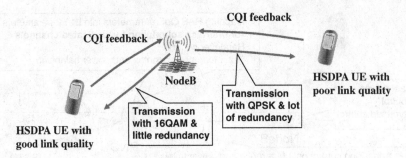

Figure 2.4   HSDPA link adaptation

detailed scheduler implementation is always left for the network implementation and is not part of the 3GPP specifications.

Transmission of a packet is anyway not always successful. In Release 99 a packet error in the decoding would need a retransmission all the way from the RNC, assuming the Acknowledged Mode (AM) of RLC is being used. Release 5 introduced physical layer retransmission enabling rapid recovery from an error in the physical layer packet decoding.

The operation procedure for the physical layer retransmission is as follows:

- After transmission a packet is still kept in the BTS memory and removed only after positive acknowledgement (Ack) of the reception is received.
- In case the packet is not correctly received, a negative acknowledgement (NAck) is received from the UE via the uplink physical layer feedback. This initiates retransmission, with the base station scheduler selecting a suitable moment to schedule a retransmission for the UE.
- The UE receiver has kept the earlier packet in the receiver soft buffer and once the retransmission arrives, the receiver combines the original transmission and retransmission. This allows utilization of the energy of both transmissions for the turbo decoder process.
- The Hybrid Adaptive Repeat and reQuest (HARQ) operation is of stop and wait type. Once a packet decoding has failed, the receiver stops to wait for retransmission. For this reason one needs multiple HARQ processes to enable continuous operation. With HSDPA the number of processes being configurable, but with the UE memory being limited, the maximum data rate is achieved with the use of six processes which seems to be adapted widely in practical HSDPA network implementations.

The HSDPA HARQ principle is shown in Figure 2.5. From the network side, the extra functionality is to store the packets in the memory after first transmission and then be able to decode the necessary feedback from the UE and retransmit. Retransmission could be identical, or the transmitter could alter which bits are being punctured after Turbo encoding. Such a way of operating is called incremental redundancy. There are also smaller optimizations, such as varying the way the bits are mapped to the 16QAM constellation points between retransmissions. The modulation aspects were further enhanced in Release 7 when 64QAM was added. While the use of 16QAM modulation was originally a separate UE capability, today basically all new UEs support 16QAM, many even 64QAM reception as well.

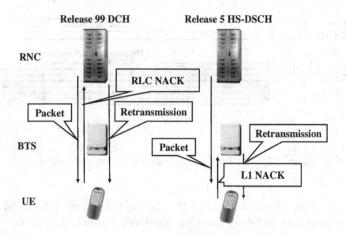

Figure 2.5 HSDPA HARQ operation

To address the practicality of handling high bit rates, dynamic code resource sharing was introduced. Release 99 UE was allocated a specific downlink channelization code(s) dimensioned to carry the maximum data rate possible for the connection. When the data rate was smaller, the transmission was then discontinuous to reduce interference generated to the network. Thus, with the lower data rates, the code was then used only part of the time, as shown in Figure 2.6. It is not possible to share such a resource with any other user, which causes the orthogonal code resource to run out quickly, especially with the case of variable rate packet data connections. Had one used 2 Mbps in the field, it would have basically enabled only a single user to have been configured in a cell at the time.

The dynamic code resource utilization with HSDPA means that for each packet there is separate control information being transmitted (on the High Speed Shared Control Channel, HS-SCCH), which identifies which UE is to receive the transmission and which codes are being used. When a UE does not have downlink data to receive on HSDPA, the codes are used

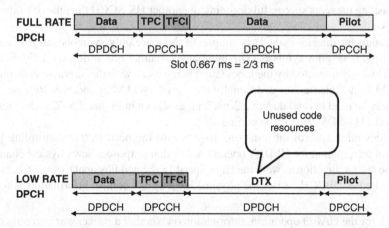

Figure 2.6 Code resource usage with Release 99 DPCH

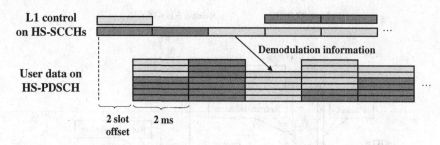

**L1 control
on HS-SCCHs**

Demodulation information

**User data on
HS-PDSCH**

2 slot   2 ms
offset

Figure 2.7   HSDPA physical channels for data and control

by another user or users. There is a pool of 15 codes with a spreading factor of 16 available for HSDPA use, assuming that there is not too much other traffic, such as Release 99 based voice, requiring codes for DCH usage. One cell can have up to four parallel HS-SCCHs which UE could monitor, but for maximum performance it is usually better to send to fewer users at a time in order to minimize the overhead needed for the signaling. From a network point of view, one could configure even more than four HS-SCCHs for one cell, giving different UEs a different set of the channels, but there is no real benefit from such a configuration. The principle of the use of HS-SCCH to point out to the UE which High Speed Physical Shared Channel (HS-PDSCH) codes to receive is illustrated in Figure 2.7. As seen also in Figure 2.6 there is a timing offset between the control and data channel, which allows the UE to determine before the 2 ms Transmission Time Internal (TTI) on HS-PDSCH which codes are intended for the UE in question. This needs to be known beforehand, especially if the UE receiver is able to receive fewer codes than the maximum of 15.

The HS-SCCH is divided into two parts, with the first part carrying information on the codes received and also which modulation is being used, while the second part carries information on the HARQ process being used, transport block size, redundancy version or whether the transmission is new packet, or whether it should be combined with existing data in the buffer for the particular HARQ process. The spreading factor for HS-SCCH is always fixed to be 128, and UE will check from the first part whether the control data is intended for it or not (UE specific masking prevents successful decoding of another HS-SCCH than the one intended for the UE).

The structure for the HS-PDSCH is simple, as there are only symbols carrying user data, not control fields or pilot symbols. The number of channel bits fitting on a single code on HS-PDSCH is only impacted by the modulation being used, with the alternatives being QPSK and 16QAM from Release 5 or additionally the use of 64QAM as added in Release 7. Up to 15 codes may be used in total during a 2-ms TTI, as shown in Figure 2.8. The channel coding applied on the HS-PDSCH is turbo coding.

Besides downlink control information, there is also the need to transmit uplink physical layer control information for HSDPA operation. For that purpose a new physical channel was added in the uplink direction as well, the High Speed Dedicated Physical Control Channel (HS-DPCCH). The HS-DPCCH carries the physical layer uplink control information as follows:

- Feedback for the HARQ operation, information on whether a packet was correctly decoded or not, thus carrying positive and negative acknowledgements. In later phases, different

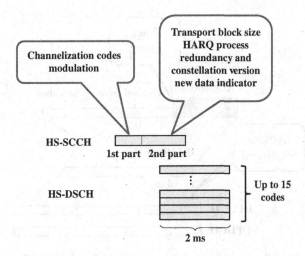

Figure 2.8   HS-SCCH structure

post/preambles have also been added to avoid NodeB having to decide between the ACK/ NACK and DTX state, thus improving the detection reliability.

- Channel Quality Information (CQI) helps the base station scheduler to set correct link adaptation parameters and decide when is a good moment to schedule the transmission for a particular user. The CQI basically gives from the UE an estimate of what kind of data rate could be received reliably in the current conditions the UE is experiencing. Besides the external factors, such as interference from other cells and relative power of the own cell (serving HSDPA cell), the CQI also covers UE internal implementation aspects, such as whether there is an advanced receiver or whether the UE has receiver antenna diversity.

The network can parameterize the HS-DPCCH to also use different power offsets for the different fields carrying CQI and ACK/NACK feedback, as shown in Figure 2.9. It is also worth noting that the HS-DPCCH is never transmitted on its own, but always in connection with the uplink DPCCH. Uplink DPCCH contains pilot symbols and thus provides the necessary phase reference for the NodeB receiver. Besides pilot symbols, the uplink DPCCH carries the downlink power control commands and the Transport Format Combination Indication (TFCI) to inform on the data rate being used on the uplink Dedicated Physical Data Channel (DPDCH). If only HSDPA is operated without HSUPA, all uplink user data is then carried on DPDCH.

The CQI information does not map directly to a data rate but actually provides an indication of the transport block size, number of codes, and level of modulation which the terminal expects it could receive correctly. This in turn could be mapped to a data rate for the assumed power available for HSDPA use. Depending on the type of UE, the CQI table covers only up to such modulation which the UE can support, after that an offset value is used. For this reason there are multiple CQI tables in [2]. In the first phase of HSDPA introduction the UEs on the market supported only QPSK modulation, but currently the generally supported modulations also include 16QAM, and in many cases also 64QAM. The values indicated by the CQI are not in any way binding for the NodeB scheduler; it does not have to use the parameter combination

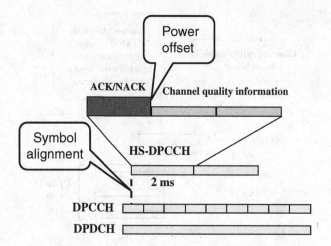

Figure 2.9   HS-DPCCH carrying ACK/NACK and CQI feedback

indicated by the CQI feedback. There are many reasons for deviation from the UE feedback, including sending to multiple UEs simultaneously or having actually different amounts of power available for HSDPA than indicated in the broadcasted value.

2.4   HSDPA Mobility

The mobility with HSDPA is handled differently from Release 99 based DCH. Since the scheduling decisions are done independently in the NodeB scheduler to allow fast reaction to the momentary channel conditions, it would be difficult to follow the Release 99 soft handover (macro diversity) principle with combining time aligned identical transmission from multiple BTS sites. Thus the approach chosen is such that HSDPA data is only provided from a single NodeB or rather by a single cell only, called the serving HSDPA cell. While the UE may still have the soft handover operation for DCH, the HSDPA transmission will take place from only a single NodeB, as shown in Figure 2.10.

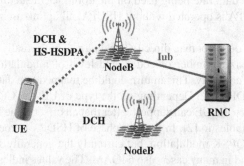

Figure 2.10   Operation of HSDPA from a single NodeB only with DCH soft handover

When operating within the active set, the UE shall signal the change of the strongest cell and then the serving HSDPA cell can be changed to another NodeB. If the strongest cell is not part of the active set, then the new cell needs to be added to the active set before HSDPA transmission can take place from that cell. When changing the serving HSDPA cell, the packets in the source NodeB buffer are thrown away once the serving cell change is complete and the RNC will forward the packets not received to the new serving HSDPA cell NodeB, in case Acknowledged Mode (AM) operation of RLC is being used.

2.5   HSDPA UE Capability

In Release 5 a total of 12 UE categories were defined supporting up to 14 Mbps downlink peak data rate. Some of the UE categories were based on the possibility of having the UE receiving data not during consecutive TTIs, but in reality that possibility has not been utilized. In the first phase the products in the market had only QPSK modulation supported, enabling only 1.8 Mbps physical layer peak data rate, but soon implementations with 16QAM also became available enabling first 3.6 Mbps and then later up to 7.2 Mbps or even up to 10 or 14 Mbps. The relevant UE categories, which actually have been introduced to the market in wider scale, are shown in Table 2.1, with all categories given in [3]. The RLC data rates are calculated with 40 bit RLC packet size. In Release 7 3GPP introduced a flexible RLC packet size which allows reduction of the RLC header overhead.

2.6   HSUPA Technology Key Characteristics

While Release 5 HSDPA introduced a significant boost to downlink packet data capabilities, the uplink operation was unchanged for user data. There were only modifications to cover the necessary uplink feedback with HSDPA operation. The problem areas in the uplink direction were not exactly the same, but fundamentally the Release 99 based design in uplink was more suited for continuous and relatively low rate data transmission than for high bit rate packet data operation. With RNC-based uplink scheduling using the uplink DCH, up to 384 kbps were only implemented in the commercial UMTS networks. If the capacity of the Release 99 network was roughly around 1 Mbps per sector, one could only have allocated roughly three 384 kbps users. With the variable data rate connection the allocated maximum data rate would basically reserve the sector capacity and could only be controlled slowly by the RNC.

Table 2.1   Release 5 HSDPA UE categories and their L1/RLC peak data rates

UE category	# of codes	Highest Modulation	Max L1 data rate (Mbps)	Max RLC data rate (Mbps)
12	5	QPSK	1.8	1.6
5/6	5	16QAM	3.6	3.36
7/8	10	16QAM	7.2	6.72
9	15	16QAM	10.7	9.6
10	15	16QAM	14.0	13.3

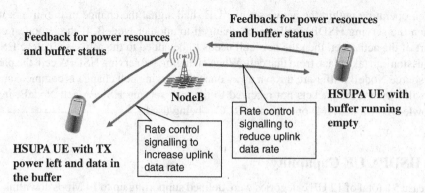

Figure 2.11   Uplink NodeB scheduling principle with HSUPA

To improve the situation, 3GPP introduced in Release 6 with the HSUPA several improvements addressing the following areas that were introduced following study [4]:

- BTS scheduling/ rate control;
- link adaptation;
- HARQ;
- shorter frame size.

With the BTS scheduling and rate control the key target was to address the burstiness of the uplink data transmission needs. The UE provides feedback for the NodeB on the power resources available as well as on the uplink transmission buffer status. This enables the NodeB to allocate more uplink transmission "allowance" for such UE which has more data in the buffer, and then the allocation can be respectively reduced from a UE that does not have that high transmission need. This principle is illustrated in Figure 2.11.

From the protocol point of view, the new intelligence in NodeB is part of the MAC-e – with "e" standing for enhanced. Since the uplink transmission is controlled, the UE has a new entity in the MAC layer, as shown in Figure 2.12.

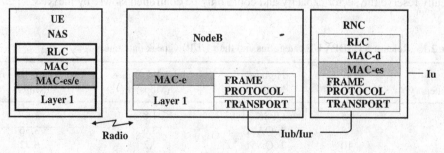

Figure 2.12   HSUPA user place protocol stack new elements

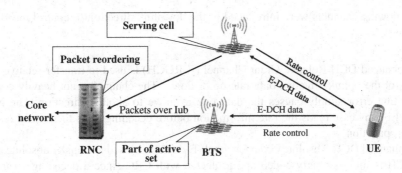

Figure 2.13   HSUPA operation in soft handover

With HSUPA there is also the new MAC entity in the RNC, called MAC-es. Since the uplink macro-diversity operation is retained with HSUPA, there can be more than a single NodeB receiving the uplink packet transmission, as shown in Figure 2.13. In this case a packet transmitted earlier may fail in the decoding in all of the NodeBs in the active set while a packet transmitter is later decoded correctly with first transmission. This, together with the physical layer retransmissions, creates the need for packet re-ordering in the RNC side to restore the order of packets before they are provided to the higher layers.

The rate control in Figure 2.13 can be implemented with two alternative methods, either by using relative grants which basically move the data rate up or down, a bit similar to power control operations. Or the other approach is the use of an absolute grant which enables moving to any possible data rate for the connection (but addressing then only a single UE at a time). Handling the rate control in the NodeB allows reaction to the bursty nature of the application, and avoids having to reserve uplink capacity according to the maximum data rate, thus effectively narrowing the uplink noise rise variance and enabling loading the uplink more fully compared to the Release 99 based uplink solution, as shown in Figure 2.14.

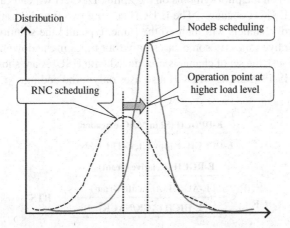

Figure 2.14   Noise rise distribution with RNC and NodeB based uplink schedulers

The following channels were introduced in the downlink direction to control the HSUPA operation:

- The Enhanced DCH Relative Grant Channel (E-RGCH) is used to transfer relative grants to control the uplink effective data rate up or down. The channel is not heavily coded as a decoding error simply causes the data rate to move to another direction. The NodeB receiver in any case reads the rate information before decoding, thus there is no resulting error propagation.
- The Enhanced DCH Absolute Grant Channel (E-AGCH) is used to transfer absolute grants. E-AGCH is more strongly coded and protected with CRC since a decoding error could cause a UE to jump from the minimum data rate to the maximum data rate. Respectively only one UE can be controlled at a time per E-AGCH.
- The Enhanced DCH HARQ Indicator channel (E-HICH) is used to inform the UE whether the uplink packet has been received correctly or not, since there is similar physical layer retransmission procedure introduced with HSUPA as with HSDPA.

In the uplink direction the following new channels were introduced:

- The E-DCH Dedicated Physical Control Channel (E-DPCCH) carries the control information related to the uplink transmission. The E-PDCCH has the Retransmission Sequence Number for HARQ, happy bit, and the Enhanced Transport Format Combination Indicator (E-TFCI). The happy bit indicates whether the UE could use a higher data rate in the uplink direction. If the UE is limited either by the uplink power resource or by the amount of data in the uplink transmission buffer and is thus not able to use higher data rate, then the UE is considered to be happy. The scheduling information, part of the MAC Payload Data Unit (PDU), carries then more detailed information of the buffer status and power headroom situation. E-TFCI indicates the data rate being on used the E-DPDCH.
- The E-DCH Dedicated Physical Data Channel (E-DPDCH) carries the actual data. There is no physical layer control information nor pilot symbols on E-DPDCH. The channel estimation is based on the pilot symbols on the uplink DPCCH which is always transmitted in parallel to the E-DCH channels. The E-DCH can use two frame durations, 2 ms and 10 ms. All the control channels will then do the same, typically the solution being repeating the 2 ms structure five times (in some cases only four times in the downlink from the other cells in the active set); the set of channels introduced with HSUPA are shown in Figure 2.15. Similarly to the HS-PDSCH, E-DPDCH also uses only turbo coding.

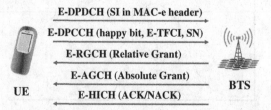

Figure 2.15 New physical channels introduced with HSUPA

The E-DPDCH carrying the actual data may use the spreading factor from 256 to as low as 2, and then also multicode transmission. Since a single code on the E-DPDCH is using basically only the I or Q branch, there are two code trees then usable under a single scrambling code. Thus the peak data (without 16QAM) of 5.7 Mbps is obtained with the use of two spreading codes with spreading factor 2 and two codes with spreading factor 4. The network may select which of the TTI values will be used, with 10 ms being better for coverage while 2 ms TTI offers larger data rates. The TTI in use may be changed with the reconfiguration.

Originally the uplink modulation was not modified at all, since the use of QPSK (Dual-channel BPSK), as covered in [2], has lower energy per bit than use of higher order modulation such as 16QAM (which was introduced in Release 7). While in downlink direction the limited power control dynamic range causes devices to receive the signal with too good SINR when close to NodeB ("high geometry" region), the use of 16QAM or 64QAM comes more or less for free. In the uplink direction the power control dynamic range can be, and has to be, much larger to avoid the uplink near-far problem. The uplink fast closed-loop power control operates as in Release 99, based on the uplink DPCCH signal received as well as typically monitoring the packet decoding performance (Block Error Rate, BLER). Therefore the uplink signal received by the NodeB from the UE does not arrive with too high a power level, and thus changing to higher order modulation only increases the interference level for other users for a given data rate. The only situation when the use of 16QAM (or higher) modulation would make sense would be a single user per cell situation. In such a case higher order modulation allows an increase in the peak data rate that is available with the uplink code resources under a single scrambling code. As discussed in Chapter 8, the use of an advanced BTS receiver can deal with the interference from other HSUPA users.

The physical layer HARQ operation with HSUPA uses a fixed number of HARQ processes, with four processes used with 10-ms TTI, as shown in Figure 2.16. With a single transmission error there is 30-ms time between the original and retransmission and thus 40 ms of extra delay involved for a single retransmission. There is 14 to 16-ms time for a NodeB to decode the uplink packet and to determine if ACK/NACK needs to be transmitted. The range is due to the downlink channel timing being set on 2-ms intervals as a function of the uplink timing. With 2-ms TTI there are eight processes in use, thus the extra delay for a single retransmission is 16 ms, with the respective NodeB and UE processing times shown in Figure 2.17. The times

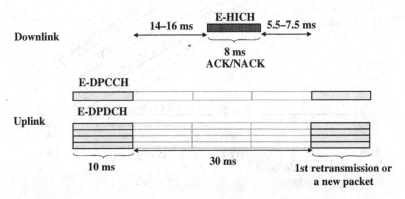

Figure 2.16 HSUPA uplink/downlink timing relationship with 10-ms TTI

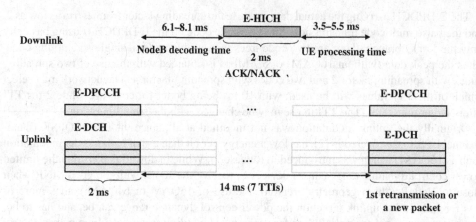

Figure 2.17   HSUPA uplink/downlink relationship with 2-ms TTI

are shortened, which is partly compensated for as the - ms TTI has less room for channel bits due to shorter duration.

2.7   HSUPA Mobility

With HSUPA the active set is operated as in Release 99, with all the base stations in the active set normally receiving the HSUPA transmission. However, there is only one serving cell (the same cell as providing the HSDPA data) which has more functionalities as part of the rate control operation for the scheduling. Only the serving HSUPA cell will send the absolute grants or relative grants which can increase the data rate. The other NodeBs in the active set will only use relative grants to reduce the data rate if experiencing interference problems, as shown in Figure 2.18. All NodeBs in the active set try to decode the packets and will then send feedback on the E-HICH if packets were decoded correctly.

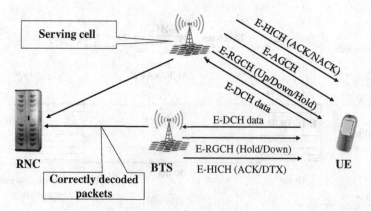

Figure 2.18   HSUPA control channel operation with mobility

Table 2.2 HSUPA UE capability in Release 6 and 7

UE category	# of codes	Max L1 data rate (Mbps)	
		10-ms TTI	2-ms TTI
1	1 × SF4	0.73	—
2	2 × SF4	1.46	1.46
3	2 × SF4	1.46	—
4	2 × SF2	2.0	2.9
5	2 × SF2	2.0	—
6	2 × SF2 + 2 × SF4	2.0	5.76
7(16QAM)	2 × SF2 + 2 × SF4	2.0	11.5

In general, the soft handover overhead, from the backhaul point of view, is also smaller with HSUPA as with Release 99. The HSUPA operation does not cause packets with decoding errors to be sent for RNC as the physical layer asks for retransmission and only correctly decoded packets are then provided for the MAC-es in the RNC for reordering.

2.8 HSUPA UE Capability

The HSUPA UE capability with the latest products typically reaches category 6, enabling 5.76 Mbps uplink peak data rate in the physical layer, as shown in Table 2.2. In the early phase of HSUPA introduction, the equipment in the field typically supported only 10-ms TTI which enabled a 2 Mbps uplink peak data rate. In later Releases new categories were then added with the introduction of 16QAM in the uplink and 64QAM in Release 11, as can be seen from Release 11 version of [3].

2.9 HSPA Architecture Evolution

HSDPA was the first step towards moving more functionality towards the NodeB, with a first step being the downlink scheduling functionality taken away from the RNC and moved to the NodeB. Still, a lot of the Release 99 based functionality remained in the RNC after Release 5 had moved the downlink packet scheduling and Release 6 the uplink packet scheduling control to the NodeB. Architecture evolution continued with 3GPP introducing in Release 7 a further fully flat radio architecture option with all the radio functionality, including RRC signaling, moved to the NodeB, as shown in Figure 2.19. This was then later also adopted to be the LTE radio protocol architecture solution as well [5]. From the core network point of view, also the user plane traffic was moved to bypass the Serving GPRS Support Node (SGSN), with the control plane still going through the SGSN.

From the control plane point of view, the signaling towards the core network goes via the SGSN, similar again to the LTE control plane solution with the Mobility Management Entity in LTE being a control plane only element in the core network side, with the LTE core network known as the Evolved Packet Core (EPC).

A further important aspect is the backhaul, since with HSDPA and HSUPA the backhaul transmission resources are not reserved on a user basis according to the peak data rate,

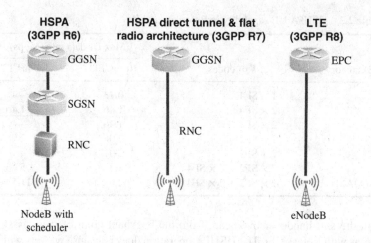

Figure 2.19   HSPA architecture evolution for user plane

thus leading to much better multiplexing gain in the HSPA backhaul network. The backhaul dimensioning for HSDPA has been studied in [6].

2.10   Conclusions

In this chapter the fundamental HSPA functionality introduced in Release 5 and 6 was covered, enabling the end users to reach the original 3G dream of 2 Mbps data rate availability. The basic HSPA features introduced have become more or less *de facto* baseline for all 3G networks, as well as even low end 3G enabled devices, with more than 1 billion users as shown in Chapter 1.

The introduction of HSUPA nicely complements the high data rates enabled in the downlink with HSDPA, so that the uplink can also provide higher data rates when there is downlink transmission burst over HSDPA. Besides the data rate increase, the network side resources are also much more efficiently utilized with HSPA technology. There is no longer a need to reserve the network resources according to the sum of the maximum peak data rates on all connections.

The latency and resulting Round-Trip Times (RTT) have also become shorter with the introduction of the 2-ms frame both in the uplink and downlink direction, with RTT values as low as 30 ms being observed in the field, as discussed in Chapter 11.

The chapters that follow will cover the key areas that have been enhanced on top of Release 6 to further improve the performance for the end user of HSPA capable handsets.

References

[1] RP-000032, Work Item Description for High Speed Downlink Packet Access, TSG RAN#7, Motorola, March 2000.
[2] Holma, H. and Toskala, A. (2010) *WCDMA for UMTS*, 5th edn, John Wiley & Sons, Ltd, Chichester.

[3] 3GPP Technical Specification, TS 25.306, "UE Radio Access capabilities", version 5.15.0, March 2009.
[4] RP-020658, 3GPP Study Item description "Uplink Enhancements for Dedicated Transport Channels", Ericsson, Motorola, Nokia, AT&T Wireless Services, September 2002.
[5] Holma, H. and Toskala, A. (2011) *LTE for UMTS*, 2nd edn, John Wiley & Sons, Ltd, Chichester.
[6] Toskala, A., Holma, H., Metsala, E. *et al.* (2005) Iub efficiency analysis for high speed downlink packet access in WCDMA. Proceedings of the WPMC, Aalborg, Denmark, September 2005.

3

Multicarrier and Multiantenna MIMO

Antti Toskala, Jeroen Wigard, Matthias Hesse, Ryszard Dokuczal, and Maciej Januszewski

3.1 Introduction

This chapter presents HSPA multicarrier and multiantenna Multiple Input Multiple Output (MIMO) capabilities introduced in Releases 7 onwards all the way to the ongoing Release 12 work on uplink enhancements involving uplink multicarrier operation. Release 99 contained support for base station transmit diversity (both closed and open loop, as covered in [1]), but Release 5 HSDPA and 6 HSUPA were basically designed around a single antenna, single frequency band, and single-carrier operation. This chapter first covers the multicarrier operation introduced with dual-carrier downlink and then extends to more carriers in downlink and also to the dual-carrier uplink operation. This chapter then continues with the multiband capabilities and is concluded with MIMO operation introduced originally in Release 7 and finalized with Release 11 uplink MIMO.

3.2 Dual-Cell Downlink and Uplink

Following the development of the baseband processing capabilities, including handling of larger bandwidths, the work in 3GPP turned towards enabling a single device to use larger bandwidth than the original WCDMA bandwidth of 5 MHz. The definition of, for example, 10 MHz bandwidth with the new chip rate was not a very attractive approach since that would have been a non-backwards compatible solution, not allowing legacy devices to access such a carrier. Thus the direction chosen was to consider using multiple 5 MHz carriers for a single device, with the first step being a downlink dual carrier. Work on the uplink dual carrier later followed. A similar approach is used in LTE as well, with the LTE-Advanced carrier

HSPA+ Evolution to Release 12: Performance and Optimization, First Edition.
Edited by Harri Holma, Antti Toskala, and Pablo Tapia.
© 2014 John Wiley & Sons, Ltd. Published 2014 by John Wiley & Sons, Ltd.

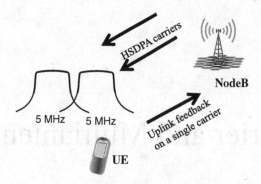

Figure 3.1   Dual-carrier HSDPA principle

aggregation combining multiple carriers of the legacy bandwidths, part of Release 10 LTE specifications, as covered in [2].

3.2.1   Dual-Cell Downlink

Dual-cell downlink was included in Release 8 specifications of HSPA. The basic dual cell HSDPA capability enables a single UE to receive two adjacent downlink 5 MHz carriers simultaneously and to decode HSDPA data on both if scheduled so by the NodeB scheduler. The UE in Release 8 uses only a single uplink to provide the feedback to the NodeB, as shown in Figure 3.1.

From the protocol point of view, the use of dual-carrier HSDPA is only visible to the MAC layer. The data is provided from the RNC Mac-d layer over the Iub to the MAC-hs. The scheduling functionality in the NodeB MAC-hs will then determine, based on the physical layer feedback provided for both downlink carriers (CQI as discussed in Chapter 2), which carrier is better suited for data transmission or if even both should be used simultaneously. Once the data is provided to one of the HS-DSCH transport channels, corresponding to one carrier, then the HARQ procedures are carrier specific, as shown in Figure 3.2.

The use of two carriers allows benefiting from the frequency domain scheduling by selecting to always use the better carrier from the frequency selective fading and interference point of view. The data rate reached can be twice that of the data rate of a single carrier, with the UEs in the field typically supporting up to 42 Mbps, double the single carrier 21 Mbps data rate when 64QAM modulation is supported.

In the downlink direction, each carrier uses its own HS-SCCH as described in Chapter 2. The UE complexity increase from receiving two carriers was limited by requiring UE to receive at most six HS-SCCHs in total while keeping the limit of a maximum of four HS-SCCHs per cell. In the uplink direction the feedback is organized with HS-PDCCH in such a way that a single HS-PDCCH may carry the necessary ACK/NACK and CQI feedback for two downlink carriers. The multiplexing of the CQI feedback for two downlink carriers on a single uplink HS-PDSCH is shown in Figure 3.3, with the solution being concatenation of two CQI feedbacks and then using (20,10) block code instead of (20,5) code. With the ACK/NACK respectively more code works are used with the 10-bit sequence. The driver for

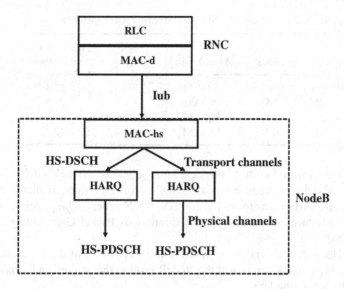

Figure 3.2   Dual-cell HSDPA impact to protocol architecture

using a single code is to avoid the further increase of peak to average ratio (or cubic metric) of the uplink transmission, which would require reduced uplink transmit power or a bigger and less power-efficient power amplifier.

For the dual-cell HSDPA operation new UE capabilities were introduced in [3], as illustrated in Table 3.1 The common UE capability implemented in the field is category 24, which supports two carriers with 21 Mbps each and thus results in a 42 Mbps peak data rate together with 64QAM support and 1/1 coding rate. Release 9 enables combining two-stream MIMO reception to reach 84 Mbps with two-carrier operation.

For terminal power consumption, it is important to avoid keeping dual-cell reception on when there is not that much data to be transmitted for a given terminal. Besides the RRC

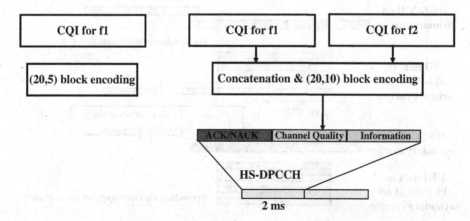

Figure 3.3   Uplink CQI feedback encoding for dual-cell HSDPA

Table 3.1   Dual-cell HSDPA in Release 8 UE capability

Cat	Codes	Modulation	Coding	Peak LI data rate (Mbps)
21	15	16QAM	5/6	23.4
22	15	16QAM	1/1	28.0
23	15	64QAM	5/6	35.3
24	15	64QAM	1/1	42.2

reconfiguration, also with the use of physical layer signaling, with HS-SCCH orders can be used to activate and deactivate reception of HSDPA on two carriers, as shown in Figure 3.4, with the secondary carrier deactivation using HS-SCCH orders. The principle of HS-SCCH orders was as introduced in Release 7 with Continuous Packet Connectivity operation, as explained in Chapter 4.

The use of HS-SCCH orders can be used both to activate and deactivate the secondary carrier, thus enabling quick reaction at the NodeB level to the amount and/or priority of data in the NodeB buffer for the UE.

The mobility is based on the primary serving HS-PDSCH cell. In the case of the Release 8 dual carrier the secondary carrier is always adjacent to the primary carrier, thus the path loss is more or less the same to NodeB from both carriers.

3.2.1.1   The Performance of Dual-Cell HSDPA

Release 8 DC-HSDPA doubles the user data rate if the number of users, or the load in general, is low because a single user can utilize two parallel frequencies. When the system load increases, the probability decreases that a single user can get access to the full capacity of both frequencies. Also with very large number of users on both carriers it is more likely that a user has good conditions and can be scheduled using low coding overhead and high

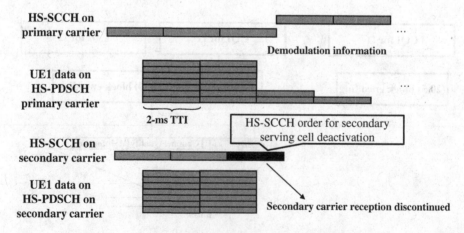

Figure 3.4   Dual-cell HSDPA deactivation with HS-SCCH order

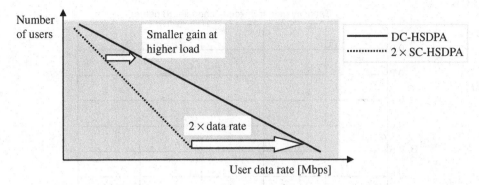

Figure 3.5 DC-HSDPA theoretical performance benefits over single-carrier HSDPA operation

order modulation. But even at high load DC-HSDPA provides some capacity benefit compared to two single carriers. The principles of the DC-HSDPA gains are illustrated in Figure 3.5 with the theoretical performance of DC-HSDPA, showing the double data rate at a low load situation and smaller gains at a high load situation. The DC-HSDPA feature is fully backwards compatible as there are no negative impacts with the legacy devices as a legacy UE does not notice any difference in the downlink waveform even if part of the devices operate in the dual cell HSDPA mode.

The capacity benefit with DC-HSDPA comes from the following sources:

- Frequency domain scheduling benefit. NodeB scheduler can now dynamically select even every 2 ms the carrier with better link conditions. There is of course only the carrier level granularity of 5 MHz available.
- Load balancing gain, as similarly the NodeB scheduler can also base the selection every 2 ms to select the carrier with less load. This provides benefits for the network as a whole since now the DC-HSDPA capable devices can be scheduled dynamically on either carrier, allowing better service for legacy devices too when compared to the situation with all devices being just single-carrier HSDPA capable.
- Multiuser diversity, as the scheduler has now larger pool of users since DC-HSDPA users are available on both carriers to choose from in the NodeB scheduler.

Figure 3.6 shows the simulated capacity gains with Pedestrian A and Pedestrian B multipath profiles. The results show that DC-HSDPA can improve the cell capacity by 20% compared to two Single-Carrier HSDPA (SC-HSDPA) carriers with ideal load balancing between the carriers. Thus the gain varies from 20% in high load cases to up to 100% gain in low load situations. An important observation from the distribution is that both average capacity and cell edge capacity are improved.

3.2.1.2 Dual-Band Dual-Cell HSDPA

Release 9 enhanced further the dual-cell HSDPA operation by defining support for the case of having the carriers on different frequency bands. This is typically beneficial when dealing

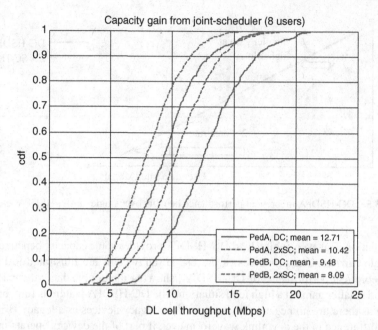

Figure 3.6   DC-HSDPA capacity gain with high load [2]

with lower frequency bands such as 900 and 2100 MHz. In this case the 900 MHz band often has room for a single 5 MHz carrier, especially if it is still partly used for GSM. The use of dual-band operation allows one to benefit from the better coverage offered by the 900 MHz band, but also when path loss is small enough to additionally benefit from the increased performance with two carriers without having to hand over to rely only on the 2100 MHz layer HSDPA coverage. The defined band combinations for dual-band dual-carrier HSPA are shown in Table 3.2.

3.2.2   Dual-Cell HSUPA

Dual-cell HSUPA was introduced in Release 9. In dual-cell HSUPA the UE may be assigned one or two uplink carriers from the same NodeB for data transmission. Compared to the downlink multicarrier concept, in the uplink the UE is power limited, which means doubling the bandwidth does not lead to doubling the power.

Table 3.2   Dual-band dual-cell HSDPA band combinations in Release 9 and 10

Band combination	3GPP Release
Band 1 (2100 MHz) and Band 5 (850 MHz)	Release 9
Band 1 (2100 MHz) and Band 8 (900 MHz)	Release 9
Band 2 (1900 MHz) and Band 4 (2100/1700 MHz)	Release 9
Band 1 (2100 MHz) and Band 11 (1450 MHz)	Release 10
Band 2 (1900 MHz) and Band 5 (850 MHz)	Release 10

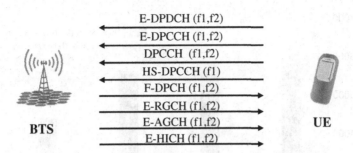

Figure 3.7   Overview of the different channels and frequencies. F1 is the primary uplink frequency

When a UE is configured into dual-cell HSUPA, a UE is at the same time configured to at least two downlink frequencies corresponding to the two uplink frequencies, that is, the UE is at the same time configured in the downlink as dual-cell HSDPA, four or eight carrier HSDPA.

Dual-cell E-DCH can only be operated with 2-ms TTI. The channels E-DPDCH, E-DPCCH, and DPCCH are sent per activated uplink frequency, while the HS-DPCCH is only transmitted on the primary uplink frequency. In the downlink the F-DPCH, E-HICH, E-RGCH, and the E-AGCH are sent per frequency on the corresponding downlink frequency. An overview of the different channels can be seen in Figure 3.7.

Power control, HARQ, and packet scheduling are performed independently per activated frequency, while also the active sets on both frequencies are independent.

The RNC configures the UE in dual-cell mode, while the NodeB can use HS-SCCH orders to activate and deactivate the Secondary Downlink Frequency and Secondary Uplink Frequency. When the frequency of the Secondary Serving HS-DSCH cell is deactivated using an HS-SCCH order, the associated Secondary Uplink Frequency is also deactivated, while the deactivation of the Secondary Uplink Frequency using an HS-SCCH order does not imply the deactivation of the Secondary Downlink Frequency. When the UE hits its maximum transmit power, the UE reduces the power used on the data channel (E-DPDCH) on the carrier with the highest DPCCH power, that is, on the carrier with the worst radio conditions.

Figure 3.8 shows the performance of DC-HSUPA and a single-carrier HSUPA network with the same load per carrier with an Inter Site Distance (ISD) of 500 meters and target Raise Over Thermal (Rot), that is, uplink interference level, increase of 6 dB. It can be seen that at low loads the throughput is doubled for UEs, whereas the gain decreases with the load. The gain also depends on the coverage. This can be seen by comparing Figure 3.8 and 3.9 where the ISD is increased from 500 to 1732 m. The gain in user throughput at low load has now decreased to 50%, while it decreases to lower numbers when the load increases. This is caused by UEs more likely being power limited and therefore unable to turn the extra carrier into extra throughput. Therefore the use of DC-HSUPA is especially beneficial in small cells or in large cells close to the base station.

3.3   Four-Carrier HSDPA and Beyond

In later releases the work has continued to define support, first up to four carriers and later even up to eight carriers. Release 10 contains the support for four-carrier HSDPA, which enables the NodeB scheduler to select dynamically from four different 5-MHz carriers, thus covering a

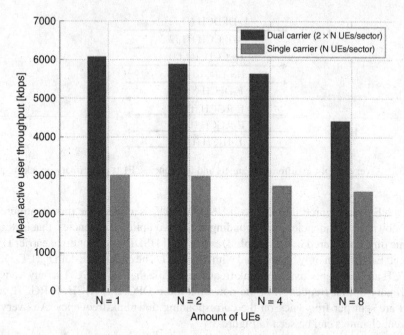

Figure 3.8 Mean active user throughput for N = [1,2,4,8] UEs/sector, ISD = 500, channel profile = PA3 and RoT target = 6.0 dB

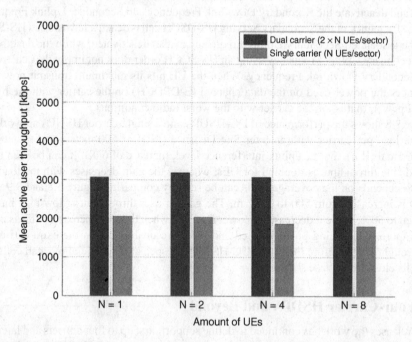

Figure 3.9 Mean active user throughput for N = [1,2,4,8] UEs/sector, ISD = 1732, channel profile = PA3 and RoT target = 6.0 dB

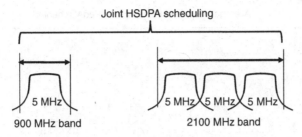

Figure 3.10   4C-HSDPA scheduling principle

total of 20 MHz of bandwidth. In many cases this is not possible for a single operator within one frequency band only, thus dual-band operation is also specified, as illustrated with the example band combination in Figure 3.10. In the case of 4C-HSDPA, the UE provides CQI feedback over four carriers and thus the scheduler can select any of the carriers, effectively using the same RF bandwidth that is used with the LTE 20 MHz carrier case. This enables the re-use of wideband RF UE components designed originally for LTE. Similarly, on the network side the activation of two or more carriers is rather simple, especially if the site already uses more than a single HSDPA carrier. The market utilization of more than two HSDPA carriers remains to be seen as currently the only dual-cell that is supported widely in the latest devices in the marketplace. It is also possible to implement the three-carrier case, which in many instances fits to a single frequency band only, such as the examples with the 2100 MHz (band 1) case.

The performance of four-carrier HSDPA offers 4 times the peak data rate in the low loaded case compared to the single HSDPA carrier case as can be seen in Figure 3.11. The frequency domain scheduling benefit is not quite as high as with the LTE since the granularity with - MHz blocks is clearly bigger than the 180 kHz granularity possible with LTE, as addressed in [4]. Similar to the dual-cell case covered earlier, the capacity benefit is reduced when the load increases. The peak data rate without MIMO is 4 times 21 Mbps, thus reaching 84 Mbps, with 2×2 MIMO up to 168 Mbps.

Release 11 has defined further support for up to eight-carrier HSDPA, enabling a peak data rate of up to 168 Mbps without MIMO and up to 334 Mbps with 2×2 MIMO and even 668 Mbps with 4×4 MIMO, which was also added in Release 11.

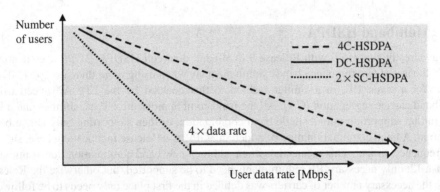

Figure 3.11   Performance of 4C-HSDPA over single-carrier HSDPA as a function of number of users

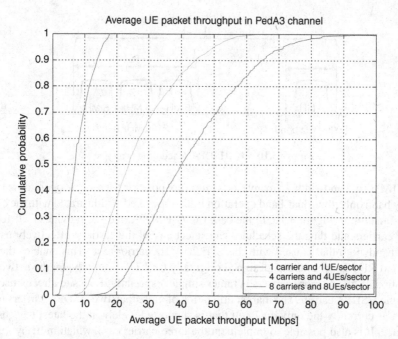

Figure 3.12   Cumulative distribution of the average UE packet throughput (Mbps) for one, four, and eight carriers at low load

The benefits of aggregating multiple carriers are significant for the end user, since often free resources are available in the other carriers. This can be seen in Figures 3.12 and 3.13, which show the cdf of the average user throughput and the mean packet throughput at low loads.

Similar to the dual-cell HSDPA, as with more than two carriers the achievable gains depend a lot on the load in the system. If the load is high, then there will be fewer free resources on the other carriers as well as the fact that scheduler is likely to find users in good channel conditions regardless of the UEs with multicarrier capability, which results in lower gains. This is illustrated in Figure 3.14, which shows the mean packet call throughput as a function of the offered load per carrier.

3.4   Multiband HSDPA

As was already addressed with Release 9 dual-band dual-cell HSDPA, 3GPP specifications enable the use of two frequency bands simultaneously when using more than a single HSDPA carrier for a given UE, in a similar fashion to that enabled by the LTE-Advanced using inter-band carrier aggregation (CA). All the band combinations are UE capabilities and a UE may end up supporting only a single band combination, or then supporting only single band operation. A band combination may be added in 3GPP as a Release independent case, similar to a frequency variant. This means that when adding a new band combination, for example in Release 12, only necessary RF requirements need to be supported, but otherwise the Release where the necessary number of carriers was defined in the first place only needs to be followed. This is similar to the LTE CA band combination principle as covered in [5].

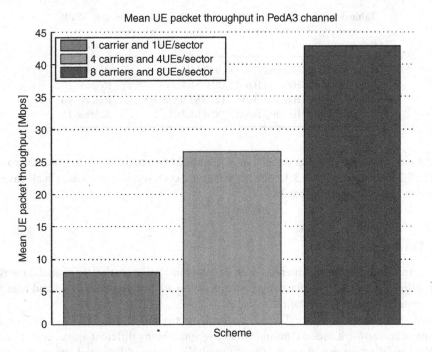

Figure 3.13 Mean UE packet throughput (Mbps) for one, four, and eight carriers at low load

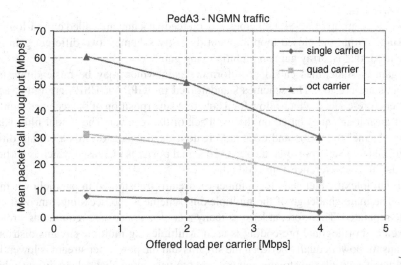

Figure 3.14 Mean UE packet throughput (Mbps) for one, four, and eight carriers as a function of the load

Table 3.3 4/8-carrier HSDPA band combinations defined in 3GPP

Bands	Release
Band 1 (2100 MHz) and Band 8 (900 MHz)	Release 10
Band 1 (2100 MHz) and Band 5 (850 MHz)	Release 10
Band 2 (1900 MHz) and Band 4 (2100/1700 MHz)	Release 10
Band 2 (1900 MHz) and Band 5 (850 MHz)	Release 11

The band combinations so far defined in 3GPP in addition to the dual-carrier cases covered in Table 3.2 are listed in Table 3.3, together with their corresponding Release. In all cases the uplink can be on either of the bands.

3.5 Downlink MIMO

Multiple transmit and receive antennas open the possibility of exploring the spatial dimension. In the following, different mechanisms for exploration are briefly summarized and then their implementation in HSPA is detailed.

Space-time coding, or Space-Time Transmit Diversity (STTD) in 3GPP, is an open-loop diversity scheme which uses a constant coding scheme. Many different space-time codes are described in literature, for example, space-time trellis codes, differential space-time codes, or space-time block codes. From a decoding perspective, orthogonal space-time block codes show very low complexity. The Alamouti code [6] was proven to be the only orthogonal space-time block code which provides full diversity and was included already in Release 5 of the HSPA standard.

An alternative group of MIMO schemes rely on feedback and are called closed-loop spatial multiplexing. Within the closed-loop transmit diversity schemes, two different gain mechanisms are used, that is, array gain and multiplexing gain.

Array gain is sometimes called beamforming gain which may be misleading since no literal beams are created. The schemes standardized in 3GPP for HSPA are rather adaptive spatial filters (pre-coders) which allow for a coherent superposition of the received signal from different transmit antennas based on the feedback of the receiver. The spatial filter adapts to the fading channel over time and does not create a fixed beam towards a certain direction. Due to the coherent superposition, the received signal power is increased which is particularly useful under high interference at the cell edge.

Spatial multiplexing increases the number of streams transmitted in parallel over multiple antennas. The main challenge of the multiplexing scheme is the decoding, since all streams have to be separated. Therefore, at least as many receive antennas are required as streams are multiplexed and orthogonal pre-coding is used. Multiplexing multiple streams distribute the overall transmit power equally between the streams and the power per stream is lower. Hence, the decoding of multiplexed streams require good to very good channel conditions which are typically found in the cell center.

With spatial filtering (beamforming) providing gains at the cell edge and spatial multiplexing in the cell center, the combination of both is natural. However, there is a natural tradeoff between array gain and multiplexing gain [7]; for example, transmitting the maximum number

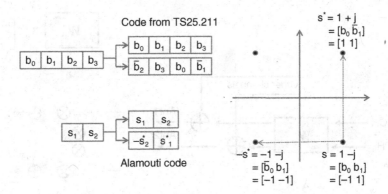

Figure 3.15 Mapping of 3GPP STTD scheme to the Alamouti code for QPSK where s* is the conjugate of s and $\bar{b} = -b$

of stream (= number of transmit antennas) leaves no array gains and when capturing the maximal array gain only one stream can be transmitted.

3.5.1 Space Time Transmit Diversity – STTD

STTD is an open-loop transmit diversity scheme and corresponds to the Alamouti code. Figure 3.15 illustrates how the bit mapping scheme defined in [8] can be translated to the Alamouti scheme by considering how the bits are mapped to complex symbols (here for simplicity QPSK).

STTD requires only one receive antenna and no additional feedback information has to be provided by NodeB. This coding scheme can be applied to all physical downlink channels and is made mandatory for the UE.

However, STTD has some technical shortcomings. First, it does not support Continuous Packet Connectivity (CPC), with CPC details as covered in Chapter 4. Second, STTD makes the channel of one symbol appear longer since it is spread over two slots. The equalizer at the UE has now to estimate the multipath components of two channels spread over two symbols. This makes the implementation of the RAKE receiver in the UE much more complex when keeping the equalization performance constant. This means that STTD is not deployed in practical systems.

3.5.2 Closed-Loop Mode 1 Transmit Diversity

Two closed-loop modes were defined in Release 99, mode 1 and mode 2. Mode 2 was later removed from Release 5 onwards because it was never implemented in the networks. In mode 1, the UE provides the NodeB with a phase offset between the first and the second transmit antenna. The phase offset is transmitted as Feedback Information (FBI) over two consecutive TTIs. Since FBI contains only 1 bit, four different phase offsets can be signaled to the NodeB. Closed-loop transmit diversity mode 1 can be applied to PDCH and HS-PDCH. In practice, mode 1 does not support continuous packet connectivity (CPC).

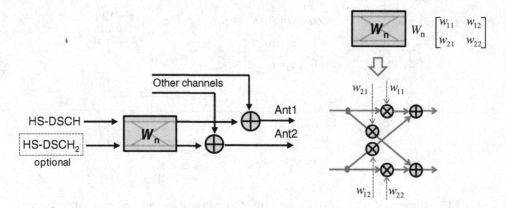

Figure 3.16   MIMO pre-coding of HS-DSCH channels

3.5.3   2 × 2 MIMO and TxAA

With Release 7 MIMO was introduced to HSDPA. The closed-loop scheme is based on feedback of the pre-coding index which defines the weights used in the spatial pre-coder. Figure 3.16 shows the pre-coding and the superposition of additional channels. As shown, MIMO can only be used for HS-DSCH. All control channels and HS-DSCH for non-MIMO do not use the MIMO pre-coder and are shown as other channels in Figure 3.16.

As shown in the right part of Figure 3.16, pre-coding is a weighted distribution of HS-DSCH to the two transmit antennas. The pre-coding weights can be combined into a pre-coding matrix W_n, with

$$
W_1 = \begin{bmatrix} \dfrac{1}{\sqrt{2}} & \dfrac{1}{\sqrt{2}} \\ \dfrac{1+j}{2} & \dfrac{-1-j}{2} \end{bmatrix} \quad
W_2 = \begin{bmatrix} \dfrac{1}{\sqrt{2}} & \dfrac{1}{\sqrt{2}} \\ \dfrac{1-j}{2} & \dfrac{-1+j}{2} \end{bmatrix}
$$

$$
W_3 = \begin{bmatrix} \dfrac{1}{\sqrt{2}} & \dfrac{1}{\sqrt{2}} \\ \dfrac{-1-j}{2} & \dfrac{1+j}{2} \end{bmatrix} \quad
W_4 = \begin{bmatrix} \dfrac{1}{\sqrt{2}} & \dfrac{1}{\sqrt{2}} \\ \dfrac{-1+j}{2} & \dfrac{1-j}{2} \end{bmatrix}
$$

as defined in Figure 3.16 and the pre-coding index n. Beside the preferred pre-coding matrix, the UE also indicates the preferred rank of the transmission, that is, how many streams would be optimal to transmit from a UE perspective. In single-stream transmission, only HS-DSCH$_1$ is present which benefits from array gains. Typically, single-stream transmission is used towards the cell edge. In dual-stream transmission, both streams are present and share the total transmit power. Considering further that multipath channels introduce additional interference between the two streams, much better reception conditions are required to decode dual-stream MIMO. Hence it is mainly used towards the cell center.

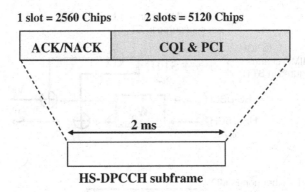

Figure 3.17   HS-DPCCH structure

Doubling the number of HS-DSCH enables a double peak data rate in downlink but also increases the feedback required in uplink. Both the hybrid-ARQ acknowledgement (HARQ-ACK) and the Channel Quality Indicator (CQI) have to be transmitted for both streams. In addition the Precoding Control Indicator (PCI) and the Rank Indicator (RI) have to be transmitted in uplink. Hence the coding of the HS-DPCCH was updated for MIMO.

Figure 3.17 shows the frame structure of the HS-DPCCH which consists of three slots, one for HARQ-ACK and two for CQI and PCI. To accommodate the additional HARQ-ACK in the first slot (10-bit), new coding with decreased Hamming distance between the code words was introduced.

For the channel quality reporting, two types of CQI report were introduced: type A and type B. The type A report is periodically sent by the UE and may contain a single 5-bit CQI value if single-stream transmission is preferred or two CQI values if dual-stream transmission is preferred. The report is 8 bit which results in 256 CQI values. The first 31 values (5 bits) are used to indicate single-stream CQI values. The remaining 225 values are used for dual-stream CQI_1 and CQI_2, where the combined CQI value is given as $CQI = 15 \times CQI_1 + CQI_2 + 31$. Note that the rank indicator is implicitly signaled by the number of preferred streams. A type B report is a 5-bit single-stream CQI report. The UE can be configured such that the report is transmitted periodically to create a reliable single-stream reference for the scheduler.

Two PCI bits are simply prepended to the 8|5 bits of the CQI report. The resulting 10|7 bits are coded by a (20,10) | (20,7) code and transmitted during the remaining two slots of the HS-DPCCH.

Figure 3.18a shows the original version of MIMO that includes the primary CPICH (P-CPICH) being transmitted in STTD mode. As mentioned earlier, STTD may have reduced performance due to channel estimation. Also, in the original MIMO scheme all non-MIMO, non-STTD signals are transmitted from antenna 1, which leads to a power difference between the two antennas. This implies that some fraction of available transmit power is not used at transmit antenna 2. The use of original MIO caused clear performance degradation with the existing devices in the field. The interference from the additional pilot caused problems for the single antenna equalizer devices and thus would cause the actual network capacity to go down if the MIMO were used in a commercial network with legacy devices.

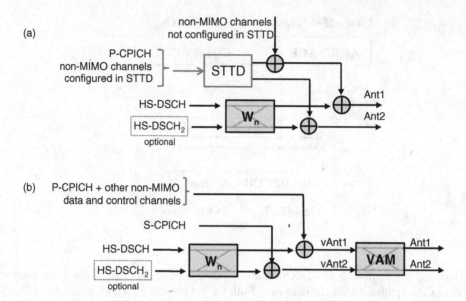

Figure 3.18 Block diagram of original Release 7 MIMO and the work-around solution

To avoid those issues, a work-around MIMO solution was standardized. The solution is shown in Figure 3.18b. Instead of using the P-CPICH transmitted in STTD mode, two different pilot signals are used where the primary CPICH is transmitted over antenna 1 and a secondary CPICH (S-CPICH) from antenna 2. This simplifies channel estimation and equalization.

To overcome the imbalanced power issue, a power balancing network can be assumed which maps the signal of the virtual antennas (vAnt1 and vAnt2) to the physical antennas (Ant1 and Ant2). This Virtual Antenna Mapping (VAM) network is not standardized and transparent for UE. The VAM appears to the UE as part of the channel since all pilot signals are also transmitted over the VAM. Depending on the used VAM matrix, the power of the first stream is not mapped equally to both transmit antennas. In the extreme case, all the power of stream 1 is mapped to one transmit antenna for two pre-coding matrices. For the other two pre-coding matrixes the power is balanced. In order to enable balanced transmit power also for single-stream transmission, this extreme case is used with the set of pre-coding weights limited to a size of 2. This reduces the performance slightly but allows for balanced output power. To enable this, the NodeB can inform the UE on how many weight sets to use. Those work-around solutions were included in Release 7 in 2012 as a late inclusion.

While Release 7 MIMO requires two receive antennas, a new transmit diversity scheme called TxAA was introduced in Release 9, which operates also with a single receive antenna in the UE. TxAA functionality is equivalent to single-stream MIMO.

3.5.4   4-Branch MIMO

Downlink 4-branch MIMO was standardized in Release 11 as a straightforward extension to 2 × 2 MIMO. The NodeB can transmit up to four parallel streams (branches) which doubles the peak data rate per carrier. However, 4-branch MIMO only supports up to four carriers and hence the HSDPA peak data rate remains at 336 Mbit/s. Since a UE supporting 4-branch MIMO is assumed to have advanced receiver, the UE has to have the capability of decoding at least two streams (two Rx antennas) and has to support at least two carriers.

In practice it is expected that there will be more terminals with two receive antennas. Four antennas will be only realistic in larger devices like tablets or laptops. Hence, the 4-branch MIMO design focuses more on optimizing 4×2 transmissions and enabling 4×4 MIMO. The main limitation of transmitting up to four streams is to feedback CQI and HARQ ACK/NACK on the uplink HS-DPCCH especially since this channel also has to carry an extended pre-coding index and rank index.

To limit the feedback in uplink, up to two transport blocks are bundled in codewords and the CQI/HARQ ACK/NACK is transmitted only per codeword. This allows for transmitting four transport blocks with the same amount of feedback as would be required for two transport blocks. The transport block bundling causes some limited performance loss since the whole codeword has to be retransmitted when the transmission of only one transport block fails. Also the feedback of combined CQI values implies that for both transport blocks the same modulation and coding is used even though they are transmitted over different layers which may have different quality. To obtain an equal performance for both transport blocks in one codeword, the transport blocks are interleaved within the codeword. If one or two transport blocks are transmitted in parallel, each transport block is mapped to one codeword and no performance reduction is observed. The codeword mapping is illustrated in Figure 3.19.

4-branch MIMO uses the same codebook for spatial pre-coding as LTE. The codebook has 16 sets of pre-coding weights where the optimal set is chosen by the UE. The NodeB is informed which set to use by the 4-bit pre-coding indicator (PCI). Figure 3.20 shows how 4-branch MIMO channels are combined with 2×2 MIMO, non-MIMO, and signaling channels. For legacy reasons, non-MIMO HS-DSCH and other signaling channels have to be transmitted together with the primary pilot (P-CPICH) over the first (virtual) antenna. The same applies for the first secondary pilot (S-CPICH$_1$) and output of the 2×2 MIMO pre-coder W_n.

The 4-branch MIMO pre-coder H_n can multiplex one to four layers (streams). If only one layer is transmitted, HS-DSCH can't be combined with a not-present channel to one codeword and is transmitted only over layer 1. The same applies for HS-DSCH$_2$.

The changes to HS-DPCCH are limited thanks to the introduction of codewords. The HARQ-ACK feedback the CQI reports remain unchanged. Only the pre-coding indicator is extended from 2 to 4 bits and 2 bits are added to indicate that the Number of Transport Blocks Preferred (NTBP) are introduced which corresponds to the rank indicator in LTE. Those four

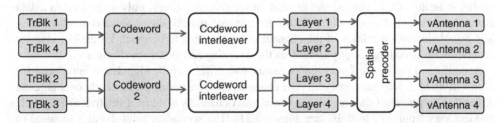

Figure 3.19 Transport block to codeword mapping introduced to 4-branch MIMO to reduce feedback

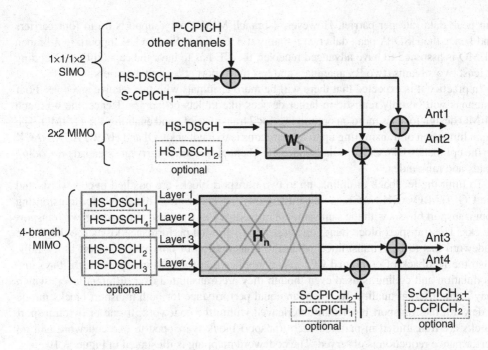

Figure 3.20 4-branch MIMO block diagram with newly introduced pilot channels and the combination with 2 × 2 MIMO and non-MIMO

new bits can be accommodated in the HS-DPCCH by using a rate 1/2 convolutional code for type A reports and rate 1/3 convolutional code for type B reports. Note that the increased error probability due to reduced coding protection has to be compensated by higher transmit power.

Since the channel between each transmit antenna and receive antenna has to be estimated separately, new pilot signals have to be introduced. Adding more pilots with similar power as the P-CPICH would imply for four transmit antennas that 40% of the total power is used for pilot signals, which is clearly too much. Hence a different solution is used.

The common and secondary pilots are required for

1. channel quality measurement and
2. channel estimation for demodulation.

While the channel quality measurement needs to be done every TTI to calculate the channel quality indicator (CQI), the channel estimation for demodulation is only needed when a data channel is scheduled in downlink. Pilots for channel quality measurements require much less power since the accuracy of the channel estimate needs to be much lower. The channel estimation for decoding directly influences the decoding and system performance.

According to the two different requirements, two different pilot signals are introduced for 4-branch MIMO. New secondary CPICH channels are transmitted permanently over antenna 3 and 4. The power of the newly introduced pilot channels is defined relative to P-CPICH and can vary from −12 to 0 dB (see Figure 3.21). The secondary pilot 2 and 3 are used to calculate periodical CQI reports. In many cases the power of S-CPICH$_{2/3}$ is also sufficient

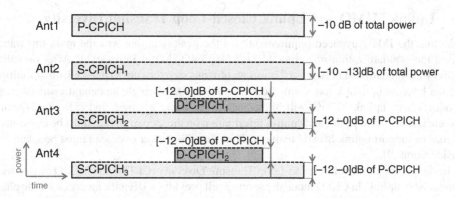

Figure 3.21 Power level of additionally introduced pilot channels for antenna 3 and 4 (D-CPICH – demodulation-CPICH)

for decoding. High pilot power similar to P-CPICH is only required for UEs at the cell edge with unfavorable channel conditions. For this worst case UEs, two additional demodulation common pilot channels (D-CPICH) can be scheduled.

Similar to S-CPICH$_{2/3}$, the power of D-CPICH$_{1/2}$ is defined by higher layers between 0 and −12 dB according to the cell size, for example. The UE is informed about the additionally scheduled D-CPICH via HS-SCCH order.

Figure 3.22 shows expected gains obtained from 3GPP conform system-level simulations with conservative pilot power settings and dynamic rank switching. In the figure all combinations of receive and transmit antennas are compared where the case of 4Tx MIMO with one receive antenna is not standardized but is given here for completeness. It can be observed that 4-branch MIMO brings clear gains with respect to 2Tx MIMO ranging from 8.5% for one receive antenna to 22% for four receive antennas.

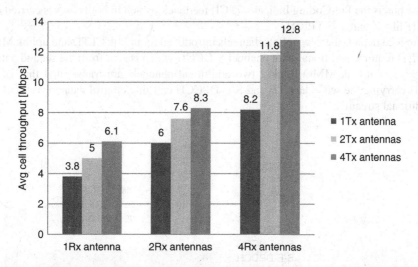

Figure 3.22 Gains for different constellations of receive and transmit antennas. Note that the 4 × 1 case is not standardized and is shown only for reference

3.6   Uplink MIMO and Uplink Closed-Loop Transmit Diversity

Following the IMT-Advanced requirements for the peak data rate was the main motivation behind the standardization of uplink MIMO (and 64QAM) in Release 11. After downlink MIMO was standardized in Release 7 it was an obvious step forward to pursue similar multiple antenna benefits in uplink too. Compared to uplink 16QAM single antenna transmission the introduction of uplink MIMO will allow increasing peak, average, and cell edge spectral efficiencies. Secondary pilot and control information on the second stream must be transmitted in order to support uplink MIMO features. Thus the pilot power overhead must be taken into consideration [9].

Similar to downlink, both Closed-Loop Transmit Diversity (CLTD) and MIMO modes were introduced in uplink. In CLTD mode the serving cell provides a UE with feedback on the phase offset between the signals from the two transmit antennas. The corresponding phase delay is then applied by the UE to the uplink data channels in order to achieve a coherent superposition of the received signal at the Base Station, which results in higher received power. Uplink MIMO allows transmission of up to two Transport Blocks on two spatial streams – this is called MIMO Rank-2. The pre-coding principle is the same as in the downlink case described in Section 3.5. MIMO Rank-1 is a transmission scheme which works technically in the same way as CLTD with a difference that CLTD is a separate feature from the standard perspective. Thus, it can be used by UEs which do not support MIMO. On the other hand, MIMO-enabled UEs utilize the benefits of both Rank-1 and Rank-2 modes and the rank selection is performed dynamically depending on the channel conditions. What differentiates uplink multiple antenna techniques from the similar approaches introduced in downlink is the possibility of switching the transmit antennas. The antenna switching mode allows the UE to use only one of the two available transmit antennas based on which one provides better received signal power. This is especially beneficial in a scenario where there is a substantial imbalance in the signal levels received from those antennas because one of the antennas suffers from strong shielding.

Both CLTD and MIMO are enabled by the HS-SCCH orders from the serving cell. What is also common is the Pre-Coding Indicator (PCI) feedback, which in both cases is carried on an F-DPCH-like channel, F-TPICH.

Figure 3.23 depicts the new uplink channels introduced for uplink CLTD and uplink MIMO. The CLTD requires additional pilot channel S-DCCH which is sent from the second antenna. In the case of uplink MMO Rank-2 two additional channels are transmitted that is, S-E-DPDCH carrying the secondary TB and S-E-DPCCH carrying control data corresponding to the additional stream.

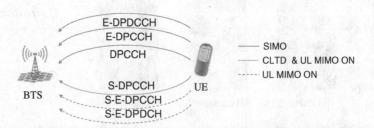

Figure 3.23   Overview of the different channels in case of SIMO, UL CLTD, and UL MIMO

During the uplink MIMO Rank-2 studies it was decided to transmit dual TB independently over the spatial streams. It was also decided that each spatial stream use independent E-TFC selection (rate adaptation). This operation provides more flexibility and maximizes the achievable throughput. The drawback of this solution is the additional overhead of the scheduling and HARQ control information which is associated to each stream in uplink. In downlink for each stream the E-TFC indicator (E-TFCI) and the ACK/NACK must be sent. The two transport blocks are acknowledged independently using the E-HICH channel, where two different signatures are configured on the same E-HICH code channel. There is only one inner and outer power control loop for both streams.

However it was not the only scheme that was evaluated during the Rel-11 studies. Two others had some benefits and drawbacks compared to the chosen one but based on the performance results were not selected. Nevertheless, in order to give some background short descriptions are provided below [10]:

- *Design #1 – single Transport Block Rank-2 transmission.* Due to the fact that in this case it is sufficient to indicate a single E-TFC and transmission rank in uplink, there is low control signaling overhead. The main disadvantage of this approach is the low channel adaptation ability, in a situation where there is strong imbalance between MIMO spatial streams, due to the fact that all parameters for both spatial streams should be the same, that is, modulation, OVSF code allocation, bit mapping and power allocation.
- *Design #2 – dual TB Rank-2 transmission,* TBs interleaved between the two spatial channels. This option can be described as a hybrid between design #1 and the standardized design which was described in the paragraph above. Interleaving allows transmission of each TB over the same channel conditions. In terms of signaling, the HARQ overhead is the same as in the standardized solution. It was discarded because, similarly to design #1, this approach facilitates less efficient scheduling.

It is worth noting that if the transmission of the TB on the secondary stream fails for whatever reason, uplink MIMO Rank-2 mode will be operated as it was uplink CLTD (or uplink MIMO Rank-1). More details about differences between uplink MIMO and beamforming CLTD are presented in the figures in Section 3.6.1 below.

3.6.1   Uplink MIMO Channel Architecture

In order to benefit from uplink MIMO transmission there is a need to introduce new uplink physical channels that is, Secondary Dedicated Physical Control Channel (S-DPCCH), Secondary Dedicated Physical Control Channel for E-DCH (S-E-DPCCH) and Secondary Dedicated Physical Data Channel for E-DCH (S-E-DPDCH) as they are shown in the Figure 3.24 below.

Details about channel configuration of UL CLTD are presented below in Figure 3.25.

As presented in the figures below, a UE is able to transmit data in an additional two instances (depending on network conditions and UE capability):

1. UL-CLTD (Uplink MIMO Rank-1) – DPCCH, E-DPDCH, E-DPCCH and S-DPCCH
2. Uplink MIMO (Rank-2) – DPCCH, E-DPDCH, E-DPCCH, S-DPCCH, S-E-DPCCH and S-E-DPDCH.

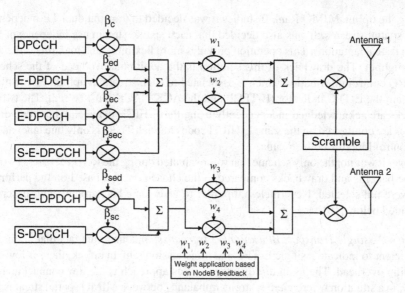

Figure 3.24   Channel configuration in case of UL MIMO Rank-2 transmission

It is worth noting that all control and data channels use the same pre-coding vector ([w1 w2]) as DPCCH, only secondary DPCCH and the secondary E-DPDCH use orthogonal pre-coding vector ([w3 w4]).

S-DPCCH has a mandatory role in both uplink MIMO transmission modes. In Rank-1, that is, uplink CLTD, it was introduced for channel sounding of the secondary stream. By this we are able to choose the optimal pre-coding vector which will be used for future transmissions. It can also be supportive for data demodulation.

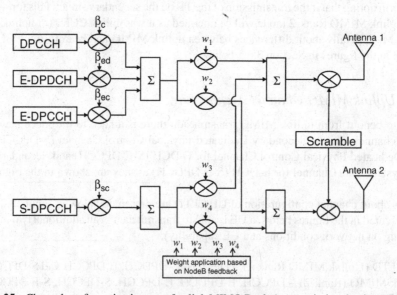

Figure 3.25   Channel configuration in case of uplink MIMO Rank-1 transmission that is, uplink CLTD

In the case of uplink MIMO Rank-2 it is used not only for the best pre-coder selection but also as a phase reference for data demodulation on the secondary stream.

An additional two channels are needed in the case of uplink MIMO Rank-2 transmission. The S-E-DPCCH channel is responsible for carrying E-TFCI and the Retransmission Sequence Number (RSN) for the secondary data stream and S-E-DPDCH channels are used to carry transport blocks on the secondary stream.

3.6.2   Scheduling and Rank Selection with Uplink MIMO

The agreed MIMO architecture includes transmission of two (for Rank-2) or one (for Rank-1) TB(s) independently over the two spatial channels. A scheduling and rank-selection algorithm has to simultaneously select the transmission rank and either two (for Rank-2) or one (Rank-1) E-TFCs (or equivalent transmission grants). The traditional scheduling procedure is applied to the primary spatial stream and the E-TFC is selected based on the transmission grant provided by the NodeB that is, the serving grant. The secondary stream uses the virtual serving grant. Grant for the second stream is calculated based on the combination of transmit power which is chosen for the primary stream and offset value which is provided by the NodeB over the E-DCH Rank and Offset Channel (E-ROCH) together with rank indication. When calculating E-TFC for the secondary stream based on the virtual SG the same legacy gain factor tables are used [11, 12].

The block-diagram of the power-based scheduling and rank selection algorithm is provided in Figure 3.26.

According to the block diagram, retransmissions are considered first and, if they happen, are done in compliance with the used MIMO H-ARQ protocol. When no retransmissions are expected, the scheduling is performed independently for Rank-1 and Rank-2 transmissions and the rank providing the maximum throughput is selected.

The Rank-1 scheduling and the primary spatial channel scheduling for Rank-2 use the legacy scheduling procedure as for the SIMO. This is possible because of the MIMO transmit power control implementation where one Inner-Loop Power Control (ILPC) and one Open-loop Power Control (OLPC) operate over the primary spatial stream. The E-TFC to be sent over the primary spatial channel for both Rank-1 and Rank-2 is selected to provide the maximum throughput and for the total predicted received power to be below the available received power budget.

For the secondary stream of a Rank-2 transmission, the scheduling approach is different and the following principles are applied:

- Transmit power used for the secondary stream is equal to the power of the primary stream.
- No independent fast power control ILPC/OLPC is applied for the secondary stream (one ILPC and OLPC for both streams).
- Data rate adaptation is used for secondary stream meaning that the E-TFC is selected based on the post-receiver SINR of the secondary stream to provide the needed BLER performance.

The used procedure actually decouples data rate control and the transmit power control for the secondary spatial stream (that are coupled in SIMO, Rank-1 HSUPA, and Rank-2 primary stream scheduling).

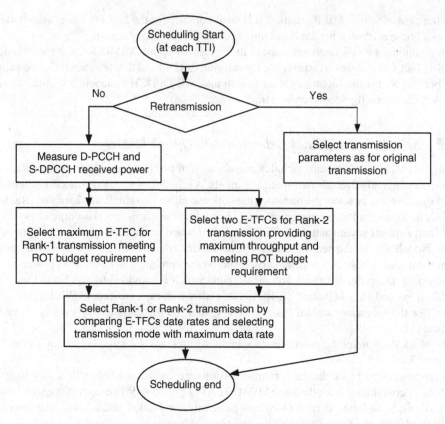

Figure 3.26   Block diagram of power-based scheduling and rank selection algorithm for HSUPA MIMO

3.6.3   Uplink MIMO Performance Evaluation

This section presents the estimated gains coming from the uplink multiple antenna schemes obtained in the 3GPP link and system level simulations. For the sake of simplicity the results are presented only for the Pedestrian A 3km/h channel model [13] and 2 × 2 antenna configuration. The conclusions drawn from the presented evaluation are valid also for other channel models with minor differences in the absolute numbers.

As can be observed from the simulation results in Figure 3.27, the uplink MIMO scheme, as defined in Release 11, was found to have the best performance for high SINR values, which was to be expected. The apparent lack of CLTD gains comes from the fact that this mode provides benefits in a form of lower UE transmit power and what follows – less interference. This does not manifest in the link level results presented in Figure 3.27 but is responsible for the better CLTD performance in the system-level simulations shown in Figure 3.28.

The system-level evaluation presented below assumes adaptive rank adaptation in the case of MIMO mode. As mentioned before, the Rank-2 transmission is beneficial only for the very high SINR conditions. That is why the system level gains are observed only for very low UE densities. In other scenarios, the overhead of additional control channels takes away the marginal gains of spatial multiplexing. The benefit of the CLTD method comes from its ability

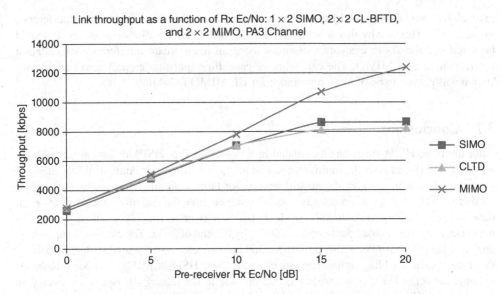

Figure 3.27 Link throughputs of different transmission modes, PedA 3 km/h, 2 × 2 (1 × 2) antenna configuration

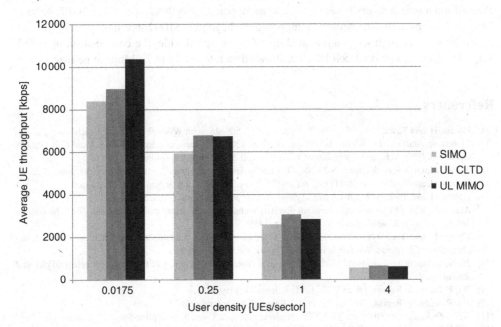

Figure 3.28 System level throughputs for different transmission modes, PedA 3 km/h, 2 × 2 (1 × 2) antenna configuration

to reach the same throughput by using less transmit power, therefore causing less interference to other UEs. This is why this scheme seems to be beneficial for all user densities. It should be noted that the above evaluation assumes using an inter-stream interference cancellation receiver in case of MIMO. The capability of cancelling the interference from the primary stream was proven to be of high importance for UL MIMO feasibility.

3.7 Conclusions

After the basic HSPA functionality created in Release 5 and 6, HSPA evolution has added a lot of capabilities in both the multiantenna and also the multicarrier domain. While many of the other features in this chapter are still waiting for large scale implementation, the support for dual-cell HSDPA has been already widely deployed globally, and offers the 42 Mbps peak data rate quite commonly. Globally, from all HSPA networks, roughly 1 out of 4 of HSPA networks will have enabled dual-carrier HSDPA by the end of 2013. The chips sets supporting dual-cell HSUPA have been also published, and are expected to be available during 2014 in the marketplace, enabling uplink support of 11.5 Mbps. HSDPA MIMO has been deployed in some networks but was not widely rolled out due to the legacy UE issues as covered in Section 3.5.3. Combining downlink MIMO with dual-cell HSDPA would enable pushing the data rates to 84 Mbps, doubling the currently achievable HSPA peak data rate in the market by many handsets today. Similarly, the HSUPA MIMO has not attracted huge interest in the market due to the highly limited use case with virtually no interfering users in a cell. Uplink closed-loop transmit diversity seems to have more potential in this respect. The 3GPP Release 11 specifications define the necessary support to go beyond the 100 Mbps limit with HSPA too, assuming large enough spectrum availability. On the uplink side, the combination of uplink 16QAM together with dual-cell HSUPA allows data rates of up to 22 Mbps to be reached.

References

[1] Holma, H. and Toskala, A. (2010) *WCDMA for UMTS*, 5th edn, John Wiley & Sons, Ltd, Chichester.
[2] Morais de Andrade, D., Klein, A., Holma, H. et al. (2009) Performance Evaluation on Dual-Cell HSDPA Operation. IEEE Vehicular Technology Conference Fall 2009, Anchorage, Alaska, 2009.
[3] 3GPP Technical Specification, TS 25.306, "UE Radio Access capabilities", v 8.12.0, Release 8, June 2012.
[4] Holma, H. and Toskala, A. (2011) *LTE for UMTS*, 2nd edn, John Wiley & Sons, Ltd, Chichester.
[5] Holma, H. and Toskala, A. (2012) *LTE-Advanced*, John Wiley & Sons, Ltd, Chichester.
[6] Alamouti, S.M. (1998) A simple transmit diversity technique for wireless communications. *IEEE Journal on Selected Areas in Communications*, **16**(8), 1451–1458.
[7] Zheng, L. and Tse, D.N.C. (2003) Diversity and multiplexing: a fundamental tradeoff in multiple-antenna channels. *IEEE Transactions on Information Theory*, **49**(5), 1073–1096.
[8] 3GPP Technical Specification, TS25.211, "Physical channels and mapping of transport channels onto physical channels".
[9] 3GPP Technical Report, TR 25.871 v.1.0.1, Release 11, May 2011.
[10] 3GPP Technical Report, TR 25.871 v11.0.0, Release 11, September 2011.
[11] 3GPP Technical Specification, TS 25.321, "Enhanced uplink; Overall description Stage 2", Release 11.
[12] 3GPP Technical Document, R1-121733 Scheduling and Rank Selection for HSUPA MIMO, RAN1 #68bis.
[13] 3GPP Technical Document, R1-112632 UL MIMO Link Level Evaluation, Nokia Siemens Networks, RAN1 #66.

4

Continuous Packet Connectivity and High Speed Common Channels

Harri Holma and Karri Ranta-aho

4.1 Introduction

WCDMA Release 99 was designed for large file transfers, not for frequent and bursty transmission of small packets such as those created by smartphones, see Chapter 13. The WCDMA air interface design is circuit switched like dedicated channels that are not optimized for efficiency, for latency, nor for terminal power consumption. High Speed Downlink Packet Access (HSDPA) in Release 5 and High Speed Uplink Packet Access (HSUPA) in Release 6 improved the efficiency, data rates, and latency considerably by providing a true packet channel-based radio interface. The limitation was still that a Dedicated Physical Control Channel (DPCCH) was required and setting up and releasing DPCCH takes time, adding to the connection setup latency and limiting the number of users with simultaneously active radio connections. We could say that the Release 6 solution was still circuit-switched from the control channel point of view.

Release 7 brings Continuous Packet Connectivity (CPC) which enables Discontinuous Transmission (DTX) for the control channel. Release 8 brings for the first time a true packet-switched channel type in HSPA with High Speed Random Access Channel (HS-RACH) for the uplink direction, called Enhanced Cell_FACH in 3GPP. The corresponding downlink channel High Speed Forward Access Channel (HS-FACH) was defined in Release 7. The evolution of the packet transmission in HSPA improves the network efficiency, terminal power consumption, and end user latency and reduces signaling load. CPC and HS-FACH complement each other nicely: CPC gives benefits also together with HS-FACH during Cell_DCH state. Additionally, CPC terminal penetration is far higher than HS-FACH penetration. These packet transmission improvements have changed WCDMA radio access from circuit-switched to packet-switched, which is important for smartphones. These solutions are illustrated in

HSPA+ Evolution to Release 12: Performance and Optimization, First Edition.
Edited by Harri Holma, Antti Toskala, and Pablo Tapia.
© 2014 John Wiley & Sons, Ltd. Published 2014 by John Wiley & Sons, Ltd.

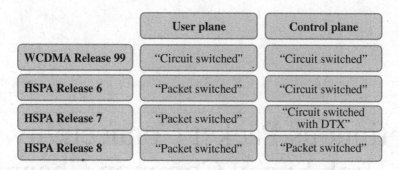

Figure 4.1   Evolution of packet transmission in WCDMA/HSPA

Figure 4.1 and described in this chapter. Section 4.2 explains CPC, Section 4.3 HS-FACH, Section 4.4 HS-RACH, and Section 4.5 further enhancements to HS-FACH and HS-RACH. Sections 4.6, 4.7, 4.8, and 4.9 show the benefit of these solutions in interference reduction, terminal power consumption improvement, signaling reduction, and latency optimization. The chapter is summarized in Section 4.10.

The relevant 3GPP references for CPC and HS-FACH are as follows: [1] is the technical report of CPC and gives an overview while [2–4] include detailed procedures. Simulation results are shown in [5–9].

4.2   Continuous Packet Connectivity (CPC)

The data channel in HSDPA and HSUPA is active only when data transmission is happening but the control channel (DPCCH) is active continuously. Continuous Packet Connectivity (CPC) brings the following improvements:

– Discontinuous transmission (DTX) of DPCCH in uplink.
– Discontinuous reception (DRX) of High Speed Shared Control Channel (HS-SCCH) in downlink.
– Downlink transmission without a High-Speed Shared Control Channel (HS-SCCH).

These features are applicable in Cell_DCH state to improve the link efficiency and reduce the UE battery consumption when the data transmission is not contiguous.

Note that CPC requires also one underlying feature called the Fractional Dedicated Physical Control Channel (F-DPCH) where the downlink DPCH can be shared by ten users to minimize the control channel overhead. F-DPCH was defined in Release 6 but it was implemented in practice only together with CPC in Release 7. The Release 7 functionality also brought enhancements to F-DPCH including more efficient operation in the case of soft handover and it is called enhanced F-DPCH (eF-DPCH).

CPC was defined already in Release 7 but it took a relatively long time to get the CPC feature working properly in the commercial networks. The main reason was that there were many parameters to be defined and to be tested in CPC, and it turned out to be a complex task to get all the chip sets and networks performing well together and achieving very low call drop rates. CPC terminal penetration in many networks is 10–15% as of 2013 and it is expected to increase rapidly with many new phones supporting CPC.

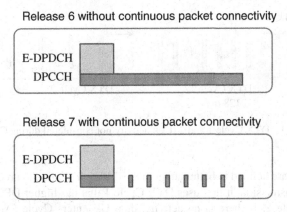

Figure 4.2 CPC concept in uplink

4.2.1 Uplink DTX

The CPC concept in uplink is illustrated in Figure 4.2. The uplink Enhanced Dedicated Physical Data Channel (E-DPDCH) runs only when there is data transmission in uplink while DPCCH runs also during data inactivity in Release 6. The typical inactivity timer before the channel is released and the UE moved to Cell_FACH or to Cell_PCH is several seconds. CPC allows using discontinuous DPCCH transmission during data inactivity when DPCCH is transmitted only in short bursts. These short bursts of DPCCH are needed to maintain the link synchronization and power control for instantaneous resumption of the data traffic. The discontinuous transmission brings two main benefits: less interference is created in the network and less battery power is used by the terminal.

Typical CPC parameter values are shown in Table 4.1 both for 2-ms TTI and 10-ms TTI. Normally, 2-ms TTI is used as long as the uplink link budget allows 2-ms TTI transmission. If the signal level gets low, then the connection is reconfigured to 10-ms TTI.

Table 4.1 Typical CPC parameters (subframe = 2 ms)

Name	2-ms TTI	10-ms TTI
UE DTX Cycle 1	8 subframes	10 subframes
UE DTX Cycle 2	16 subframes	20 subframes
Inactivity threshold for UE DTX Cycle 2	64 subframes	80 subframes
UE DPCCH burst 1	1 subframe	1 subframe
UE DPCCH burst 2	1 subframe	1 subframe
UE DTX long preamble	4 slots	4 slots
Inactivity threshold for UE DRX cycle	64 subframes	64 subframes
UE DRX cycle	8 subframes	8 subframes
MAC DTX cycle	8 subframes	10 subframes
CQI feedback cycle with CPC	4 subframes	5 subframes

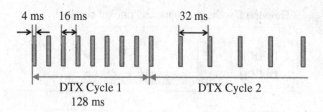

Figure 4.3   DTX Cycle 1 and 2 (CQI activity not considered during Cycle 1)

Two DTX cycles are defined to better adapt to the different application requirements. When UE enters DTX transmission, it first uses DTX Cycle 1 that has higher DPCCH transmission frequency than Cycle 2. If there is no activity, then UE enters Cycle 2 after the inactivity threshold. This concept with typical parameters is shown in Figure 4.3.

The transmission burst of a DTX cycle consists of a preamble, a burst of DPCCH, and a postamble, see Figure 4.4. Preamble and postamble are pilots before and after the DPCCH burst and they are used for channel estimation and uplink power control convergence. The preamble length is two slots and postamble length one slot. The actual DPCCH transmission burst is of configurable length and is typically the same with 2-ms and 10-ms TTI. A long preamble of 4 or 15 slots is also allowed in DTX Cycle 2 if E-DCH data transmission is started for better power control convergence after a longer transmission gap.

The HSUPA uplink must maintain the correct received power levels with DTX operation too in order to prevent near–far problems when the data transmission on E-DCH is started again. The NodeB can estimate the uplink Signal-to-Interference Ratio (SIR) during the transmission burst. The power control command is sent on downlink Fractional DPCH (F-DPCH) to UE. There is no need to send any power control commands unless NodeB has updated the SIR estimate, in other words, if there is no transmission in the uplink in a particular slot, then there is no need to send power control commands in the corresponding downlink slot either. The power control process is illustrated in Figure 4.5.

The uplink HS-DPCCH transmission is needed to carry the control information related to the downlink transmission: HARQ acknowledgements (ACK/NACK) for retransmissions and CQI for the link adaptation. ACKs are transmitted in the same positions as without DTX; that means ACKs have higher priority than DTX and ACKs will cause additional uplink DPCCH activity. NodeB knows when to expect ACK because the offset between downlink transmission

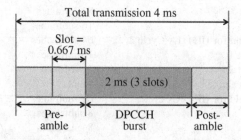

Figure 4.4   DPCCH transmission burst

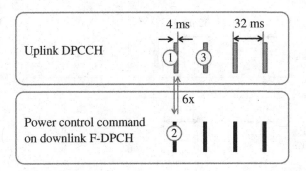

① = NodeB estimates signal-to-interference ratio (SIR) for each slot

② = Power control command is sent on downlink F-DPCH for each slot
Total 6 commands per transmission burst

③ = UE transmission resumes at the power prior to the gap

Figure 4.5   Uplink power control with DTX

and uplink ACK is fixed. CQI is transmitted with normal periodicity if there has been HS-DSCH transmission during the last "CQI DTX Timer" subframes. If there has not been any recent HS-DSCH activity, then CQI is transmitted only when DPCCH is transmitted anyway. The idea is that CQI transmission does not increase the uplink activity unless it is needed due to simultaneous downlink activity. The CQI transmission with uplink DTX is illustrated in Figure 4.6.

The DTX operation helps to reduce the uplink interference levels. The lower the transmission activity, the less interference is created to the network. The activity factor in DTX Cycle 2 is illustrated in Figure 4.7. The transmission burst length is 4 ms and the DTX cycle 32 ms with 2-ms TTI and 40 ms with 10-ms TTI leading to activity of 12.5% with 2-ms TTI and 10% with 10-ms TTI. These activity factors can lead to a substantial decrease in the uplink interference levels as shown in Section 4.6.

NodeB can configure UE specific offsets for DTX to spread the transmission interference equally in the time domain.

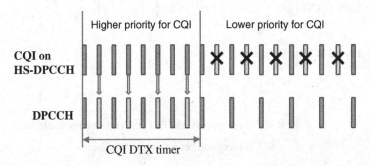

Figure 4.6   CQI transmission with uplink DTX

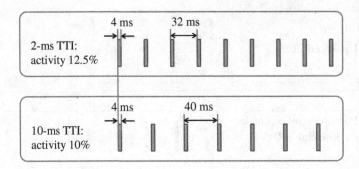

Figure 4.7 Activity factor in DTX Cycle 2

4.2.2 Downlink DRX

UE must monitor continuously up to four HS-SCCHs, E-AGCH, and E-RGCH in normal operation in the Cell_DCH state. The continuous reception keeps UE power consumption relatively high. Release 7 enables downlink discontinuous reception (DRX) to save UE batteries. If there has been no data to UE within the "Inactivity threshold for UE DRX cycle" number of subframes, the UE can start to use DRX. The "UE DRX cycle" defines how frequently UE must wake-up to check the HS-SCCH again to see if there is data scheduled to it. When there is again data transmission on HS-PDSCH, the UE needs to monitor the HS-SCCHs continuously in subsequent TTIs until the inactivity timer has expired. The DRX concept is shown in Figure 4.8.

Downlink DRX only works together with uplink DTX because uplink power control is needed for uplink transmission. If uplink is running continuously, then the power control commands need to be delivered in downlink and DRX is not possible. That means uplink transmission also prevents downlink DRX. The concept is shown in Figure 4.9. It makes sense to harmonize the UE DTX cycle and UE DRX cycle parameters to maximize the benefits from downlink DRX power savings.

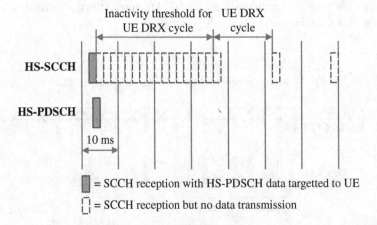

Figure 4.8 Downlink DRX concept

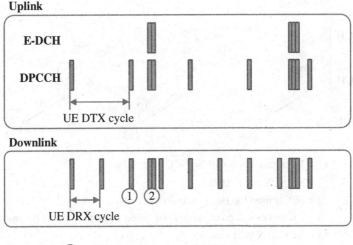

<table>
(1) = UE reception due to DRX cycle
(2) = UE reception due to uplink transmission
</table>

Figure 4.9   Downlink DRX with uplink activity

4.2.3 HS-SCCH-Less Transmission

The downlink control channel overhead was reduced considerably in Release 6 with Fractional DPCH (F-DPCH) where 10 downlink connections can share a single spreading code of length 256. F-DPCH needs to carry only the power control bits for the uplink. Release 7 allows further downlink optimization with HS-SCCH-less transmission of HS-DSCH data. The HS-SCCH overhead is low for large packet sizes but for small packets HS-SCCH-less transmission gives benefits. Such a service is, for example, Voice-over-IP (VoIP) with packet sizes typically below 50 bytes.

Transmission without HS-SCCH requires blind decoding of the HS-PDSCH by UE. A 24-bit CRC check masked with UE identifier is used on the HS-PDSCH transport block in HS-SCCH-less transmission. The identifier is called the HS-DSCH Radio Network Temporary Identifier (H-RNTI), which is normally used on HS-SCCH to identify the UE for which the transmission is intended. In order to keep the blind decoding complexity low, a few constraints have been defined for HS-SCCH-less transmission:

– A maximum of four possible transport block sizes can be used.
– A maximum of two HS-PDSCH codes can be used for data delivery.
– Only QPSK modulation is allowed.

When HS-SCCH is used, then the UE knows when to expect data on HS-PDSCH. UE can then send either ACK or NACK in uplink depending on the decoding success of the data block on HS-PDSCH(s). The same approach cannot be used with HS-SCCH-less transmission: if UE decoding of HS-PDSCH fails, the reason could be that there was an error in the transmission, or that there was no data sent to that UE at all. Therefore, the UE cannot send NACK

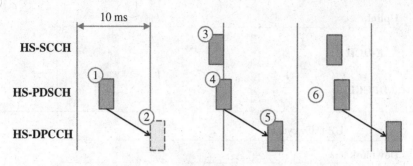

① = 1st transmission without SCCH but it fails
② = No ACK nor NACK
③ = SCCH used for 1st retransmission
④ = UE makes soft combining of 1st failed transmission and retransmission
⑤ = ACK or NACK sent
⑥ = 2nd retransmission is allowed but not more

Figure 4.10   HS-SCCH-less transmission and retransmissions

in HS-SCCH-less transmission at all. With failed decoding, the UE stores soft information in the memory to be used with potential retransmissions. The retransmission must always use HS-SCCH. HS-SCCH-less transmission is only possible for the first transmission. A maximum of two retransmissions are allowed together with HS-SCCH-less first transmission. The retransmissions with HS-SCCH are shown in Figure 4.10.

The UE decodes the signal as follows in the case that HS-SCCH-less transmission is configured: first the UE checks if it can find HS-SCCH. If no HS-SCCH is found, the UE tries to decode HS-PDSCH with four different MCS hypothesis on the preconfigured HS-PDSCH codes. If HS-PDSCH decoding fails, the UE buffers these predefined HS-PDSCH codes to be combined with the potential retransmission. The retransmission must use HS-SCCH and it points to the first transmission, which allows the UE to make correct HARQ combining.

The HS-SCCH-less mode has a few benefits:

– An increase in voice over HSDPA capacity because of the HS-SCCH overhead reduction.
– More HS-SCCH capacity available for other simultaneous services.

These benefits help especially in downlink voice capacity but also in mixed voice and data traffic. Downlink VoIP simulation results with HS-SCCH and with HS-SCCH-less transmission are shown in Figure 4.11. The simulations assume 40 bytes of voice frame arriving every 20 milliseconds during active speech phase and a silence indicator packet arriving every 160 milliseconds during speech inactive phase with 50% voice activity. A single antenna Rake receiver and a maximum of six code multiplexed users are assumed in both cases. HS-SCCH is power controlled based on UE feedback. The results show that HS-SCCH-less transmission can increase VoIP capacity by 10%.

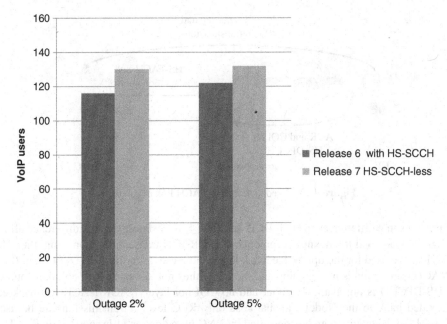

Figure 4.11   Capacity benefit of HS-SCCH-less transmission for voice case. Data from [8]

4.3   High Speed FACH

Release 99 RACH and FACH have very limited data rates: FACH 32 kbps and RACH in practice even less than 10 kbps when time spent on power ramping is considered. RACH and FACH are typically used to carry data volumes that are smaller than 512 bytes. HSDPA and HSUPA, on the other hand, require channel setup signaling and are better suited for larger packet sizes in excess of tens of kilobytes. HS-FACH and HS-RACH are designed to carry medium packet sizes and fill the gap between Release 99 common channels and HSDPA/HSUPA.

HS-FACH has many similarities with HSDPA with both using the HS-DSCH channel. The idea in HS-FACH is to take advantage of the advanced solutions already defined for HSDPA, like shared channel usage, NodeB based scheduling, link adaptation, and discontinuous reception. Figure 4.12 illustrates HS-DSCH with HS-FACH.

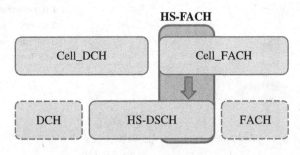

Figure 4.12   HS-DSCH usage in Cell_FACH state

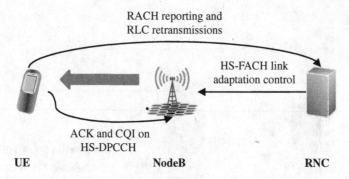

RACH reporting and
RLC retransmissions

HS-FACH link
adaptation control

ACK and CQI on
HS-DPCCH

UE   NodeB   RNC

Figure 4.13   Feedback for HS-FACH link adaptation

There are some differences in Cell_DCH and Cell_FACH states: the first difference affects the CQI feedback and the acknowledgments. If HS-RACH transaction is ongoing, then HS-DPCCH can be used for the uplink feedback in the same way as in the Cell_DCH state. But if HS-RACH transaction is not ongoing and the UE is thus not allocated with uplink resources, then HS-DPCCH is not available either and no CQI nor layer 1 ACK/NACK feedback can be provided back to the NodeB. In that case only RLC level retransmission can be used. HS-FACH link adaptation can be controlled by RNC based on the UE reporting on RACH or based on the RLC retransmission frequency. The feedback solutions are shown in Figure 4.13. It is obvious that HS-FACH performance will be better if layer 1 feedback with HS-RACH is available. Fortunately, most of the packet sessions are uplink initiated and, therefore, the lack of uplink fast feedback is typically not a major issue, and even without uplink feedback, using shared HSDPA code and power resources the HS-FACH is a much more capable data delivery vehicle than FACH over S-CCPCH.

HS-FACH performance with different retransmission approaches is illustrated in Figure 4.14. The file size of 100 kbytes is assumed to be transferred on HS-FACH, the

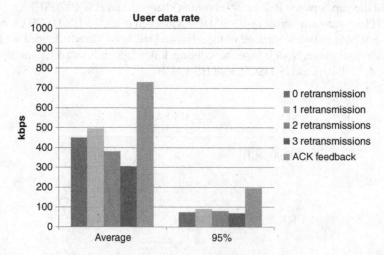

Figure 4.14   HS-FACH performance. Data from [9]

link adaptation is based on the initial RACH measurement report and is not adjusted during the transfer of the file. To model the case with no HS-RACH allocation and no fast uplink feedback, a different number of fixed retransmissions (0... 3) are investigated. An upper bound of the no-feedback case is modeled with ACK feedback.

The Release 99 FACH throughput is typically 32 kbps and in practice even less when considering RLC retransmissions. With HS-FACH and no uplink feedback and always blindly retransmitting each packet once, the average throughput is close to 500 kbps already considering RLC retransmissions. A HS-RACH capability of 1.8 Mbps peak rate is assumed with five HS-PDSCH codes, QPSK-only modulation, and a single-antenna Rake UE. When the UE is allocated to HS-RACH, then fully-fledged HS-DPCCH feedback, with both HARQ-ACK and CQI, is available and the HS-FACH performance equals that of the Cell_DCH HSDPA performance.

HS-FACH brings a few important benefits:

- Less interference by the small or medium size packet transmission than using FACH (no power control) or using HSDPA (setup signaling and DPCCH interference).
- Less signaling than with HSDPA.
- Lower latency for the end user.
- Lower terminal power consumption.
- Sharing of the same code and power resources between the Cell_FACH and Cell_DCH state users.

We also note that paging messages can be sent over HS-DSCH. This concept is called the High Speed Paging Channel (HS-PCH) and it increases the paging channel capacity considerably. The paging channel capacity in Release 99 is limited to below 100 paging messages per second in the case of 8 kbps PCH, and below 400 paging messages per second in the case of 24 kbps PCH. HS-PCH increases the paging channel capacity beyond these limits.

4.4 High Speed RACH

HS-FACH in Release 7 increased the downlink data rates in the Cell_FACH state but the uplink was still based on Release 99 RACH. It was obvious that a corresponding uplink enhancement was required: HS-RACH was included in Release 8. In practice, HS-FACH and HS-RACH are implemented together. HS-RACH follows the same logic as HS-FACH by using the already defined HSPA transport channels also in Cell_FACH state. HS-RACH uses the Enhanced Dedicated Channel (E-DCH) from HSUPA Release 6 which allows it to take advantage of link adaptation, NodeB based scheduling, and layer 1 retransmissions. The concept is shown in Figure 4.15.

HS-RACH signaling is shown in Figure 4.15.

1. UE starts by sending Random Access Channel (RACH) preambles with increasing power levels. UE selects one of the preambles reserved for requesting E-DCH capacity.
2. When the received power of the preamble exceeds the predefined NodeB internal threshold, NodeB sends layer 1 acknowledgement on Acquisition Indication Channel (AICH) pointing UE to the E-DCH resources with Extended Acquisition Indication (E-AI).

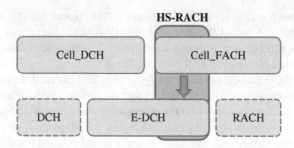

Figure 4.15   E-DCH usage in Cell_FACH state

3. UE starts transmission of control channels DPCCH and E-DPCCH and data channel E-DPDCH with initial data rate.
4. UE is power controlled with downlink F-DPCH that is reserved for the corresponding E-DCH resources.
5. NodeB commands UE to increase its data rate by using the E-DCH Absolute Grant Channel (E-AGCH). With E-AGCH the NodeB also confirms that the UE has passed collision resolution and may proceed.
6. UE increases its data rate.
7. NodeB sends layer 1 acknowledgement on E-HICH.
8. Layer 1 retransmission can be utilized.

In short, HS-RACH transmission is the combination of the RACH access procedure with E-DCH transmission and rate control, and with power control on F-DPCH, as shown in Figure 4.16. With Release 8 the HS-RACH can be configured to either 2-ms or 10-ms TTI operation in a particular cell. Release 11 enhancement to HS-RACH enables the coexistence

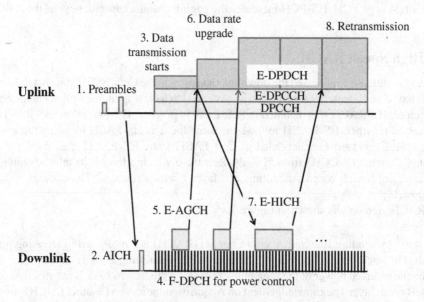

Figure 4.16   HS-RACH transmission

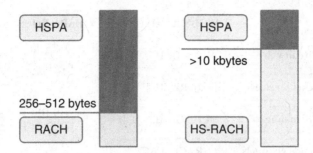

Figure 4.17   HS-RACH extends common channel usage

of both 2- and 10-ms TTI HS-RACH in the same cell and the TTI selection is taken based on the path loss at the beginning of the channel setup.

HS-RACH can enhance the common channel usage area. The maximum length of HS-RACH allocation is 640 ms. The allocation duration is limited because no handover is supported with HS-RACH and HS-FACH. If we assume a 0.5–2.0 Mbps data rate with HS-RACH and HS-FACH, the maximum data volume is 40 to 160 kB. That indicates that HS-RACH and HS-FACH are best suited for data volumes up to tens of kilobytes which extends common channel capabilities considerably compared to Release 99. Many smartphone applications create packet sizes that are higher than 512 bytes but less than tens of kB. The average data volumes per HS-DSCH allocation in smartphone dominated networks is 50–100 kbytes and the median values are far lower. The maximum data volume for Release 99 RACH is typically 256 or 512 bytes. Figure 4.17 illustrates the extended use of common channels with HS-RACH. Therefore, HS-RACH and HS-FACH are nicely suited to smartphone traffic.

The main benefits of HS-FACH and HS-RACH can be summarized as follows:

* The same HS-DSCH and E-DCH channels are used as already defined for Cell_DCH state.
* Minimized uplink interference for small and medium size packets compared to Cell_DCH.
* Minimized signaling load compared to Cell_DCH.
* Very fast access to high data rates without setup latency.
* Reduced terminal power consumption.

WCDMA Release 99 defined a number of different transport channels for voice, data, signaling, common channels, and paging. HSPA Release 8 makes it possible to run all services on top of HSPA: Circuit Switched (CS) voice over HSPA is defined, packet data and VoIP can use HSPA, SRB can be mapped on HSPA, common channels can utilize HSPA channels with HS-FACH and HS-RACH, and also paging is supported on HSPA. The benchmarking of Release 99 and Release 8 is shown in Figure 4.18. It would be possible to get rid of all Release 99 channels and run just HSPA in the network if there were no legacy devices. The complete 3GPP specifications have been re-written by Release 8 with just a few remaining physical channels such as the Synchronization Channel (SCH), Common Pilot Channel (CPICH), Paging Indicator Channel (PICH), and AICH. These major changes also explain why HSPA evolution shows substantially higher performance than WCDMA Release 99. The major advantage is still that all these new HSPA features can be utilized on the same frequency as any legacy Release 99 terminal.

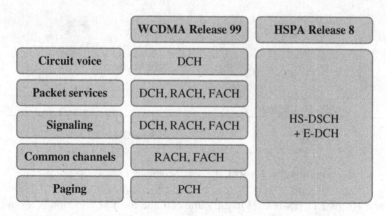

4.5   High Speed FACH and RACH Enhancements

3GPP Release 11 brought a number of enhancement options on top of the Release 7/8 HS-FACH/HS-RACH. These features are collectively known as Further Enhanced FACH (FE-FACH).

NodeB triggered HS-DPCCH feedback for HS-FACH allows for the NodeB to send a HS-SCCH order triggering the UE to initiate HS-RACH procedure and establish a HS-RACH connection without waiting for the data in the uplink transmit buffer to trigger this. This feature can be used to obtain HS-DPCCH feedback for the HS-FACH transmissions as well as preemptively setting up the HS-RACH link before the uplink data transmission need occurs.

Concurrent 2-ms and 10-ms TTI HS-RACH deployment enables the coexistence of 2-ms and 10-ms TTI HS-RACH in the same cell. The TTI selection is done in the PRACH preamble ramp-up phase, if the preamble power required for acquisition is below a threshold a 2-ms TTI can be used, otherwise 10-ms TTI is selected.

Fallback to Release 99 RACH was introduced to help with HS-RACH resource blocking. The NodeB needs to allocate separate resource pools for Release 99 RACH and HS-RACH access, and if all the HS-RACH resources are already allocated it may be beneficial to direct a new access attempt to Release 99 RACH rather than reject it.

TTI alignment and per-HARQ process grants for 2-ms TTI HS-RACH enable time-aligning of the TTIs of different HS-RACH UE transmissions, and time-aligning HS-RACH transmissions to Cell_DCH HSUPA transmissions, and also controlling the activity states of the HARQ processes independently in the Cell_FACH state. These modifications make the 2-ms TTI handling of Cell_FACH and Cell_DCH UEs equal and makes it easier to manage the UEs' transmissions in time domain.

A second, longer DRX cycle in Cell_FACH allows for the UE to fall back to a longer DRX cycle in the Cell_FACH state after a period of inactivity. This makes the battery efficiency of the Cell_FACH state comparable to that of Cell_PCH state making it possible to keep the UE longer in the Cell_FACH state without penalizing the UE standby time.

PRACH preamble delay reduction redefined the time-domain randomization algorithm for the first PRACH preamble to minimize the latency. This optimization relates to UE implementation only and does not require any changes in the network.

Common E-RGCH based interference control enables the UEs to monitor E-RGCH channels during HS-RACH transmissions in exactly the same way as is done with HSUPA transmissions in Cell_DCH state during soft handover.

Network controlled mobility to LTE in Cell_FACH enables the network to configure the UE to measure the LTE frequency layer and report the results to the network in Cell_FACH and leave the decision to move the UE to LTE to the network's algorithms. The interworking between HSPA and LTE is presented in detail in Chapter 15.

Absolute priority cell reselection to LTE and inter-frequency UTRAN enables triggering of the inter-frequency and inter-RAT (to LTE only) mobility measurements and cell reselection in Cell_FACH state even if the current cell quality is good.

4.6 Fast Dormancy

Before the networks activated all power saving features, like Cell_PCH and CPC, some terminals implemented a proprietary solution to minimize power consumption. The idea was to release the RRC connection and go to idle state without any network control. Such a solution is referred to as proprietary fast dormancy. Proprietary implementation means that it was not defined in 3GPP specifications. Fast dormancy was used particularly if the inactivity timer in Cell_DCH was configured for too long, for example more than 5 seconds. When using such a fast dormancy, the mobile application informs the radio layers when the data transmission is over, and the UE can then send a Signaling Connection Release Indication (SCRI) to the RNC, simulating a failure in the signaling connection. Consequently, the UE releases the RRC connection and moves to idle state. This approach keeps the UE power consumption low, but it causes frequent setups of packet connections, increasing unnecessarily the signaling load. In addition to the high signaling load, the network counters indicate a large number of signaling connection failures as this battery saving method cannot be distinguished from a genuine signaling connection failure in the network.

3GPP Release 8 specified a fast dormancy functionality clarifying the UE behavior and providing the network with information on what the UE actually wants to do, but leaving the network in control of the UE RRC state. That is, the UE is not allowed to release the RRC connection and move to an idle state on its own without network control. When the RNC receives SCRI from a UE with a special cause value indicating a packet data session end, the RNC can command the UE to Cell_PCH state instead of releasing the RRC connection and dropping the UE to idle state. This approach avoids unnecessary Radio Access Bearer (RAB) setups and enables the network to separate signaling connection failures from fast dormancy related signaling connection release indications. The Release 8 fast dormancy feature helps to save UE battery when the radio layers get information directly from the application layer when the data transmission is over. The fast dormancy functionality is illustrated in Figure 4.19.

The signaling messages in the state transitions are shown in Figure 4.20. The messages include Radio Resource Control (RRC), NodeB Application Part (NBAP), Access Link Control Application Part (ALCAP), and Radio Access Network Application Part (RANAP). RRC messages run between RNC and UE. NBAP and ALCAP messages are transmitted between RNC and NodeB, and RANAP between RNC and packet core. If UE is idle, a total of 30 messages are needed to get to Cell_DCH state. If UE has a RRC connection and is in Cell_PCH state, 12 messages are required to get to DCH state. If UE is in Cell_PCH state and has a

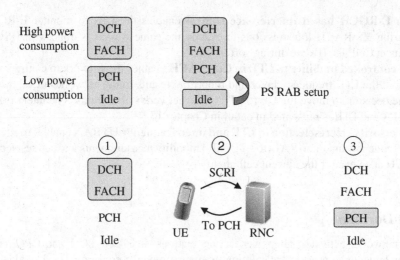

(1) = UE is in high power consumption state = DCH or FACH

(2) = UE sends Signaling Connection Release Indication (SCRI) to RNC

(3) = RNC moves UE to PCH state for low power consumption. No need to
move UE to idle

Figure 4.19   Release 7 fast dormancy functionality

low amount of data that can be carried over FACH, then the transition to Cell_FACH takes
only three messages. There is a major benefit from the signaling point of view to remain RRC
connected instead of moving to idle.

The network broadcasts an inhibit timer (T323), setting a minimum delay between the two
SCRI messages for fast dormancy a UE is allowed to send. This is to prevent a UE from sending
a constant flow of SCRI messages if for some reason the network is temporarily unable to move
the UE to a battery saving state. The presence of T323 in the broadcast can also be considered
an indication of network support of the fast dormancy functionality, triggering the UE to use
the SCRI messages with the cause value indicating the packet data session end. 3GPP standard
based fast dormancy was defined to be early implementable, that is, even though the behavior
and changes in the signaling are defined in the Release 8 specifications it was possible to build
these extensions to a UE and network that is not Release 8 compatible. That approach allowed
the quick introduction of the fast dormancy feature.

4.7   Uplink Interference Reduction

Smartphone applications tend to transmit small packets very frequently. The traffic behavior
is analyzed in Chapter 13. The average data volume per HS-DSCH allocation can be below
100 kB but the allocations can happen 20 times per hour for an average user and with even
higher frequency for active social networker users. If we assume a 5 Mbps data rate and 100 kB

From idle	Large packet from PCH	Small packet from PCH
From idle to DCH	**From PCH to DCH**	**From PCH to FACH**

From idle to DCH

1. RRC Connection Request (RRC)
2. Radio Link Setup (NBAP)
3. Radio Link Setup Response (NBAP)
4. Establish Requests (ALCAP)
5. Establish Confirm (ALCAP)
6. RRC Connection Setup (RRC)
7. NBAP Synchronization (NBAP)
8. RRC Connection Setup Complete (RRC)
9. Initial Direct Transfer RRC)
10. Initial UE Message (RANAP)
11. Common Id (RANAP)
12. Security Mode Command (RANAP)
13. Security Mode Command (RRC)
14. Security Mode Complete (RRC)
15. Security Mode Complete (RANAP)
16. Downlink Direct Transfer (RRC)
17. Uplink Direct Transfer (RRC)
18. RAB Assignment Req (RANAP)
19. GTP Tunnel Setup (RANAP)
20. RB Setup (RRC)
21. Radio Link Setup (NBAP)
22. Radio Link Setup Response (NBAP)
23. Establish Requests (ALCAP)
24. Establish Confirm (ALCAP)
25. Establish Requests (ALCAP)
26. Establish Confirm (ALCAP)
27. RB Setup Complete (RRC)
28. RANAP:RAB Assignment Resp (RANP)
29. Measurement Control (HO) (RRC)
30. Measurement Control (TVM) (RRC)

From PCH to DCH

1. Cell update with TVM (RRC)
2. Radio Link Setup (NBAP)
3. Radio Link Setup Response (NBAP)
4. Establish Requests (ALCAP)
5. Establish Confirm (ALCAP)
6. Establish Requests (ALCAP)
7. Establish Confirm (ALCAP)
8. CU confirm with RB reconfig (RRC)
9. NBAP Synchronization (NBAP)
10. RB reconfiguration complete (RRC)
11. Measurement Control (HO) (RRC)
12. Measurement Control (TVM) (RRC)

From PCH to FACH

1. Cell update (RRC)
2. Cell update confirm (RRC)
3. UMI confirm (RRC)

ALCAP = Access Link Control Application Part
CU = Cell Update
GTP = GPRS Tunneling Protocol
HO = Handover
NBAP = NodeB Application Part
RAB = Radio Access Bearer
RANAP = Radio Access Network Application Part
RB = Radio Bearer
RRC = Radio Resource Control
UMI = UTRAN Mobility Information
TVM = Traffic Volume Measurement

Figure 4.20 Signaling messages in state transitions

packet size, the transmission time is just 0.16 s. The typical inactivity timers are set to 2–10 s which means that more than 90% of the time the uplink DPCCH is running without any data activity (Figure 4.21). This explains why networks with high smartphone penetration suffer from uplink interference effects: smartphone devices keep transmitting unnecessarily in the uplink direction while the channel is up.

One of the main benefits of CPC is the reduced uplink interference due to DTX. The activity factors in Figure 4.7 are 12.5 and 10% for 2- and 10-ms TTIs. These activities would indicate in theory 8–10 times higher uplink capacity when the radio link is inactive and can fully utilize DTX. In practice, the expected uplink capacity gain is lower because power control accuracy is sacrificed during DTX. In order to find out the realistic gains, field measurements were run with a large number of terminals in Cell_DCH state but with very low activity. The maximum number of simultaneous users per cell is evaluated with 3 and 6 dB noise rises corresponding to 50 and 75% loadings. The testing was done with CPC off for all users and CPC on for all users. The 2-ms case is shown in Figure 4.22 and 10-ms case in Figure 4.23. The maximum

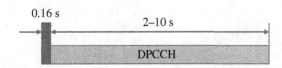

0.16 s

2–10 s

DPCCH

Figure 4.21 DPCCH causes uplink interference with small packet sizes

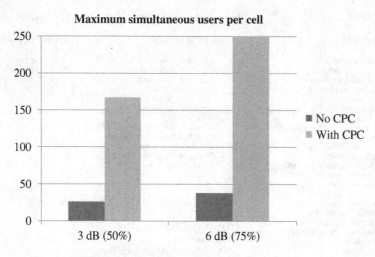

Figure 4.22   Measurements with 2-ms TTI

number of simultaneous users per cell without CPC is only 25–50 even if the users are inactive. That means the uplink is fully congested by the control channels even if the cell throughput is zero. DTX improves the uplink capacity by a factor of 4 to 6 in the measurements. More than 200 simultaneous users can be supported per cell with DTX. The lower interference levels with DTX may allow extension of the inactivity timer for CPC users to keep the UE in the Cell_DCH state for a longer period. A longer inactivity timer could be beneficial from the end user performance point of view if the state transition can be avoided.

Next, we show how to turn these measurements into the estimated number of subscribers per cell depending on the application activity and packet size. The Release 6 case assumes that data is transmitted on HSUPA without CPC. Release 7 assumes DTX during the inactivity

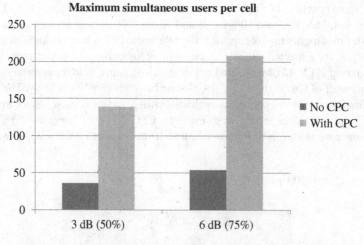

Figure 4.23   Measurements with 10-ms TTI

timer. Release 8 assumes that the data is carried by HS-RACH. We use the following further assumptions in the calculations:

– Maximum load factor 75% = 6 dB noise rise.
– Application packet arrival period 60 s.
– Application packet size 0.5–64 kB.
– HSUPA inactivity timer 4 s for Release 6 and Release 7.
– HS-RACH inactivity timer 40 ms for Release 8.
– Average HSUPA data rate 384 kbps.
– Average HS-RACH data rate 384 kbps.
– HSUPA E-DPDCH (data) vs. DPCCH (control) offset 9 dB. E-DPDCH Eb/N0 requirement 2 dB.
– DPCCH interference reduction with DTX -75%, that means 4 times more capacity.
– All terminals support Release 7 CPC and Release 8 HS-RACH, which is naturally optimistic in the short term.

The assumptions correspond to approximately 37 simultaneous users without CPC and 140 with CPC if they are all in Cell_DCH state, which is similar to the field measurements. Figure 4.24 shows the maximum number of subscribers per cell when the traffic model is considered. The maximum number of subscribers per cell with Release 6 is approximately 500. The number is not sensitive to the transmitted packet size because most of the interference is caused by the control channels, not by the data transmission. It is easy to understand that the network can get congested in mass events with tens of thousands of people each with smartphones. Assume 80 000 people in a large sports stadium. The operators would need a minimum 80 000/500 = 160 cells in the stadium to provide enough capacity. Since the

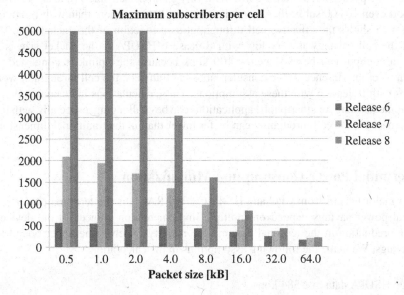

Figure 4.24 Maximum simultaneous subscribers per cell

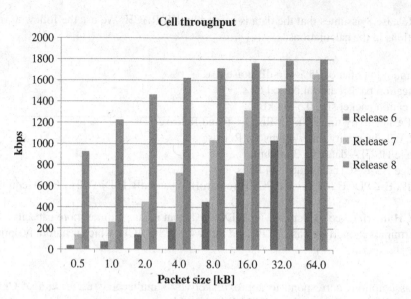

Figure 4.25   Effective uplink cell throughput

spectrum is limited, they would need multiple sectors which again would increase the inter-cell interference and further reduce the capacity. Release 7 CPC brings nearly four times more capacity because most of the interference comes from inactive users and DTX helps a lot. The CPC gain is still two times with a packet size of ten kB. The gains with Release 8 HS-RACH are even more massive, since the transmission can be stopped very quickly when the data transmission is over. The relative gain of HS-RACH compared to Release 7 CPC is more than two times if the packet size is below 5 kB. With very small packet sizes, Release 8 HS-RACH can support even 10 000 subscribers per cell, which is 20 times more than with Release 6.

Figure 4.25 shows the effective cell throughput as a function of the packet size. We can note that the cell capacity is very low with Release 6 HSUPA if the packet size is small. The cell throughput can be even below 100 kbps because the uplink is congested by the control channel interference. If we consider large file transfers, the cell throughput would be 1800 kbps with Release 6 with these assumptions. These calculations indicate that the uplink network dimensioning for smartphone applications can be challenging, especially with Release 6 devices, because the cell throughput can be far lower than in the traditional dimensioning.

4.8   Terminal Power Consumption Minimization

The other major benefit from CPC and HS-FACH/HS-RACH is the terminal power savings. The actual power savings depend on multiple implementation choices in the devices. We present the reduction in the terminal transmit RF activity which is the upper bound for the power savings. We make the following assumptions in the calculations

– Average HSUPA data rate 384 kbps.
– Transmit RF activity during HSPA setup 200 ms.

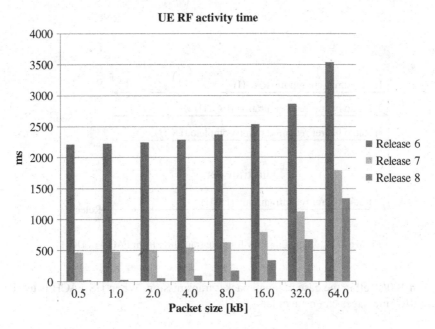

Figure 4.26 UE transmit RF activity time

- Transmit RF activity during HS-RACH setup 10 ms.
- HSPA inactivity timer 2 s.
- DTX activity factor with Release 7 CPC 13%.

Figure 4.26 shows the RF active time for the transmission of small packets. Most of the RF activity with Release 6 happens during the 2-s inactivity timer. Release 7 RF activity is reduced considerably with DTX during the inactive time. Release 8 is still substantially more efficient especially for small packet transmission. The gains in terms of power consumption are lower because there are many other components in the terminal taking power. Power saving measures are shown in Chapter 14.

4.9 Signaling Reduction

The number of RRC signaling messages can be reduced with HS-FACH and HS-RACH. Setting up and releasing the HSPA channel requires a minimum of four RRC messages over the air and two NBAP messages over Iub. In the case of HS-FACH and HS-RACH no RRC messages are required. The reduction in the signaling messages is generally beneficial from the network efficiency point of view. The signaling flow for the state transition to Cell_DCH is shown in Figure 4.27. All this signaling can be avoided when HS-FACH and HS-RACH are used for the data transmission.

Figure 4.28 shows the number of RRC messages per cell per second in a high loaded cell if every packet transmission requires HSPA channel allocation. The number of messages can be

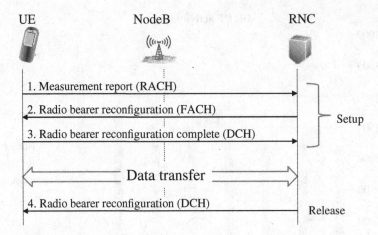

Figure 4.27 Signaling flow for state transition from FACH to DCH

more than 100/s/cell in the case of small packet transmission. When HS-FACH is used, most of these RRC messages can be avoided.

4.10 Latency Optimization

HS-FACH and HS-RACH also improve the end user experienced response times. HSDPA/HSUPA make the download time very short but the channel setup time still increases the total response time. Assuming a HSDPA data rate of 10 Mbps, even the average web page

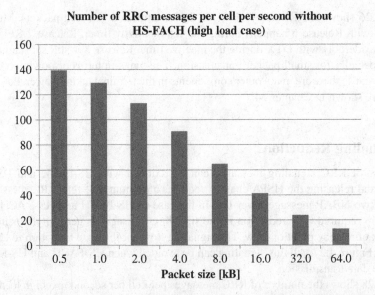

Figure 4.28 Number of RRC messages for HSPA channel allocation without HS-FACH

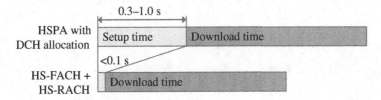

Figure 4.29   End user latency improvement with HS-FACH and HS-RACH

of 1–2 Mbytes can be downloaded in 1–2 s without considering any TCP impacts. The state transition from Cell_FACH or from Cell_PCH to Cell_DCH takes typically 0.5 s. In practice, the application layer delay can be up to 2 s due to the state transition. HS-FACH and HS-RACH allow access to high speed data without any state transition delay. The end user latency benefit is illustrated in Figure 4.29.

4.11   Summary

3GPP Releases 7 and 8 bring CPC and HS-FACH/HS-RACH functionalities that are important for the growing number of smartphone applications that tend to transmit small packets frequently. CPC includes uplink DTX, downlink DRX, and SCCH-less transmission. Uplink DTX reduces uplink interference up to a factor of four with small packet transmission. Uplink DTX together with downlink DRX helps in minimizing terminal power consumption. SCCH-less transmission optimizes downlink control channel usage and increases the maximum capacity. HS-FACH and HS-RACH further improve the efficiency of small packet transmission and the end user experienced latency. All these functionalities were introduced commercially to the networks and first devices by 2014. Further enhancements on top of HS-FACH and HS-RACH have been included in Release 11 specifications.

References

[1] 3GPP Technical Report 25.903 "Continuous connectivity for packet data users", v. 7.0.0, 2007.
[2] 3GPP Technical Specification 25.214 "Physical layer procedures (FDD)", v. 11.7.0, 2013.
[3] 3GPP Technical Specification 25.308 "High Speed Downlink Packet Access (HSDPA); Overall description; Stage 2", v.10.6.0, 2013.
[4] 3GPP Technical Specification 25.319 "Enhanced uplink; Overall description; Stage 2", v.11.7.0, 2013.
[5] 3GPP R1-061476 "TP on uplink 10 ms TTI VoIP capacity with UL DPCCH gating", RAN1 meeting #45, 2006.
[6] 3GPP R1-061576 "UE tx and rx parts power consumption improvement with DTX and DRX", RAN1 meeting #45, 2006.
[7] 3GPP R1-062249 "Further link simulation results for UL DPCCH gating, HARQ on", RAN1 meeting #46, 2006.
[8] 3GPP R1-062251 "HSDPA VoIP capacity", RAN1 meeting #46, 2006.
[9] 3GPP R1-063339 "Further analysis of HSDPA in CELL_FACH state", RAN1 meeting #47, 2006.

5

HSDPA Multiflow

Thomas Höhne, Karri Ranta-aho, Alexander Sayenko, and Antti Toskala

5.1 Introduction

This chapter presents the Release 11 HSDPA Multiflow, which was introduced to address and improve the cell edge performance of the HSDPA system. As described in Chapter 2, the Release 5 based HSDPA connection is provided from a single cell only, even though the active set may contain more cells. With HSDPA Multiflow, there is the possibility of obtaining additional diversity and boosting the cell edge data rate by receiving several data streams from neighboring cells. In the following, we first describe the Multiflow principles, before highlighting specific architectural and implementation aspects. The chapter is then concluded with a Multiflow performance evaluation.

5.2 Multiflow Overview

Multiflow is designed to improve the throughput of cell edge users and network-wide capacity by RRM load balancing. It belongs to the category of cell-aggregation features, a number of which were studied and presented in [1] as part of the study item that led to the Multiflow work item effort [2]. Multiflow is very similar to multicarrier HSDPA, with the generalization that cells do not need to have the same timing and can be on the same frequency. Multiflow also has a similarity with the LTE-Advanced dual connectivity, which at the point of writing is being studied as a feature for Release 12 [3]. A comparison to other multipoint and aggregation techniques is provided at the end of this chapter.

In a Multiflow transmission a UE, which is situated at the border between two cells and has those cells in its active set, may receive data independently and concurrently from several cells. The cells may belong to the same site, leading to intra-site Multiflow, or may belong to different sites, leading to inter-site Multiflow, as shown in Figure 5.1. From a UE perspective, gains are visible because twice as many transmission resources can be put at its disposal. On

HSPA+ Evolution to Release 12: Performance and Optimization, First Edition.
Edited by Harri Holma, Antti Toskala, and Pablo Tapia.
© 2014 John Wiley & Sons, Ltd. Published 2014 by John Wiley & Sons, Ltd.

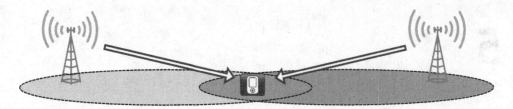

Figure 5.1   NodeBs schedule and transmit data independently to a UE at the cell edge

a system level, gains materialize when either of the cells has free resources that are put to use with Multiflow. This works best in low–medium load situations, and Multiflow gains are thus best understood as short-term load balancing gains between cells.

Unlike LTE Rel-11 CoMP feature, Multiflow does not require coordination among sites and allows for independent cell schedulers, enabling easy inter-site deployment and comparatively low implementation complexity in the network. In fact, no new hardware requirements arise on the network side. From the UE implementation perspective, the Multiflow functionality is based on existing HSDPA multicarrier architecture.

In summary, it is the following aspects of Multiflow that stand out and have led to a speedy adoption in specifications and implementation:

- operator-friendly network capacity improvements by better utilization of unused resources;
- cell-edge user throughput benefits (of up to 50%, as shown towards the end of this chapter);
- software-only network implementation without a requirement to coordinate or synchronize NodeBs;
- reliance on existing UE hardware building blocks.

5.2.1   Multiflow Principle

In intra-site Multiflow, RNC sends all data to one NodeB, which distributes packets between the cells, making the overall functionality almost identical to multicarrier HSDPA. In inter-site Multiflow the source data coming to the RNC is split at the level of RLC PDUs and distributed between two independent flows sent to different sites, as shown in Figure 5.2. NodeBs of both sites schedule and transmit the RLC PDUs independently of each other. In turn, the UE merges PDUs received from the two NodeBs and re-orders them into a single stream of PDUs with consecutive sequence numbers before delivering them to higher layers.

Multiflow has been designed to support transmission from only two different sites. The reason is that transmissions from three or more sites improve performance only marginally. Furthermore, more than two sites will result in having potentially more than two cells with different timing, as will be explained later in Section 5.6.3.

5.2.2   Multiflow Configurations

In multicarrier HSDPA, the NodeB transmits content to the UE on several carriers: up to eight carriers may be used for transmission to a UE as per 3GPP Release 11 functionality. Multiflow

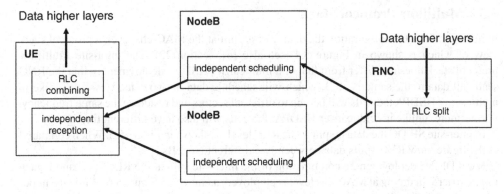

Figure 5.2 Data flow split, independent scheduling, and combining at the UE for inter-site Multiflow

configurations are limited to two cells on the same frequency and two NodeBs transmitting on at most two frequencies, leading to a total amount of aggregated cells of four. The naming convention for the different configurations is:

$$< \text{number of frequencies} > - < \text{number of cells}>.$$

Of all conceivable combinations three configurations have been agreed in Release 11:

- SF-DC single frequency – dual cell.
- DF-3C dual frequency – three cell.
- DF-4C dual frequency – four cell.

The SF-DC configuration has already been illustrated in Figure 5.1, and Figure 5.3 shows the example of DF-3C with the assisting secondary cell on the second frequency (an alternative DF-3C scenario is to configure the cell on the second frequency at the serving NodeB). The group of cells that do not belong to the serving NodeB are designated as "assisting."

It is worth noting that both DF-3C and DF-4C Multiflow configurations can be applied to non-adjacent and dual-band deployments. Referring to Figure 5.3, the cell on frequency 1 and on frequency 2 can be either adjacent in frequency or there can be a gap between them, if an operator does not have a contiguous spectrum. The same logic can be applied to dual-band, when the aforementioned cells can be in different bands.

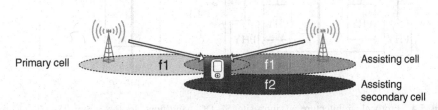

Figure 5.3 Multiflow terminology for the example of DF-3C

5.3 Multiflow Protocol Stack

In intra-site SF-DC configuration the data flow is split at the MAC-ehs level, a protocol stack view of which is shown in Figure 5.4 (see also [4], section 22). The intra-site Multiflow protocol stack does not differ from that of a multicarrier HSDPA deployment in that the RNC sends all data to the same NodeB, which will schedule data in two cells. However unlike in multicarrier HSDPA the cells will belong to different sectors and reside on the same frequency. Furthermore, unlike in multicarrier HSDPA, the cells will also have different timing.

In inter-site SF-DC the data is split at the RLC level, as shown in Figure 5.5. When compared to the figure, now RNC splits data between two different NodeBs.

For a DF-3C deployment, a combination of the inter-site split at the RLC level and legacy multicarrier scheduling at MAC-ehs level is employed, as shown in Figure 5.6. While the upper part of the figure looks the same as that of intra-site Multiflow in Figure 5.4, the difference is that two cells belong to the same sector and thus have the same timing.

In inter-site Multiflow, as shown in Figures 5.5 and 5.6, the data can be forwarded over the Iur interface using the frame protocol if the second NodeB is connected to a different RNC.

5.4 Multiflow Impacts on UE Architecture

A Multiflow UE must be able to receive, equalize, and decode data concurrently from both cells that are configured for Multiflow operation. First, the UE must demodulate the received signals from several cells on the same frequency, imposing a new requirement compared to earlier HSDPA receivers. Just as in multicarrier, the cells will have different channel responses, but unlike in multicarrier HSDPA the cells also will have different frame timing, and for inter-site potentially different frequency error because of Doppler and because of clock drift. The reception of two Multiflow cells on the same frequency but with different timing and different frequency error and so on could in principle be carried out with one receiver chain only.

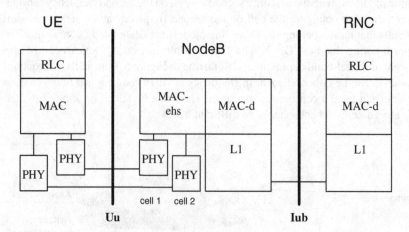

Figure 5.4 Protocol architecture of intra-site SF-DC Multiflow. In the figure, data flows from RNC to UE

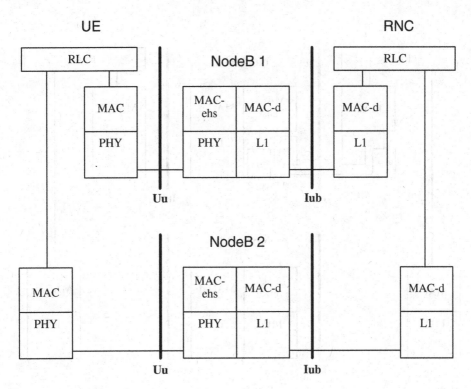

Figure 5.5   Protocol architecture of inter-site SF-DC Multiflow. The RNC splits the RLC PDUs into two MAC-d flows

However, in practice re-using multicarrier receiver chains for that purpose is easier, and in fact specifications have been written to mandate a multicarrier capability for a Multiflow UE. Figure 5.7 exemplifies a high-level physical layer block diagram for a dual-carrier dual-antenna UE receiver (a more extensive review of multicarrier receiver architectures is provided in Chapter 14). For Multiflow, the RF front-end of the second receiver chain of the multicarrier receiver blocks has been tuned to the carrier frequency of the neighboring cell. In the example, the neighboring cell has the same frequency, but of course a different scrambling code and a different frame timing.

The decoding of transmissions that arrive simultaneously with about equal receiver strength on the same carrier frequency make it necessary to employ at least an interference suppressing receiver, such as a type3i receiver, discussed in Chapter 8.2.

5.5   Uplink Feedback for Multiflow

5.5.1   HS-DPCCH Structure with Multiflow

As background, we recall that for multicarrier HSDPA the uplink feedback for all cells is carried on a single HS-DPCCH feedback channel, on the paired uplink frequency of the primary downlink carrier. The uplink feedback consists of the downlink CQI and the acknowledgements

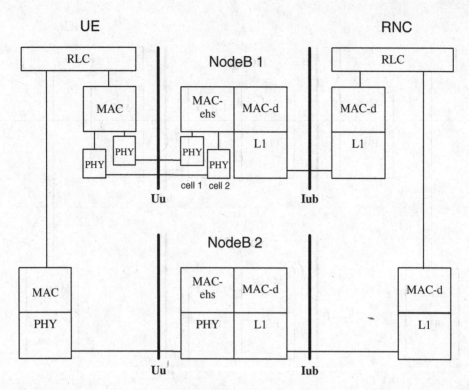

Figure 5.6 Protocol architecture of DF-3C Multiflow. The RNC splits the RLC PDUs into two MAC-d flows and one NodeB schedules data over two cells

(A/N) for downlink HARQ transmissions, and its format depends on the number of configured cells and whether MIMO is enabled. For instance, a four-carrier transmission with no MIMO could have a HS-DPCCH as shown in Figure 5.8.

In the chosen example, every TTI carries the HARQ acknowledgments (Ai) for all four cells, while the four CQI values (CQIi) are time multiplexed over two TTIs. We further note that in multicarrier HSDPA, except for the primary cell, the position of a secondary cell's feedback within the HS-DPCCH can be configured freely and is referenced with an index carried in the RRC signaling.

For Multiflow, the feedback structure is almost identical with the difference that the position of a HARQ acknowledgement and a CQI within the HS-DPCCH depends on which site (or

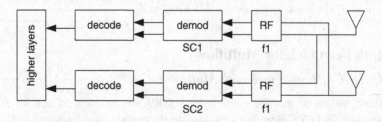

Figure 5.7 Typical UE receiver architecture for a Multiflow UE, supporting dual-carrier and SF-DC

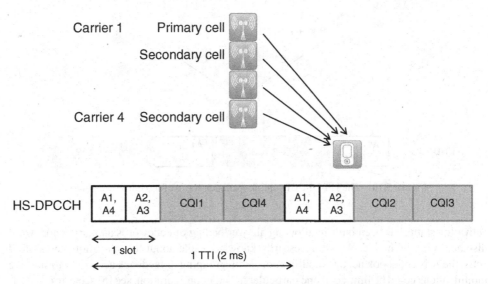

Figure 5.8 The HS-DPCCH uplink feedback format for 4C-HSDPA

cell group in 3GPP specifications) it belongs to. Thus, the position of the feedback for a cell within the HS-DPCCH is fixed and is governed by whether a cell belongs to the serving or the assisting NodeB and whether it is the primary serving/assisting cell or the secondary serving/assisting cell. As an example, the DF-4C non-MIMO Multiflow feedback format is shown in Figure 5.9.

With this design in DF-3C or DF-4C Multiflow where the feedback spans over two TTIs, both serving and assisting NodeB will have to read CQI information from only one TTI and decode the CQI only of cells belonging to the corresponding NodeB. Furthermore, dynamic carrier activation/deactivation becomes possible without the requirement of cumbersome inter-site coordination. A third advantage of the feedback grouping approach lies in the relative A/N codebook performance, compared to a design where a NodeB has less *a priori* knowledge of the possible values that it is about to receive: in some code designs the awareness of a cell's

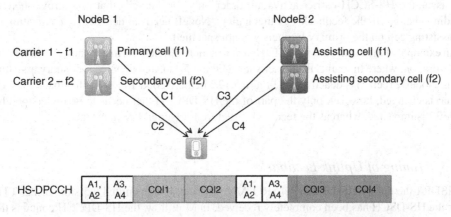

Figure 5.9 The HS-DPCCH uplink feedback format for Multiflow DF-4C

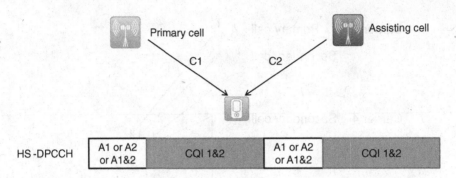

Figure 5.10   The HS-DPCCH uplink feedback format for Multiflow SF-DC

activation status has been built-in allowing an overloading of codewords and larger codeword distance. With both NodeBs not necessarily knowledgeable about the activation status of all cells, the A/N codebook needed modification. With grouping of feedback according to sites the modifications could be limited to one particular MIMO-configuration, see [5] section 4.7.3D.1.

For SF-DC the HS-DPCCH structure follows the dual-carrier format, as shown in Figure 5.10, allowing a direct reuse of the feedback design implemented for dual-carrier HSDPA.

A complete list of the configuration options and their feedback formats has been tabulated in [5] section 4.7.4.3.

5.5.2   Dynamic Carrier Activation

Multicarrier HSDPA allows for the activation and deactivation of already configured cells by means of HS-SCCH orders. This functionality was introduced to enable greater battery savings, as well as better uplink spectral efficiency, as the UE does not need to report CQIs for deactivated cells, and in dual-band receivers may also switch off one receive chain. Dynamic carrier (de-)activation can also be applied to Multiflow, with some restrictions. While in multi-carrier HSDPA the serving NodeB can deactivate any carrier, in inter-site Multiflow a NodeB may issue the HS-SCCH carrier activation order only for the cells that it controls, thereby avoiding changes to the feedback TTI that a given NodeB needs to decode. A deactivation of the assisting cell on the primary frequency is not specified.

An example of the change of uplink feedback to a more efficient format is shown for a DF-3C configuration, where first all cells are activated (Figure 5.11), and then the secondary assisting cell in NodeB2 (cell 3) is deactivated (Figure 5.12): in the second TTI of the feedback the CQI format is changed, however, only the part of the HS-DPCCH that needs to be understood by NodeB2 is impacted, whereas the feedback structure for NodeB1 remains the same.

5.5.3   Timing of Uplink Feedback

In HSDPA the uplink HS-DPCCH HARQ A/N feedback is transmitted 7.5 slots or 2.5 TTIs after the HS-DSCH has been completely received. In Multiflow, the HS-DPCCH contains the feedback of up to four HS-DSCH cells, which may belong to two different sites, which in

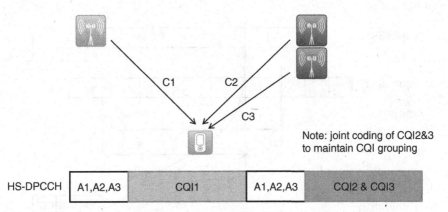

Figure 5.11 DF-3C feedback with all carriers activated. In this Multiflow configuration the CQIs of the second TTI are jointly coded, as they belong to the same NodeB

turn in general will be unsynchronized. Furthermore, even in the intra-site Multiflow, cells belonging to different sectors may have a different timing, as configured by the so-called T_cell parameter. An increase in the latency of the feedback transmission would result in a reduction of the throughput or would require larger numbers of HARQ processes, neither of which is desirable. Instead of increasing latency one could also speed up processing of the HARQ acknowledgments (and CQIs) for the cells that appear later. In the worst case, one cell lags the other by 1.5 slots, and hence a new requirement was introduced for the UE to process the *downlink* transmissions within 6 slots, instead of within 7.5 slots.

As the feedback for two different sites (or cell groups) is being bundled into one feedback message, one of the cells can be defined as leading, and is referred to as the *time reference* cell, see Figure 5.13. In case the timing difference between cells becomes larger than 1.5 slots, for instance because of clock drift, the network will reconfigure the other cell to take the role of time reference. To avoid a reconfiguration ping-pong effect when the timing difference is

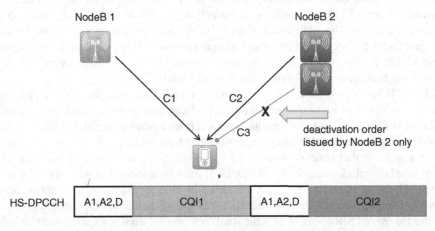

Figure 5.12 DF-3C feedback with one carrier deactivated

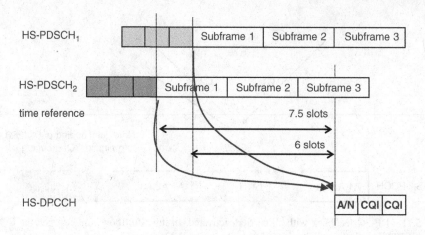

Figure 5.13   HS-DPCCH uplink timing for inter-site Multiflow

close to exactly 1.5 slots, an additional hysteresis of ± 20 chips was agreed, see Section 7.7.2 in [12]. It is worthwhile pointing out that the role of time reference cell is independent of whether a cell is assisting or not.

5.5.4 HS-DPCCH Power Levels

In the chosen design for relaying feedback in the uplink direction, both NodeBs receive and decode one and the same HS-DPCCH sent by the UE. This is possible because even in non-Multiflow HSDPA operation the UE's HS-DPCCH power is dimensioned such that it can be received by any cell in the active set, that is, also by inter-site neighbor cells. This safety margin is required as a UE may have already transitioned into a new cell's area, but still needs to maintain connectivity with the source cell before a serving cell change can be carried out. Nevertheless, in order to allow for a wider Multiflow area, which requires a better uplink control channel coverage, it was decided that the HS-DPCCH amplitude boosting factor β_{hs} be increased from a factor of 3 relative to the uplink DPCCH to a factor of close to 5, see section 4.2.1.2 in [6]. The additional boosting factor translates into increasing the maximal power gain factor from 9.5 to 14 dB. The additional boosting factor is specified as a separate Release 11 UE capability, which can be applied outside of the Multiflow operation. A UE implementing Multiflow must always indicate this capability.

HS-DPCCH boosting is especially useful in HetNet co-channel deployments (see Chapter 7), where a small cell of lower transmittal power is placed within the coverage area of a macro cell. At the border between the macro and small cell, at a point of equal downlink strength, the uplink signal strength at the macro and small cell NodeBs receivers will be very uneven, creating a problem that is referred to as the uplink/downlink imbalance. Thus, even at the point of equal downlink strength, the HS-DPCCH must be allocated more power than would be required if the small cell were to receive it alone, in order for the macro to be able to decode it. A way of helping the HS-DPCCH reception beyond a possible uplink imbalance of 14dB is to raise the overall SIR target level at the small cell beyond what is required there, to satisfy the needs at the macro.

The high HS-DPCCH power levels that may be needed for a larger Multiflow area can be reduced by using repetition and combining. For instance, the same CQI can be repeated over one or more consecutive HS-DPCCH sub-frames and the energy of a single CQI message is thus spread over multiple sub-frames reducing the peak power requirement in proportion to the amount of repetition. The downside of CQI repetition is that it limits the frequency of fresh CQI reports, but may nevertheless be useful. ACK/NACK repetition is also possible, but in inter-site Multiflow, where the main usefulness of repetition lies, it is not possible to coordinate the transmissions of the two NodeBs making dynamic repetition difficult. Furthermore, ACK/NACK repetition cuts the achievable data rates as the UE cannot be scheduled with data in consecutive sub-frames.

5.6   RLC Impact

5.6.1   RLC Timer_Reordering

One of the challenges of inter-site Multiflow transmission is that data is split by RNC and sent over two different NodeBs, who may experience different delays, link speeds, loads, and so on. Differing channel and load conditions may lead quite easily to a situation where a UE receives a set of packets without having received a set of preceding packets from another NodeB, which can be either under transmission or can be still in the NodeB buffer waiting for their scheduling turn. This situation, where RLC PDUs may arrive to a UE not in the same order as they were sent, creates a so-called skew problem. In legacy functionality the RLC layer perceives absence of a particular RLC PDU as indication that its transmission has failed over the radio link. This is a safe assumption, because the underlying HARQ layer has its own sequence numbering and re-ordering and in the non-Multiflow case it will ensure in-sequence delivery. Thus, if a certain RLC PDU is missing at the RLC layer, then a NACK is triggered and sent in a corresponding RLC STATUS PDU message. In inter-site Multiflow, however, the absence of an RLC PDU can be either due to a true transmission failure or due to the fact that it is still waiting in a buffer inside another NodeB. If a UE configured with inter-site Multiflow reported any missing RLC PDU as NACK back to the RNC, as in the legacy case, performance would decline because the RNC would re-transmit PDUs that are still in the NodeB buffer, and the same PDUs might be transmitted over the Uu interface twice. This behavior is illustrated in an example in Figure 5.14.

In the example, NodeB1 has a buffering delay $t5-t1$, but NodeB2 is able to transmit earlier, leading to skewed reception at the UE. While RLC PDU 0 is genuinely lost, and PDUs 2 and 3 are still in the NodeB1 buffer, the UE counts PDUs 0, 2, and 3 as missing once it receives PDU 4. Even though there is Timer_Status_Prohibit that prevents a UE from sending RLC STATUS messages too frequently, there is always a chance that upon its expiry a UE will have gaps in RLC sequence numbers. Referring to the example above, the RLC STAUS message will contain NACKs for PDUs 0, 2, and 3, which in turn will trigger the RNC to resend them. Hence PDUs 2 and 3 are transmitted twice over the air interface.

To address the problem that a UE cannot know the exact reason for a missing PDU, and sending back NACKs for all the missing PDUs is not optimal, a new RLC timer "Timer_Reordering" was adopted. Its purpose is to achieve a tradeoff between how long a UE waits for a missing PDU versus the time when the missing PDU is finally reported as NACK in the status report.

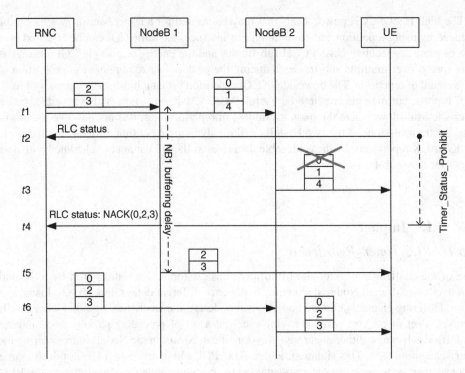

Figure 5.14   RLC ACK/NACK with the reception of out-of-sequence PDUs. PDUs 2 and 3 are transmitted to the UE twice

Figure 5.15 illustrates how, for the above example, the sending of the NACK is delayed, and unnecessary retransmission of PDUs 2 and 3 is avoided. Once a UE receives PDU 4 and detects that PDUs 0, 2, and 3 are missing, it starts the Timer_Reordering timer for those PDU thus delaying the transmission of NACKs. In the example, PDUs 2 and 3 arrive before the expiry of that timer, and the NACK will include only PDU 0.

Timer_Reordering is mandatory for a UE and is optional for the network in the sense that the network provides the exact value for this timer and, if it is omitted, then a UE does not activate the timer at all. Thus the network may deactivate this timer at the UE if it has other mechanisms to overcome the problem when NACK is received for some PDUs that were still under transmission. The value for Timer_Reordering may vary from 50 ms to 1 s. The exact value is chosen by the network based on its anticipation of how fast the flow control can work and the NodeB scheduling strategy that has a direct impact on the delay that packets may experience in the NodeB buffer.

5.6.2   RLC Reset

Along with a new RLC timer described in the previous section, another RLC enhancement introduced for Multiflow is a different handling of the RLC STATUS PDU. Before the introduction of the Multiflow operation, whenever a UE received the RLC STATUS PDU with a sequence number outside of the RLC transmission window, it initiated immediately the RLC

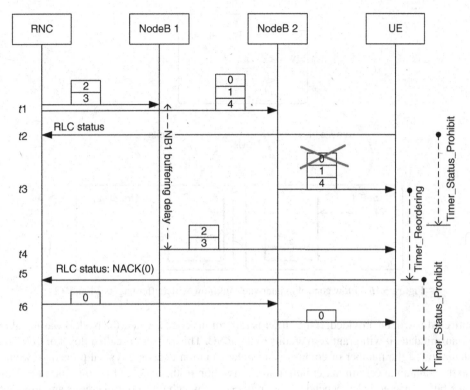

Figure 5.15 RLC ACK/NACK with the reception of out-of-sequence PDUs. Timer_Reordering eliminates unnecessary re-transmissions

RESET procedure. This functionality was needed to achieve robust behavior: if a UE received an out-of-sequence status PDU it might be a sign of severe RLC state machine mismatches. However, with Multiflow the RNC may reply to the UE uplink traffic by sending the STATUS PDU message either over the serving, or assisting, or even both the serving and assisting links. Such RNC behavior may easily lead to a situation when RLC STATUS PDU messages arrive at the UE in a different order, thus triggering unnecessarily the RLC RESET operation. It was decided that the RNC should have the possibility of deciding flexibly how to send RLC feedbacks and, as a result, it was agreed that for the inter-site Multiflow operation, no RLC RESET operation should be initiated under these circumstances. At the same time, a UE still initiates the RLC RESET procedure in case of intra-site Multiflow to be functionally close to the multicarrier HSDPA.

5.7 Iub/Iur Enhancements

5.7.1 Flow Control

When the RNC provides data to a NodeB, it does so based on "credits" that the NodeB has given to the RNC. The NodeB's credits are based on its buffer status. If the NodeB buffer is empty,

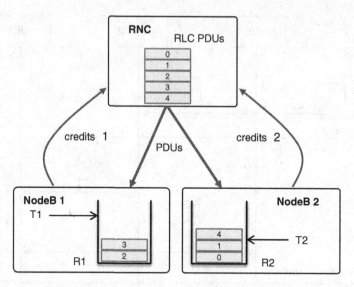

Figure 5.16   Flow control in inter-site Multiflow with buffer targets T1 and T2

more credits will be provided. If the buffer is full, for instance because the NodeB was not able to transmit data, it will grant zero credits to the RNC. This process is called flow control. The calculation of the number of credits is an implementation choice; a typical procedure would seek to maintain a certain target buffer size in relation to the average transmission rate. This target buffer size will also depend on the frequency at which flow control credits are provided to the RNC and the speed at which the RNC can hence react to the NodeB having already emptied its buffers. A faster flow control loop allows for smaller NodeB buffers and delays, a longer flow control period requires larger buffers to avoid a situation where a buffer underrun occurs and transmission opportunities are missed.

In inter-site Multiflow – as there are two NodeB buffers to fill – the RNC maintains two independent flow control loops, as shown in Figure 5.16. The total burst rate and the user experience achieved at the UE will be governed not only by the sum of the individual link speeds, but also by when data from each link arrives to the UE. We note that the overall duration of a data burst is smallest when both NodeBs finish their transmissions at the same time. Thus, the RNC should run an efficient flow control mechanism to ensure an even data split in the sense that one cell does not transmit for longer than the other, which is important in order to achieve maximal throughput gains for the UE.

5.7.2   Multiflow Extensions

As mentioned in the previous subsection, an efficient flow control running in RNC should aim at ensuring that NodeB buffers will be emptying at the same moment thus avoiding skew as much as possible, skew for which the UE RLC Timer_Reordering would hold on sending back the RLC STATUS PDUs. However, changing load conditions in the serving and the neighboring cells may lead to the situation when data in the NodeB buffer will stall for a noticeable moment of time. Since Multiflow allows for sending data over two paths to a UE, one can consider several options on how to tackle the stalled data and potentially resend it over another link. All these considerations have led to the introduction of several new mechanisms for the Iub/Iur interface: a possibility for the RNC to discard data in the NodeB buffer, and a possibility to let the RNC know that certain PDUs were dropped from the NodeB buffer.

Both new mechanisms rely upon a new sequence number field inside the HS-DSCH DATA FRAME that carries data from the RNC towards the NodeB. If RNC wants to remove data from the NodeB buffer, it uses the aforementioned sequence number to indicate which packets must be dropped. Similarly, a new control frame HS-DSCH PDU DROP INDICATION was introduced to signal back to the RNC which PDUs were dropped from the NodeB buffer. It is worth noting that while HS-DSCH DROP INDICATION allows indication that an individual RLC PDU was dropped by NodeB, the RNC cannot ask NodeB to remove an individual RLC PDU. All the PDUs, which were previously sent in the HS-DSCH DATA FRAME by the RNC, would be discarded from the NodeB buffer.

5.8   Multiflow Combined with Other Features

5.8.1   Downlink MIMO

Multiflow data transmission can be combined with downlink 2×2 MIMO in single-stream and dual-stream modes. MIMO functional behavior with Multiflow remains unchanged and only HS-DPCCH formats have been modified as explained in Section 5.5.1.

It should be noted that the combination of Multiflow and dual-stream downlink MIMO will result in four MIMO streams, which requires a UE to have four Rx antennas to allow for a rank-four channel reception. At the same time, combination of single-stream MIMO and Multiflow can still be achieved with two Rx antennas. These considerations led to a design where a UE can signal independently whether it supports MIMO with Multiflow or not, and in the case that it does, whether it is limited only to the single-stream MIMO or both single-stream and dual-stream modes.

The downlink 4×4 MIMO HSDPA transmission introduced in Release 11 is not supported in combination with Multiflow.

5.8.2   Uplink Closed-Loop Transmit Diversity and Uplink MIMO

Along with the Multiflow feature aiming at improving the downlink performance, 3GPP Release 11 has introduced Closed-Loop Transmit Diversity (CLTD), which improves system performance in the uplink direction by allowing NodeB to steer the UE uplink transmission beam in such a way that the received power is maximized. It facilitates improving the cell edge performance as a UE can transmit at higher data rates, where it otherwise would be power limited. Of course, neither Multiflow is dependent on uplink CLTD, nor does uplink CLTD require a UE to support Multiflow. However, since both features improve the cell edge performance, a combination of these features was discussed and agreed in 3GPP. Thus, the network can configure them simultaneously if a UE supports these features.

Uplink MIMO is another Release 11 feature that was introduced to boost the uplink peak rates (see Chapter 3). Even though uplink MIMO is not likely to be used at the cell edge, there is a scenario when Multiflow is configured between two cells belonging to the same site. In this case, a UE can be close to the NodeB, and thus Multiflow can be also combined with uplink MIMO.

5.8.3 DTX/DRX

3GPP Release 7 introduced a set of features under the Continuous Packet Connectivity (CPC) umbrella, where the UE DTX and DRX were specifically aimed at reducing the UE power consumption and uplink capacity consumption. For the sake of technical clarity, we recall that the main CPC features are downlink DRX and uplink DTX. The uplink DTX allows a UE to avoid transmitting continuously uplink control and pilot channels, if it does not have any data or HS-DPCCH feedback to send. Uplink DTX can be configured alone without downlink DRX. In turn, the downlink DRX allows a UE not to listen to the downlink control channels, whereupon the Uplink DTX must be also configured because a UE cannot transmit without listening to the uplink power control commands in the downlink. The CPC feature is explained in more detail also in Chapter 4.2.

In the case of intra-site Multiflow, both the uplink DTX and downlink DRX can function in exactly the same way as for the normal multicarrier HSDPA with no changes or limitation in the functionality. For inter-site Multiflow, even though there are two independent NodeBs with different state machines, uplink DTX can work as before, because there still is only one uplink even if the downlink has multiple data flows. The downlink DRX operation with Multiflow may seem to be challenging due to the fact that a UE enters the DRX state based on the downlink HS-SCCH channel activity of one NodeB, which is not visible to the other NodeB configured for Multiflow. More specifically, if one NodeB schedules the UE on the HS-SCCH channel, the UE will be active for Inactivity_Threshold_for_UE_DRX_cycle and will be monitoring both cells. However the second NodeB is not aware of the activity of the first NodeB and therefore will not be able to use these periods to schedule the UE, as it still assumes that the UE is inactive until the next scheduling opportunity defined by the DRX pattern. However, this case does not create any problems, because even if in the example the second NodeB assumes that the UE is in downlink DRX, even though it is not, the NodeB will just transmit on the HS-SCCH to the UE at the next scheduling opportunity. The most important aspect in this scenario is the fact that the UE will be able to have idle opportunities for battery saving purposes and there will be no risk of data loss. If neither of the NodeBs transmits in the DL direction and the UE is entering its DRX cycle, there still will be no state mismatch issues because each site will follow DRX rules independently and will not transmit data to the UE until the next scheduling opportunity according to each respective DRX cycle.

From the UE point of view, there is a single DRX state machine common across the configured Multiflow cells, as in the legacy behavior. As an example, referring to Figure 5.17, NodeB2 may stop scheduling data at $t0$. Once the UE DRX inactivity timeout expires, it will enter the DRX mode at $t1$, thus listening to the downlink HS-SCCH channel only at particular moments of time. When NodeB1 wants to send downlink data, it has to wait for the next UE scheduling opportunity which happens in the example at $t2$. Note that NodeB2 is not aware of the fact that the UE has quit DRX. However, if NodeB2 wants to schedule some data, it just has to wait for the moment of time, $t3$, where the UE would listen for the downlink channel as if it were in the DRX cycle. The delay caused by this fact can be bounded by the configured DRX cycle length.

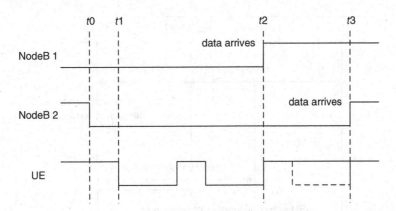

Figure 5.17 DRX with inter-site Multiflow with two independent DRX state machines in the NodeBs

5.9 Setting Up Multiflow

As mentioned earlier, from a specification perspective Multiflow can be understood as an extension of multicarrier HSDPA. Thus, setting up a Multiflow connection is similar to configuring a multicarrier radio bearer (RB): in addition to the typical RB information of carrier frequency, and other cell configurations, a few new information elements were added to the configuration.

In a typical sequence of events a UE would first carry out measurements and send reports on the strength of neighboring cells to the RNC. The RNC can – based on the measurement report and load information of the neighboring cells that it has available – identify which UEs would benefit from the Multiflow transmission. From there on, the RNC would proceed to perform radio link (re-)configurations of the involved NodeBs, as well as the UE. In particular the RNC will configure the NodeB with a "RL Addition" or "RL Reconfiguration" (or "RL Reconfiguration Prepare") message, which carries the relevant parameters. The UE will be configured by the RNC as part of an active set update or as part of a separate RB (re-)configuration. A schematic overview of the sequence of steps is shown in Figure 5.18, a more detailed view on RL and RB reconfiguration signaling flows applicable also to Multiflow is shown for instance in [4] Figure 9.5-1.

The new parameters required to configure a UE are contained within the new information element "Multiflow configuration info" (see section 10.3.6.31a in [7]). Those additional Multiflow specific parameters are:

- Indication of the "cell group," which is effectively an indication of whether the cell belongs to the serving or the assisting NodeB.
- Indication of which cell is a downlink time reference for the bundling of uplink feedback.

In inter-site Multiflow, the configuration sent by the RNC to a NodeB is not visible to the other NodeB. However, both NodeBs require some information, and the configuration messages were thus enhanced with following parameters (see section 9.2.2.170 in [8]):

- Indication on whether NodeB is a serving or assisting.
- Total number of cells of the overall Multiflow configuration.

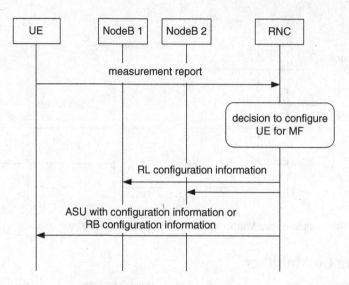

Figure 5.18   Configuring Multiflow

- Indication of which cell is the downlink time reference for the bundling of uplink feedback. For the non-time reference cell, RNC also indicates the timing offset.
- Indication of whether at least one cell in the other NodeB is configured in MIMO mode. Once at least one cell in any NodeB is configured with MIMO, the whole HS-DPCCH format changes.

5.10   Robustness

5.10.1   Robustness for RRC Signaling

Multiflow data transmission can be applied to RLC acknowledged mode (AM) and RLC unacknowledged mode (UM). When illustrating the radio bearer to transport channel mapping for inter-site Multiflow in Figure 5.19 it becomes apparent that Multiflow can be applied not only to the user RBs, but also to signaling RBs (SRBs), which are mapped to either RLC-UM or RLC-AM (for a list of all SRBs see [7] section 6.3). The advantage of doing so lies in increasing the robustness of, for example, a serving cell change: a message for RB reconfiguration can be relayed not only over the one link, but also over the other if a first RLC transmission failed. For the highest robustness the reconfiguration message could be relayed over both links from the beginning. Thus, a reconfiguration message would not reach the UE only if both configured NodeBs have lost connectivity with the UE.

5.10.2   Radio Link Failure

In the UE a radio link failure (RLF) is declared if synchronization to the serving cell is lost and the power control command carrying the F-DPCH signal level from the serving HS-DSCH cell falls below a fixed threshold for a given time duration, see [9] section 4.3.1.2. RLF leads to

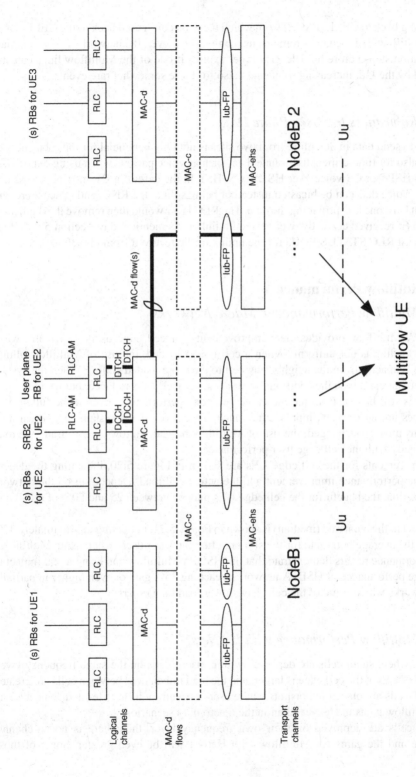

Figure 5.19 RB and SRB mapping onto logical and transport channels in Multiflow

the UE falling back to CELL_FACH, from which it can be configured back to the CELL_DCH state. In Multiflow RLF behavior remains unchanged. However, the network is able to initiate reconfiguration steps before the UE RLF timer expires if one of the Multiflow links can still be received by the UE, increasing robustness also for these somewhat rare events.

5.10.3   Robustness for User Plane Data

The ability to send data in downlink from two sites is not only beneficial for the robustness of SRBs, but also for time-critical user plane data, such as voice packets when using either VoIP mapped on HSPA or CS voice over HSPA (CSoHS). In the case of a UE moving across cell borders, the voice data can be bicasted instead of being split at the RNC, and hence there will be always at least one link providing the data. The RLC layer would then remove the duplicated RLC PDUs (if received correctly over two radio links). As mentioned in Section 5.6.2, RNC can also bicast RLC STATUS PDUs for the uplink traffic received from the UE.

5.11   Multiflow Performance

5.11.1   Multiflow Performance in Macro Networks

The HSDPA Multiflow provides clear improvements for cell edge users, especially when the system loading is not uniform. When a cell is less loaded, the use of Multiflow allows "borrowing" capacity from the neighboring cell to the more loaded cell(s). When the system load starts to increase, then the additional system capacity effect is reduced, but for individual users there is still benefit from the increased cell edge performance, even if the full system capacity does not necessarily improve that much. Thus, even if the system capacity at high load remains mostly unchanged, the use of Multiflow basically allows one to trade the total system capacity with the cell edge user performance.

The improvements for the cell edge UEs are shown in Figure 5.20, indicating the dependency of the performance improvement and the actual cell load. Depending on the network parameterization, the benefit for the cell edge UEs may be between 25 and 50% of all UEs in the network.

The impact to the burst rate (median) is shown in Figure 5.21, indicating approximately 35% increase to the average burst rate for the soft or softer handover users when using Multiflow.

The performance results demonstrate that the HSDPA Multiflow can bring a real impact to the cell edge performance of HSDPA networks, reaching 35% gain or even higher in partially loaded networks where some of the cells have lower load than others.

5.11.2   Multiflow Performance with HetNets

In HetNets, where small cells are deployed within macro cells on the same frequency layer, the amount of UEs at the cell edge is larger, and the load imbalances between cells are greater, as the small cells are placed in hotspots where user concentrations are elevated. This leads to higher Multiflow gains in HetNets than in the macro only scenario.

If small cells are deployed on their own frequency layer, then there is no co-channel interference and the gains for Multiflow with HetNet will be even greater. Some of those

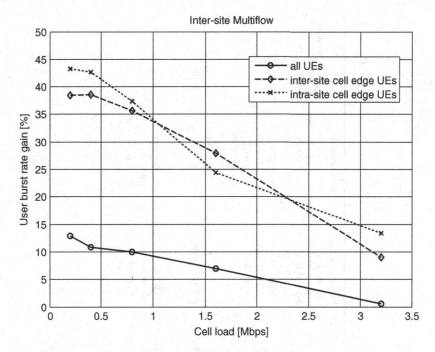

Figure 5.20 Performance of Multiflow

scenarios require, however, that the NodeBs are able to receive the uplink of the UE on a frequency which is not paired to their downlink carrier. Even more importantly, it remains to be seen whether operators will reserve a dedicated carrier for small cells only, rather than configure macro stations as multicarrier cells.

A more detailed account of gains in Multiflow with HetNets is provided in Chapter 7.

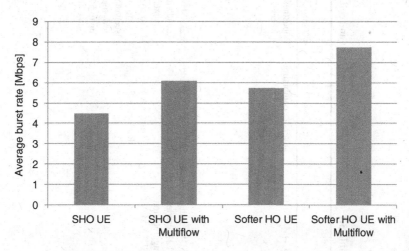

Figure 5.21 Median burst rate improvement with Multiflow

Table 5.1 Overview of downlink multipoint and cell aggregation techniques

	Gain principle	Range of gain	Data flow	Inter-site network operation	Other network requirements	3GPP Release
WCDMA SHO	Diversity combining	Robustness diversity benefit	Same data	Symbol level synchronized transmission from two cells on individual PSC	Asynchronous network, SHO user-specific transmission however synchronized. Slightly increased backhaul traffic	R99
HSDPA multicarrier, LTE carrier aggregation	Increased bandwidth, load balancing	Scales with bandwidth, see Chapter 3	Independent data and schedulers	Not applicable	Same cell timing	Rel8–11
HSDPA Multiflow	Cell aggregation, load balancing	Macro co-channel (same bandwidth) scenario: low–medium load: up to 50% for cell edge, 10% average gain	Independent data and schedulers	No NodeB–NodeB coordination requirement, data flow split in the RNC	Asynchronous network, no increase in backhaul traffic	Rel11

	Technique	Gains	Data/scheduling	Requirements	Network/other	Release
LTE CoMP	Dynamic multisite coordination and MIMO (dynamic point selection, coordinated beamforming), up to three involved sites	Depending on mode and scenario, also for high loads: up to 30% for cell edge [10]	Depending on mode independent data, but coordinated transmission	Very demanding inter-site coordination requirements, however intra-site CoMP already provides most of the gains	Synchronized network, joint transmission mode requires centralized baseband	Rel11
LTE feICIC	Interference coordination	HetNet co-channel scenario: up to 100% cell edge gains	Independent data but coordinated scheduling	HetNet feature: Coordination of transmit resources	Synchronous network	Rel11
LTE dual connectivity	Cell aggregation, load balancing	HetNet inter-frequency scenario: low–medium load: up to 50% cell edge, 20% average gain with realistic backhaul	Independent data and schedulers	HetNet feature: data flow is split in RLC layer of macro eNB and forwarded to low power eNB. Independent schedulers	Increase in backhaul or X2 traffic because of data forwarding	Rel12 planned

5.12   Multiflow and Other Multipoint Transmission Techniques

The feature most similar to Multiflow is, as mentioned earlier, LTE dual connectivity, which is planned for 3GPP Release 12. The similarity comes with the usage of independent schedulers and a data flow split across two sites, and that transmissions can occur on different carriers. The LTE Release 11 interference coordination scheme, feICIC, is benefitting specifically cell edge users just as Multiflow does.

LTE CoMP – coordinated multipoint transmission/reception – has some similarity with Multiflow, in that several nodes may transmit to one receiver. LTE CoMP comprises a number of schemes, ranging from dynamic coordinated transmission point selection to coordinated joint transmission from up to three sites. The gains for CoMP materialize from exploiting the large diversity of the channel as well as interference coordination. One of the prerequisites for LTE CoMP is a perfect backhaul that limits its practical use. Also the amount and the quality of the feedback available from the UE diminishes the gains that can be achieved in reality.

In a comparison with other multipoint and cell aggregation techniques, HSDPA Multiflow is positioned as a very low complexity technique without additional spectrum requirements, providing gains for low–medium loads. LTE CoMP, on the other hand, takes the place of maximal complexity with benefits also for high loads. An overview of the techniques and their requirements and gains is shown in Table 5.1.

5.13   Conclusions

This chapter has introduced HSDPA Multiflow protocol and physical layer operation principles as well as demonstrated the achievable capacity benefits. As the impacts to the UE and network implementation are reasonable and resulting performance benefits accrue especially to the cell-edge users, it is foreseen that HSDPA Multiflow will be an attractive feature to be adopted in the marketplace.

As use of HSDPA Multiflow is fully backwards compatible, that is, it can be operated in a network with both Release 11 UEs and older ones, it can be rolled out gradually to the network to boost the cell edge performance.

The LTE-Advanced carrier aggregation [11] has some similarity with the HSDPA Multiflow, only that the Release 10 carrier aggregation is done as intra-site aggregation and only as part of the Release 12 small cell evolution studies [3] the inter-site operation is expected to be introduced as part of the dual connectivity in Release 12 specification, due by the end of 2014.

References

[1] 3GPP Technical Report, TR 25.872, "High Speed Packet Access (HSDPA) multipoint transmission", Release 11, 2011.
[2] 3GPP Tdoc, RP-111375, WI description "HSDPA Multiflow data transmission", 2011.
[3] 3GPP Technical Report, TR 36.842 "Study on small cell enhancements for E-UTRA and E-UTRAN - Higher-layer aspects", Release 12, December 2013.
[4] 3GPP Technical Specification, TS 25.308, "HSDPA overall description", Release 11. 11, March 2013.
[5] 3GPP Technical Specification, TS 25.212, "Multiplexing and channel coding (FDD)", Release 11, March 2013.
[6] 3GPP Technical Specification, TS 25.213, "Spreading and Modulation (FDD), Release 11, March 2013.
[7] 3GPP Technical Specification, TS 25.331, "Radio resource control protocol specification", Release 11, March 2013.

[8]   3GPP Technical Specification, TS 25.433, "UTRAN Iub interface Node B Application Part (NBAP) signalling", Release 11, September 2013.

[9]   3GPP Technical Specification, TS 25.214, "Physical layer procedures (FDD)", Release 11, March 2013.

[10] 3GPP Technical Report, TR 36.819, "Coordinated multi-point operation for LTE physical layer aspects", Release 11, September 2013.

[11] Holma, H. and Toskala, A. (2012) *LTE-Advanced*, John Wiley & Sons, Ltd, Chichester.

[12] 3GPP Technical Specification, TS 25.211, "Physical channels and mapping of transport channels onto physical channels (FDD)", Release 11, June 2013.

6

Voice Evolution

Harri Holma and Karri Ranta-aho

6.1 Introduction

Circuit Switched (CS) voice was part of the WCDMA offering from Day 1 together with the Packet Switched (PS) data service. WCDMA voice uses Adaptive Multirate (AMR) codec with a data rate of 12.2 kbps. In many developed countries more than 50% of mobile voice minutes are now carried by WCDMA CS voice. The voice is still important from the operator revenue point of view even if data traffic is far higher than voice from the volume point of view. The voice service keeps improving further: AMR wideband (AMR-WB) offers High Definition (HD) quality, lower rate AMR improves voice capacity, voice-over HSPA provides both VoIP and CS voice capability, CS fallback from LTE to WCDMA CS voice enables voice services for LTE smartphones, and Single Radio Voice Call Continuity (SRVCC) brings handover from VoIP to CS voice. These evolution steps are summarized in Figure 6.1 and described in more detail in this chapter.

6.2 Voice Quality with AMR Wideband

AMR Wideband (AMR-WB) enhances voice quality by a higher voice sampling rate in voice encoding. AMR Narrowband (AMR-NB) uses an 8 kHz sampling rate which enables an audio bandwidth of 300–3400 Hz, similar to traditional landline phones. AMR-WB uses 16 kHz which increases the audio bandwidth to 50–7000 Hz. Voice sounds substantially better when the low and high frequencies are reproduced. The audio bandwidth is illustrated in Figure 6.2.

AMR-WB codec can use data rates between 6.6 kbps and 23.85 kbps depending on the compression level. The higher data rate gives slightly better voice quality while the low data rates provide higher voice capacity. The typical data rate is 12.65 kbps, which implies that AMR-WB does not consume more capacity than AMR-NB: wideband audio with narrowband radio. AMR-WB also supports voice activity detection and discontinuous transmission to minimize the average data rate. All AMR-WB data rates and the subset with Spreading Factor

HSPA+ Evolution to Release 12: Performance and Optimization, First Edition.
Edited by Harri Holma, Antti Toskala, and Pablo Tapia.
© 2014 John Wiley & Sons, Ltd. Published 2014 by John Wiley & Sons, Ltd.

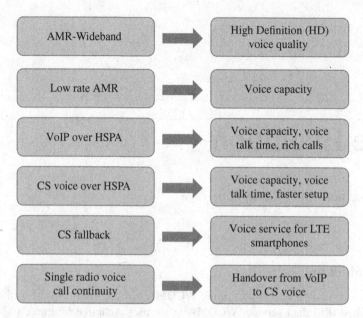

Figure 6.1   Voice enhancement options in WCDMA/HSPA

(SF) 128 are shown in Figure 6.3. If we use higher than 12.65 kbps, the spreading factor needs to be 64, which means double spreading code consumption compared to SF 128.

Voice quality with AMR-NB and AMR-WB with different data rates is shown in Figure 6.4. The quality is shown as the Mean Opinion Score where a high number indicates better voice quality. Comparing narrowband and wideband codecs with the same MOS scale is controversial, but the aim is to illustrate the major benefit of the wideband codec from the customer perspective. AMR-WB can provide MOS of 4.0 while AMR-NB has a maximum MOS of 3.2. AMR-WB with 8.85 kbps offers higher MOS than AMR-NB 12.2 kbps. AMR-WB can provide both quality and capacity benefits. AMR-WB brings very good quality already with a data rate of 12.65 kbps, which means no additional radio capacity utilized compared to the typical AMR-NB 12.2 kbps.

The better voice quality with AMR-WB requires that both parties have an AMR-WB-capable phone. Therefore, the higher quality will only come when AMR-WB terminal penetration increases. AMR-WB terminal penetration is increasing rapidly since most of the new smartphones have AMR-WB capability.

The AMR adaptation algorithms and benefits are described in more detail in [1].

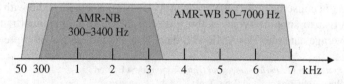

Figure 6.2   AMR-NB and AMR-WB audio bandwidth

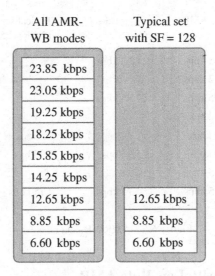

Figure 6.3 AMR-WB data rates

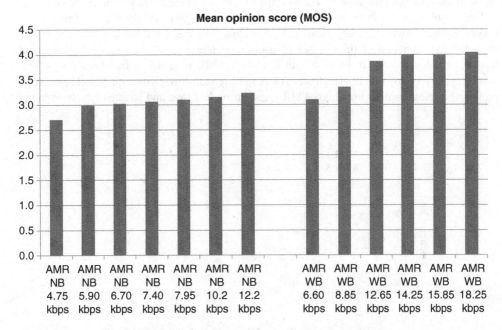

Figure 6.4 Voice quality with AMR-NB and AMR-WB

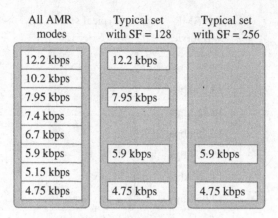

Figure 6.5   AMR-NB data rates

6.3   Voice Capacity with Low Rate AMR

The AMR data rate can be modified by the radio network according to the capacity requirements. AMR-NB data rate adaptation is supported by all terminals. The typical data rate options are shown in Figure 6.5. There is a total of eight modes defined while four modes are used with SF 128 and two modes with SF 256. The number of available spreading codes is equal to the spreading factor and SF 256 helps to increase the spreading code capacity.

The uplink capacity benefit in the measurements with lower AMR rate is illustrated in Figure 6.6 with AMR 12.2 kbps and AMR 5.9 kbps. The fractional loading of a single cell is shown as a function of the number of users. For example, with 30 users the fractional loading is reduced from 20 to 13% with the lower AMR rate, which translates into +54% voice capacity improvement. The lower AMR rates lead to lower voice quality. Therefore, a preferred implementation uses high AMR rates during low load and adaptively selects lower

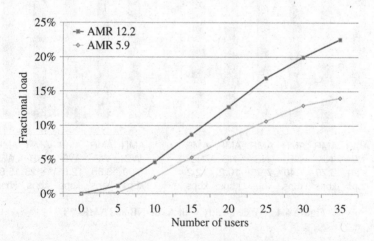

Figure 6.6   Measured uplink fractional loading with AMR12.2 and 5.9 kbps

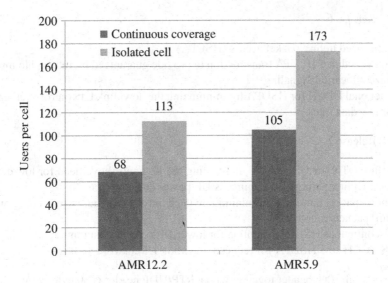

Figure 6.7 Estimated capacity with 75% uplink loading

AMR rates when the cell is congested. Similar adaptation and capacity benefit can be applied for AMR-WB as well.

The measurements from Figure 6.7 can be used to estimate the maximum cell capacity. Figure 6.7 shows the estimated capacity per cell both in an isolated cell and in the case of continuous coverage by assuming an inter-cell interference ratio of 0.65. The capacity of AMR5.9 can be more than 100 simultaneous users. The practical capacity may be lower due to simultaneous data and signaling traffic.

6.4 VoIP Over HSPA

Voice traffic, both AMR-NB and AMR-WB, has been carried on a Release 99 dedicated channel (DCH). Another option is to run voice over High Speed Downlink Packet Access (HSDPA) and High Speed Uplink Packet Access (HSUPA). Release 7 introduced features that make VoIP practical over HSPA. The main features required for efficient VoIP over HSPA are as follows:

3GPP Release 4

– Robust Header Compression (ROHC).

3GPP Release 5

– High speed downlink packet access (HSDPA).
– Code multiplexing of several parallel users, thus supporting multiple simultaneous low data rate connections.
– Quality of service (QoS) differentiation parameters.

3GPP Release 6

- High speed uplink packet access (HSUPA).
- Non-scheduled HSUPA transmission to provide guaranteed bit rate and to minimize allocation signaling.
- Fractional DPCH for HSDPA for minimizing the downlink L1 control overhead for low data rates.

3GPP Release 7

- Uplink DTX for HSUPA for minimizing the L1 control overhead for low data rates. Uplink gating also minimizes UE power consumption.
- Uplink packet bundling for minimizing the control overhead when sending two VoIP packets together.
- Discontinuous HSDPA reception for lower UE power consumption.
- HS-SCCH-less HSDPA for reduced downlink L1 control overhead.

The size of a full IPv6 header together with a RTP/UDP header is 60 bytes while the size of a typical voice packet is 30 bytes. Without header compression 2/3 of the transmission would be just headers. IP header compression can be applied to considerably improve the efficiency of VoIP traffic in HSPA. We assume robust header compression (ROHC) which is able to minimize the size of the headers down to a few bytes. Figure 6.8 illustrates the required data rate with full headers and with compressed headers. The required data rate is reduced from close to 40 kbps down to below 16 kbps.

The header compression with HSPA is done on Layer 2 PDCP (Packet Data Convergence Protocol) in UE and in RNC, therefore it saves not only the air interface capacity but also the Iub transmission capacity. The header compression location is illustrated in Figure 6.9.

The main motivations for using VoIP instead of CS voice are:

- New services. When voice runs on IP, it makes the integration of rich call services simpler.
- LTE interworking: voice in LTE must be VoIP. If also 3G networks support VoIP, the interworking between 3G and LTE is simpler.

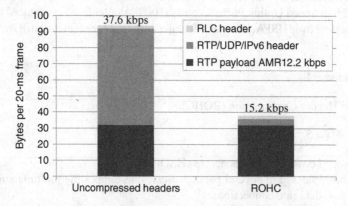

Figure 6.8 Impact of Robust Header Compression (ROHC)

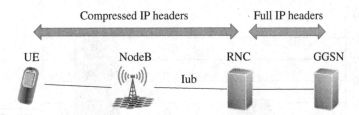

Figure 6.9   IP header compression between UE and RNC

- Higher spectral efficiency: voice over HSPA can support more users than voice over Release 99 DCH.
- Longer talk time: discontinuous transmission and reception reduces UE power consumption.
- Faster call setup: when signaling runs on top of HSPA, it becomes faster.

We will consider two topics in more detail: spectral efficiency and talk time of voice over HSPA. HSPA improves the efficiency of data transmission considerably due to advanced features. The same features can also help in voice efficiency:

- UE equalizer can remove intra-cell interference. An equalizer is included in practice in all HSPA receivers but not in WCDMA receivers.
- UE inter-cell interference cancellation also known as the Type 3i receiver.
- Layer 1 retransmissions can also be applied to voice, even if the delay budget is limited, since the retransmission delay is only 16 ms.
- HSDPA fast scheduling allows improvement in efficiency even with tough delay requirements.
- Optimized Layer 1 control channel reduces control channel overhead. The downlink solution is fractional DPCH and HS-SCCH-less transmission, and the uplink solution is discontinuous transmission.
- Uplink interference cancellation, which is typically implemented only for HSUPA, not for WCDMA.

The results of the system-level simulations are shown in Figure 6.10. Voice over HSPA can provide up to 80–90% capacity benefit when all HSPA optimization features are included. These capacities are assumed to be uplink limited. The detailed simulation results are presented in [2–4].

UE power consumption will benefit from running voice over HSPA. WCDMA uses continuous transmission and reception regardless of the voice activity. Voice over HSPA can utilize discontinuous transmission (DTX) and discontinuous reception (DRX). These functionalities are explained in more detail in Chapter 4. The most efficient approach to minimize power consumption is DTX and DRX. That also explains why GSM terminals tend to have longer talk times than WCDMA terminals: GSM can use DTX and DRX while WCDMA cannot.

The power consumption improvements can be estimated from the required transmission and reception activities. The estimated RF activity factors are shown in Table 6.1 for three different cases: uplink voice while downlink is silent, downlink voice while uplink is silent, and silence in both directions. The uplink activity is lower than the downlink because the downlink

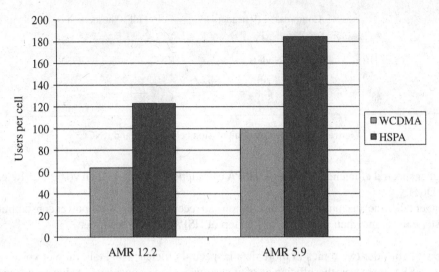

Figure 6.10 Capacity benefit of voice over HSPA

reception is needed for the reception of the uplink power control commands. We assume that voice transmission in downlink has 45% probability, voice in uplink 45% probability, and silence 10% probability. The average activity factors are 24% in uplink and 47% in downlink. The radio modem can utilize the inactivity periods for power savings. These activity factors assume that each voice packet is transmitted separately. The RF activity can be further reduced by using packet bundling, where two voice packets are sent together over the air interface. The activity factors with power bundling are shown in Table 6.2. If voice over HSPA were used, the power consumption could potentially be 50% lower, providing up to two times longer talk time.

Table 6.1 Voice over HSPA RF activity factors. Adapted from Holma and Toskala, 2010 [5]

HSPA	Uplink (%)	Downlink (%)	Probability (%)
Uplink voice, downlink silent	26	46	45
Downlink voice, uplink silent	26	52	45
Silent both directions	6	26	10
Average activity	24	47	

Table 6.2 Voice over HSPA RF activity factors with packet bundling. Adapted from Holma and Toskala, 2010 [5]

HSPA	Uplink (%)	Downlink (%)	Probability (%)
Uplink voice, downlink silent	14	40	45
Downlink voice, uplink silent	15	27	45
Silent both directions	6	26	10
Average activity	14	33	

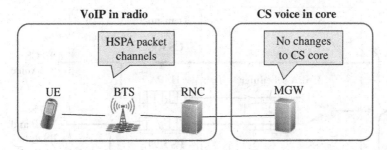

Figure 6.11 CS over HSPA uses VoIP in the radio and CS voice in the core network

6.5   Circuit-Switched Voice Over HSPA

Voice over HSPA was defined in 3GPP Release 7 for Voice over IP (VoIP). The rollout of VoIP service in the mobile networks happened only later, together with LTE. Therefore, Release 8 included also Circuit-Switched (CS) voice over HSPA. The CS over HSPA was a simple definition in 3GPP because it combines VoIP over HSPA in the radio network and CS voice in the core network. The high level view is shown in Figure 6.11. From the core network point of view, there is no difference between CS over WCDMA and CS over HSPA. Actually, the core network cannot even see if the radio network uses WCDMA or HSPA channels.

The CS over HSPA radio network solution is similar to VoIP over HSPA. Both use HS-DSCH in downlink, E-DCH in uplink for carrying the voice packets, and Unacknowledged RLC mode. The minor differences come in the Packet Data Convergence Protocol (PDCP) layer where VoIP uses header compression which is not needed in CS voice. On the other hand, CS voice uses a dejitter buffer in the RNC to keep the packet transmission timings to the core network fixed. The dejitter buffer in the case of VoIP is located in the other terminal, not in the radio network. The large similarities between CS and VoIP over HSPA also make practical implementation simpler. The three different voice solutions are compared in Figure 6.12.

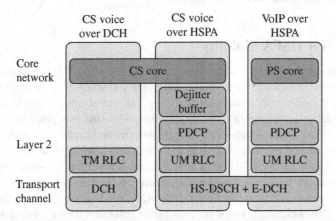

Figure 6.12 CS over HSPA in the radio network (TM = Transparent Mode. UM = Unacknowledged Mode)

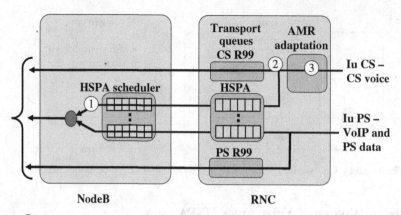

① = HSPA scheduler prioritizes voice packets
② = CS voice can be mapped on DCH or HSPA
③ = AMR bitrate adaptation according to the system load

Figure 6.13 Multiplexing of voice and data transmissions

The multiplexing of data and voice transmissions in the radio network is illustrated in Figure 6.13. The CS voice is carried over Iu-CS while VoIP and packet data is carried over Iu-PS. AMR data rate adaptation is selected in RNC according to the radio network loading. Release 99 channels are allocated in RNC. All the HSPA transmissions including CS voice, VoIP, and packet data, are multiplexed by NodeB taking into account the latency requirements. Voice packets are transmitted with higher priority compared to the data packets.

CS voice over HSPA has not yet been commercially deployed in 2014. One reason is that CPC functionality is needed first and that was only properly tested and optimized during 2013. Another reason is that the data traffic takes most of the capacity and the voice traffic eats only small fraction of the network capacity. Therefore, the capacity gain provided by CS voice over HSPA is less relevant when data traffic is booming.

6.6   Voice Over HSPA Mobility

Mobility on dedicated channels relies on soft handover, while HSDPA does not use soft handover. This section explains the mobility solution for HSDPA. The procedure is shown in Figure 6.14. We consider both procedural delay and voice interruption time in the user plane. The procedural delay is relevant for mobility reliability while the user plane break is important for voice quality. First, the new target cell emerges and enters the reporting range in UE at time $t1$. Some delay is caused by the UE measurement averaging, which is at least 200 ms. When the reporting trigger is fulfilled, UE sends a measurement report on the signaling radio bearer at time $t2$. The serving RNC reserves the base station and Iub resources to the target NodeB #2. Once the resources are ready at time $t3$, the RNC sends a radio bearer reconfiguration message to the UE, which still keeps receiving data from the source NodeB #1. When the UE has decoded the reconfiguration message and the activation time has expired at time $t4$, the UE will move the reception from the source cell to the target cell. This case is called synchronous procedure. Another option is asynchronous procedure, where the activation of

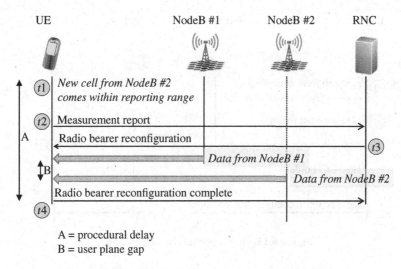

A = procedural delay
B = user plane gap

Figure 6.14   Mobility procedure on HSDPA

the reconfiguration message is immediate and the activation time is now. The UE specific MAC-hs in the source cell is reset at the time of the cell change and the buffered PDUs are deleted. At the same time, the flow control unit in the MAC-hs in the target cell starts to request PDUs from the serving RNC, so that it can start to transmit data on the HS-DSCH to the user. It is also possible for the RNC to send duplicate transmissions of the packet to both NodeBs during the cell change. When the RNC receives the Reconfiguration Complete message from the UE, it can release the resources from the source cell. The total procedural delay from $t1$ to $t4$ is mainly dominated by the measurement delay from $t1$ to $t2$. The critical factor is that UE must receive the reconfiguration command before the downlink radio link deteriorates too much due to signal fading or inter-cell interference.

The transmission gap, denoted as the time B in Figure 6.14, can be very low since the UE makes the cell change synchronously with the network, switching the transmission from the source cell to the target cell. The break can be squeezed down to a few tens of ms and almost completely eliminated with bicasting the voice packets to both NodeBs during the mobility procedure.

A few enhancements can be applied to further improve the mobility: re-establishment and enhanced serving cell change with bicasting. Re-establishment works as follows: if the radio link fails, UE sends a cell update message to the new target cell. RNC will then re-establish a new radio link to the UE and continue the voice connection. There is a break of a few seconds in the voice communication in this procedure. The voice break is mainly caused by the radio link time-out timer. The major benefit of re-establishment is that call drops can be avoided. Re-establishment is widely used for Release 99 voice channels.

Enhanced serving cell change can improve the mobility reliability because UE can switch to the new target cell immediately after sending the measurement reporting without waiting for the reconfiguration command from RNC. The fast switching can avoid radio link failure in downlink that may happen while UE waits for the reconfiguration command. The procedure is shown in Figure 6.15. UE starts to monitor HS-SCCH from NodeB #2 after sending the measurement report. UE still receives data from NodeB #1. The HS-SCCH code has been

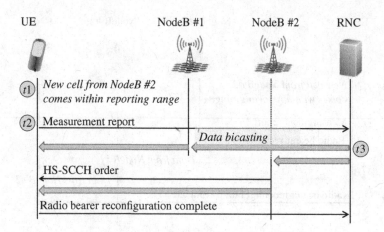

Figure 6.15 Enhanced serving cell change

indicated to UE earlier for all those cells that are in the active set. When the NodeB #2 gets the handover information from RNC, it indicates the servicing cell change to UE on HS-SCCH order. When UE receives the cell change order on HS-SCCH, it changes the data reception to NodeB #2 and sends a reconfiguration complete message to RNC. The user plane break can be minimized by using bicasting where RNC sends the same data to both NodeBs after receiving the measurement report.

6.7   Circuit-Switched Fallback

All LTE smartphones need a voice solution. Voice over LTE (VoLTE) is the long-term solution but has not yet been widely deployed. Most of the initial LTE smartphones use WCDMA or GSM networks for carrying the voice as traditional CS voice while LTE is used only for data transmission. This solution is called CS fallback handover. UE normally camps in the LTE network in order to provide access to LTE data rates. The paging message of the mobile terminating call is transmitted from the Mobile Switching Center Server (MSC-S) to the Mobility Management Entity (MME) and further to UE. When UE responds to the paging message, the LTE network commands UE to move to WCDMA or GSM network for the voice call. The UE can be moved by redirection or by handover. When the voice call is over, UE returns to the LTE network by reselection or by redirection. If there is a data connection running in LTE, the data is handed over to WCDMA or GSM during the voice call. That means data connection also continues during the voice call but the LTE data rates are not available, only HSPA or EDGE data rates. The overview of CS fallback handover is shown in Figure 6.16. For more details about LTE interworking see Chapter 15.

CS fallback has turned out to perform well in live networks. Some optimization steps have been implemented in order to minimize the call setup time and to minimize the return time back to LTE. The main solutions for minimizing the call setup time are

– Skipping the reading of the System Information Block (SIB) in the target system.
– Avoiding Location Area Update (LAU) in the target system.

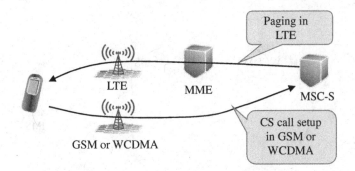

Figure 6.16   CS fallback handover from LTE to GSM/WCDMA

SIB reading can be avoided by providing the target cell SIB via LTE the network or providing the SIB only after the call setup in the target system. UE may also store the target system SIB if CS fallback happens multiple times to the same cell. LAU can be avoided in the call setup by configuring the core network so that the same location area code is used by the target MSC and the MSC where the paging is coming from. With optimized CS fallback procedure the additional call setup time can be below 1 s, which makes CS fallback performance in practice nearly as good as any WCDMA or GSM call.

The delay in getting UE back to LTE after the CS call can be minimized by using network controlled redirection after the voice call to push UE back to the LTE network. The initial solution used UE based reselection, which can cause 10 s or more delay when returning to LTE. When the redirection is implemented to the networks, the UE can instantaneously return to LTE after the voice call.

GSM/WCDMA operators can select the target network to be either WCDMA or GSM. Most operators use WCDMA as the CS fallback target. The call setup time is lowest when the CS fallback is to WCDMA. Many WCDMA networks also use AMR-WB for the best voice quality. All WCDMA networks and UEs also support simultaneous voice and data transmission. The network configuration for CS fallback to WCDMA is also simpler since only the target frequency needs to be given to the UE. Some operators use GSM as the target network, mainly if the WCDMA network does not provide full coverage. The target system can also be selected based on UE measurements: CSFB to WCDMA if the coverage is available.

6.8   Single Radio Voice Call Continuity

There may be a need to make a handover from VoIP to CS voice due to mobility. The typical use case is handover from Voice over LTE (VoLTE) to CS voice in WCDMA when UE runs out of LTE coverage area and there is no VoIP support in HSPA network. The handover functionality from VoIP to CS domain is referred to as Single Radio Voice Call Continuity (SRVCC). The solution does not require UE capability to simultaneously signal on two different radio access technologies – therefore it is called the single radio solution.

LTE eNodeB first triggers the UE to start inter-system measurements of the target WCDMA system and receives the measurement report. eNodeB then sends the handover request to MME, which triggers the SRVCC procedure via Sv interface to MSC-Server. Sv is a new

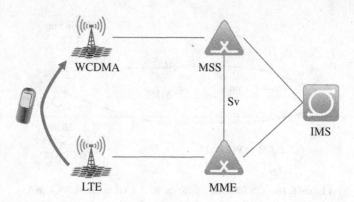

Figure 6.17   Single Radio Voice Call Continuity (SRVCC) from LTE to WCDMA

interface between MSC server and MME. The resources in the target cell are reserved and the necessary information is provided for the UE via LTE access. When the SRVCC procedure has been completed successfully, then VoIP connection remains from MSC towards the other side of the ongoing session. The CS connection exists towards the WCDMA radio access network. SRVCC architecture is shown in Figure 6.17.

In the case of simultaneous voice and non-voice data connection, the handling of the non-voice bearer is done by the bearer splitting function in the MME. The process is done in the same way as the normal inter-system handover for packet services.

In the case of roaming, the Visited PLMN controls the radio access and domain change while taking into account any related Home PLMN policies.

6.9   Summary

The WCDMA/HSPA system offers a comprehensive set of voice services. CS voice over WCDMA dedicated channels have been the baseline voice solution. Voice quality can be enhanced with AMR-Wideband codec and voice capacity can be improved with AMR-Narrowband lower codec rates. The voice service can also be provided over HSPA channels both as VoIP and as CS voice. The initial LTE smartphones typically use WCDMA CS voice with CS fallback functionality. When voice over LTE service is started, it can be complemented with the handover from VoLTE to WCDMA CS voice, called SRVCC.

References

[1] Holma, H., Melero, J., Vainio, J. *et al.* "Performance of Adaptive Multirate (AMR) Voice in GSM and WCDMA", VTC 2003.
[2] Holma, H., Kuusela, M., Malkamäki, E. *et al.* "VoIP over HSPA with 3GPP Release 7", PIMRC 2006, September 2006.
[3] 3GPP Technical Report 25.903 "Continuous connectivity for packet data users", v. 7.0.0, 2007.
[4] NSN White paper "Long Term HSPA Evolution meets ITU IMT-Advanced requirements", 2012.
[5] Holma, H. and Toskala, A. (2010) *WCDMA for UMTS – HSPA Evolution and LTE*, 5th edn, John Wiley & Sons, Ltd, Chichester.

7

Heterogeneous Networks

Harri Holma and Fernando Sanchez Moya

7.1 Introduction

Heterogeneous Network (HetNet) refers to the joint optimization of different cell layers and technologies. Current mobile networks are based on wide area macro cells that provide several hundred meters of coverage, even tens of kilometers, and normally use three sectors per site. Small cells can be added to provide more capacity and coverage. These small cells can be outdoor small cells, public indoor small cells, or home or enterprise small cells. The outdoor small cells are often termed micro cells, indoor cells pico cells, and home small cells femto cells. The HetNet concept is shown in Figure 7.1.

The HetNet concept also considers the use of the different technologies GSM, HSPA, LTE, and WLAN together. This chapter focuses on HSPA small cell deployments together with existing macro cells on WLAN interworking. Section 7.2 presents the drivers for the small cell rollouts. Section 7.3 shows the base station categories. Sections 7.4 and 7.5 present the main challenges with HSPA HetNets – interference management and link imbalance, Section 7.6 studies the data rate and capacity benefits provided by the small cells, and Section 7.7 illustrates field deployment result with HSPA small cells. Femto cell architecture and optimization are discussed in Section 7.8 and WLAN interworking in Section 7.9. The chapter is summarized in Section 7.10.

7.2 Small Cell Drivers

The introduction of small cells is driven by capacity and coverage requirements. As data traffic increases in many HSPA networks, macro cells are getting congested and operators have used all their spectrum resources in them. Capacity could be enhanced by adding new macro sites or by adding new sectors to the existing macro sites. However, it may be difficult or expensive to find a location for a new macro site or to add more antennas to the existing sites. Therefore,

HSPA+ Evolution to Release 12: Performance and Optimization, First Edition.
Edited by Harri Holma, Antti Toskala, and Pablo Tapia.
© 2014 John Wiley & Sons, Ltd. Published 2014 by John Wiley & Sons, Ltd.

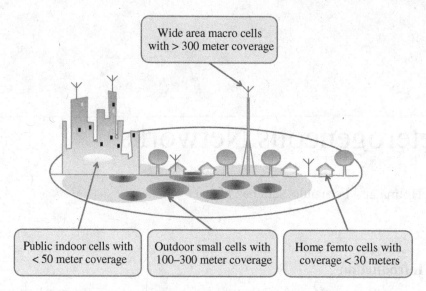

Figure 7.1 HetNet concept

the most practical way of increasing network capacity may be to add new small base stations. The small cells can also be deployed to improve the coverage and increase local data rates.

Small cells have been discussed in the industry for a long time but it is only recently that the implementation of small cells has become simpler for operators: the small base station products are very compact due to the evolution of digital and RF technologies, the network configuration and optimization is becoming automatic with Self Organizing Network (SON) solutions, and attractive wireless backhaul options are emerging.

The need for small cells may be postponed by the availability of new spectrum and LTE radio, which provide a lot more capacity with existing macro sites. The problem may still be that the LTE terminal penetration does not increase fast enough to migrate the traffic from HSPA to LTE. Operators may need to build HSPA small cells even if their LTE network is sitting empty in the same area.

7.3 Base Station Categories

3GPP has defined four base station categories with different RF requirements. The categories are defined with the maximum output powers and the Minimum Coupling Loss (MCL). MCL refers to the minimum path loss between the base station antenna and UE antenna and it defines the type of base station installation. High MCL requires that UE cannot get too close to the base station antenna, which is typical for high mounted macro cell antennas. Low MCL means that UE can get very close to the base station antenna. The idea in categories is that some of the RF specifications can be relaxed for the low power base stations to enable even lower cost implementation. The sensitivity requirements are relaxed because the link budget is anyway limited by the low power in downlink. Higher spurious emissions are allowed for the low power products because there are no co-sited base stations with high sensitivity requirements. Some

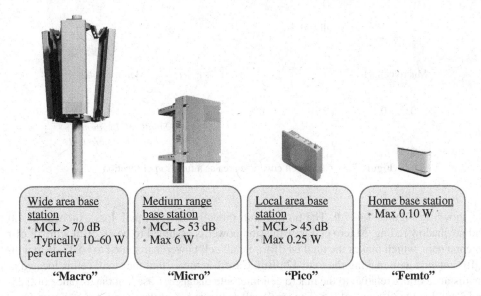

Figure 7.2   Base station categories

of the blocking requirements are higher for the low power base stations because interfering UEs can get closer to the base stations. The frequency stability requirements are relaxed for the lower power base stations because there are no high vehicular speeds used in those cells.

The four categories are shown in Figure 7.2. The wide area category applies for all high power base stations with more than 6 W of output power. The wide area category can also be termed a macro base station. The required MCL is 70 dB. The medium range base station, or micro base station, can have up to 6 W of power. The local area base station is limited to 0.25 W and the home base station to 0.10 W of power.

7.4   Small Cell Dominance Areas

When the data traffic increases, operators add more frequencies to their macro network. When all the spectrum is used by the macro cells, there is no dedicated frequency available for the small cells, which leads to the co-channel deployment of macro cells and small cells. Co-channel deployment creates some challenges: the small cell dominance area is impacted by the macro cell signal level and the power level differences between macro and small cells can create link budget imbalances. These topics are discussed in this section.

The small cell location has a major impact on the small cell coverage area in the case of co-channel deployment. The UE must be connected to the cell providing the highest received Common Pilot Channel (CPICH) signal level. It is possible to use offset values in cell reselections and handovers to favor low power small cells, but in practice the offset values can be just a few dB. An overview is illustrated in Figure 7.3. The small cell coverage area is largest when it is located in an area with low macro cell signal level between two macro cells, denoted as location 0.5. If the small cell is placed closer to the macro cell with higher macro cell signal level, the small cell coverage area becomes smaller. Three different cases

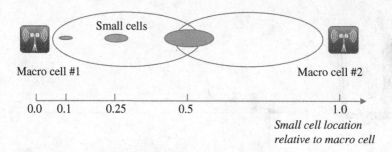

Figure 7.3   Small cell coverage area as a function of location

are shown in Figures 7.4–7.6. The propagation model assumes a path loss exponent of 4.0 and no shadow fading. Macro cell transmission power is assumed to be 46 dBm with 18 dBi antenna gain, which makes the total 64 dBm. Small cell power is assumed to be 37 dBm with 5 dBi antenna gain making the total 42 dBm. Case 1 assumes that the small cell is located at the distance of 0.1 relative to the macro cell inter-site distance, Case 2 at the distance of 0.25, and Case 3 at the distance of 0.5. The small cell radius is 6% of the macro cell radius in Case 1, which corresponds to a relative cell area of 6% x 6% = 0.36%. The relative small cell area grows by a factor of 4 in Case 2 and by a factor of 12 in Case 3. The small cell area is 4% of the macro cell area in Case 3. The coverage areas are summarized in Table 7.1. This example shows that the macro cell signal level needs to be considered when planning the small cell coverage area.

The small cell range can be estimated based on the macro cell signal level. We assumed that the small cell transmission power, including the antenna gain, is 42 dBm and the path loss 140 dB + 40*log10(distance[km]). The result is shown in Figure 7.7. If the macro cell

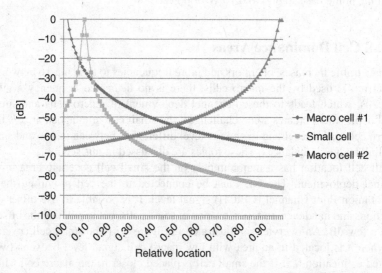

Figure 7.4   Small cell located close to macro cell #1 (Case 1)

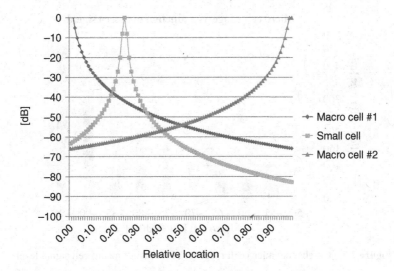

Figure 7.5   Small cell located between macro base station and cell edge (Case 2)

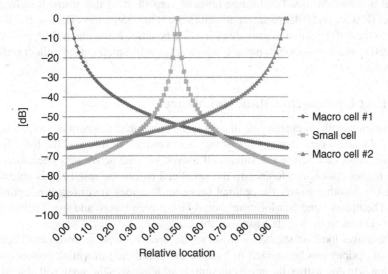

Figure 7.6   Small cell located at the edge of macro cell (Case 3)

Table 7.1   Co-channel small cell coverage area

Small cell location	Covered distance by small cell (%)	Covered area by small cell (%)	Relative coverage area
Case 1: Close to macro cell #1 (0.1)	6	0.4	
Case 2: Between macro cell center and cell edge (0.25)	15	2.3	6x larger small cell coverage than case 1
Case 3: At the edge of the macro cell (0.5)	21	4.4	12x larger small cell coverage than case 1

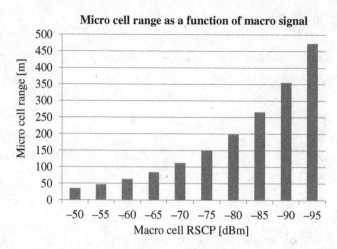

Figure 7.7   Co-channel micro cell range as a function of macro cell signal level

signal level is good, the small cell range remains very short. If the macro Received Signal Code Power (RSCP) is -50 dBm, the small cell range is just 35 meters. The small cell must be very close to the traffic source to provide offloading benefits. If the macro RSCP is -90 dBm, the small cell range becomes 350 meters, which allows the small cell to collect traffic from larger area.

7.5   HetNet Uplink–Downlink Imbalance

Another challenge with co-channel small cells is the link imbalance between macro and small cells. Since the small cell transmission power is lower than in macro cells, the UE can get closer to the small cell before the small cell moves into the active set. Therefore, the UE may cause higher uplink interference to the small cell before the small cell is able to power control the UE. In other words, the optimal handover locations are different in uplink and in downlink. The uplink handover location should be between macro and small cells where the path loss equalizes to both cells assuming the same base station uplink sensitivity. But the downlink handover location should be closer to the small cell due to lower small cell power. A number of options can be utilized to balance the uplink and downlink connections. One simple option is to desensitize the small cell uplink, which makes the small cell able to tolerate more interference. The desensitization is taken into account in 3GPP base station categories: the micro base station required sensitivity is 10 dB more relaxed compared to the macro base station case. That balances the uplink and downlink handover area. The link imbalance and the desensitization are shown in Figure 7.8. We assume macro cell transmission power of 40 W and small cell 6 W. The upper case assumes the same base station sensitivity of -121 dBm according to 3GPP while the lower plot assumes relaxed sensitivity of -111 dBm for the small cell.

Desensitization is not an ideal solution since it increases unnecessarily the UE transmission power when connected to the small cells. Another solution is to use Cell Individual Offset (CIO) for the handovers which expands the small cell coverage area. The maximum value for CIO is normally 3 dB. If a larger CIO value is used, the downlink signal quality from the small cell tends to become too weak.

Same base station sensitivity in macro and small cell

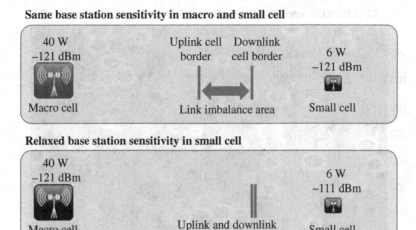

40 W
−121 dBm

Macro cell

Uplink cell   Downlink
border   cell border

Link imbalance area

6 W
−121 dBm

Small cell

Relaxed base station sensitivity in small cell

40 W
−121 dBm

Macro cell

Uplink and downlink
balanced

6 W
−111 dBm

Small cell

Figure 7.8   Balancing uplink and downlink cell borders

The area of potential interference issues is shown in Figure 7.9. When UE is in the soft handover area, UE power is controlled by both cells and there are no uplink interference issues. When UE is outside the soft handover area, UE may cause interference to the small cell. The gray color in Figure 7.9 indicates the area where UE can cause uplink interference to the small cell because it is not in soft handover and is not power controlled by the small cell. This area

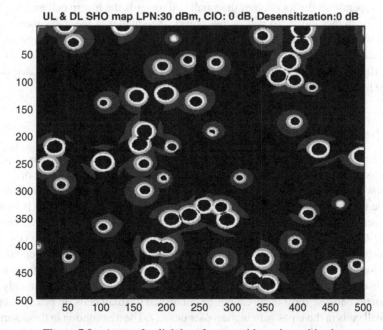

UL & DL SHO map LPN:30 dBm, CIO: 0 dB, Desensitization:0 dB

Figure 7.9   Areas of uplink interference without desensitization

UL & DL SHO map for LPN 30 dBm, CIO: 0 dB, Desensitization: 6 dB

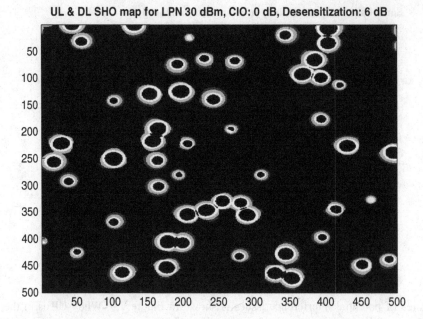

Figure 7.10   Areas of uplink interference with desensitization

can be quite significant and reach up to 14% of the total area of a HetNet network layout assuming no desensitization and a small cell power of 1 W. The shape of the interference area largely depends on the location of the small cell towards the macro cell and on the small cell power. The higher the small cell output power, the smaller is the imbalance zone. This is straightforward, as the higher the power the more similar small cells become to a macro cells and hence the imbalance problem is reduced. The area of uplink interference is reduced in Figure 7.10 by using desensitization. This and other issues and solutions related to UMTS HetNet have been studied in 3GPP document [1].

7.6   HetNet Capacity and Data Rates

HetNet capacity and data rate gain in system simulations are presented in this section. The simulations assume macro cell transmission power of 20 W with antenna gain of 14 dBi and small cell power of 0.25 to 5 W with antenna gain of 5 dBi. The macro base station uses three sectors while the small cell has omni transmission. The number of small cells is two or four for every macro cell, which makes 6 to 12 times more small base stations than macro base stations. The Cell Individual Offset (CIO) has been 0 dB or 3 dB. The macro cell inter-site distance was 500 meters. The small cells were randomly placed in the macro cell areas.

The user locations in HetNet simulations have a major impact on the benefit of the small cells. If the users are randomly located in the macro cell area, the small cells can only provide marginal benefit because most users cannot be connected to small cells. In practice, operators place the small cells in the areas with highest user density. The assumption in these simulations is that 50% of users are placed close to the small cells within 20–60 meters depending on small

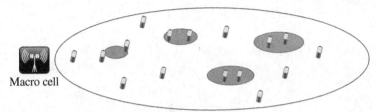

Four small cells within macro cell

Figure 7.11 User locations in HetNet simulations

cell power. The other 50% of users are placed randomly in the macro cell coverage area. An example placement of the small cells and UEs is show in Figure 7.11. The simulations assume a total of 16 UEs per macro cell.

The first set of simulations assume a full buffer traffic model. The average and the cell edge user data rates are shown in Figure 7.12 and the gains compared to the macro only case in Figure 7.13. The user data rates are relatively low because of the high number of simultaneous full buffer users. The average data rate with the macro only case is 300 kbps and the cell edge 100 kbps. Both data rates are increased considerably by the introduction of small cells. The average data rate increases by 100–150% with two small cells and 150–250% with four small cells and reaching up to 1 Mbps. The cell edge data rates increase by 50–100% with small cells. The average data rate gain from two small cells is 125% and cell edge gain 70%. The

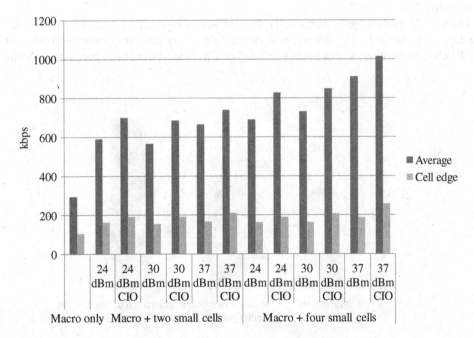

Figure 7.12 User data rates with full buffer

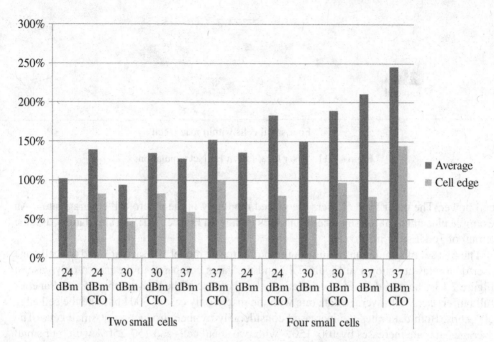

Figure 7.13   Gains compared to macro only case with full buffer

corresponding gains with four cells are 185 and 85%. The simulations also show that the gain from 3-dB CIO is 15–25%.

The network capacity gain with small cells is shown in Figure 7.14. Only one number for the capacity is shown for the different cases since the capacity was very similar with different small cell configurations with a given small cell density. The system throughput was 6 Mbps

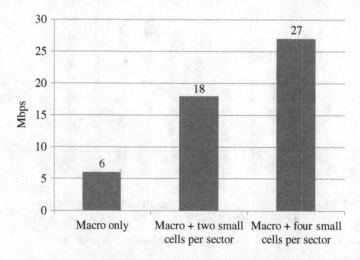

Figure 7.14   Network capacity with small cells

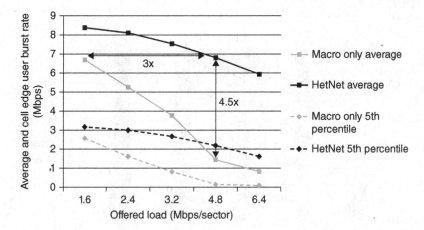

Figure 7.15   User data rates as a function of offered load per sector

without small cells, 18 Mbps with two small cells, and even 27 Mbps with four small cells. It is interesting to note that the capacity provided by the small cells is approximately the same as the capacity of the macro cell, that is 6 Mbps per cell.

In order to study the improvements introduced by HetNet in terms of user experience, it is important to analyze the relation between the user burst rate and the offered load per sector in the system. Figure 7.15 shows the average user and cell edge (5th percentile) user burst rate versus offered load for macro only and HetNet scenarios. We assume four small cells with output power of 37 dBm, CIO, and 16 UEs per macro sector area. The gain introduced by HetNet can be interpreted in different ways. One can look at a constant user experience, for instance 7 Mbps, and see that the offered load can be increased by a factor of 3 or one can see the gain in the user throughput at a given load: a factor of 4.5 for 4.8 Mbps/sector load.

The HSDPA Multiflow (see Chapter 5) [2] is characterized by simultaneous data transmission from two neighboring cells to a UE located at the border of those cells. The Multiflow solution can improve the cell edge data rates and balance the loading between adjacent cells, and that improvement is especially true for the HetNet scenarios.

In inter-site Multiflow the original data stream is split at the Radio Network Controller (RNC) and sent to each base station participating in the Multiflow transmission. Multiflow could be compared to a soft handover in Release 99: the UE must have a cell in its active set to establish Multiflow, and also in Release 99 soft handover two cells transmit to the UE. However, unlike soft handover the Multiflow data streams are different from each other, and are scheduled independently by the base stations and not by the RNC. Further, there is no automatic establishment of a Multiflow connection when a cell is taken into the active set. Instead, the usage of Multiflow transmission is determined for every UE by the RNC, and typically will not only depend on the SINR difference between the cells in the active set but also on the cell loads and UE traffic type.

There are two main reasons behind the development of Multiflow. Firstly, the problem with the reception quality at cell edge when the UE is far from the transmitter. In addition, cell edge UEs experience stronger interference from neighboring base stations. Secondly, the possibility of using transmissions from several cells provides a load balancing function where UE can

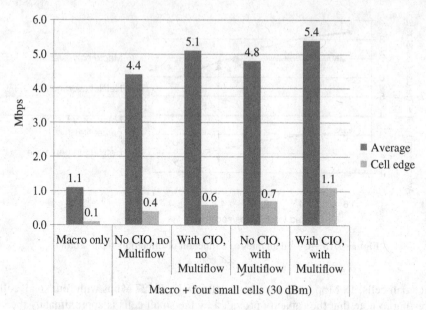

Figure 7.16   Downlink user burst rate gains introduced by HetNet and Multiflow

utilize the resources from less loaded base stations. The load balancing is especially interesting in HetNet scenarios where an offloading of UEs to small cells is desired, but at the same time due to the smaller size of the cells more frequent handovers can be expected. With Multiflow, a higher level of robustness and smoother transitions between cells can be accomplished.

Simulation results with a bursty traffic model and HSDPA Multiflow are considered in Figure 7.16. The average file size is 750 kbits, inter-burst arrival time 5, and 16 UEs per macro cell, which makes the offered load 2.4 Mbps per sector. That leads to a loaded macro that still operates under a stable regime, that is, the offered load per UE is lower than the UE burst rate. The feedback channel for Multiflow operation was assumed to be ideal, assuming HS-DPCCH boosting can overcome UL/DL imbalances. The gains are expressed over the macro only case, with Multiflow contributing to improve the cell edge experience and HetNet pushing both the average and cell edge burst rates. Multiflow on top of HetNet provides extra gain in the cell edge, leading to 110% gain for a cell edge user burst rate over that which could be achieved by a user in the macro only scenario.

7.7   HetNet Field Measurements

This section presents a summary of small cell deployment in a large European city. The analysis focuses in more detail on the impact of two small cells: one deployed for coverage and another one deployed for capacity. The impact of small cell output power and small cell frequency usage is studied. The small cell is deployed with two different approaches: on the shared frequency with the macro cells and on the dedicated small cell frequency. The two small cell locations are illustrated in Figure 7.17.

The traffic collected by the small cells was compared to the traffic collected by the adjacent macro cells in Figure 7.18. We can notice that the small cell collects more traffic when

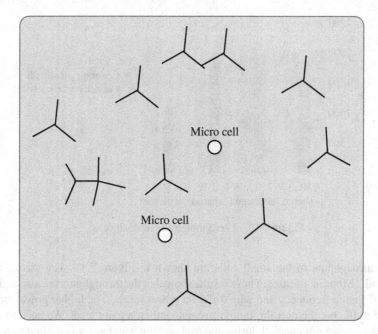

Figure 7.17 Two small cells (=circles) in the middle of three-sector macro cells

the small cell power is increased and when a dedicated small cell frequency is used. That behavior is expected since the small cell coverage area increases with higher power and with dedicated frequency. Also, the coverage driven small cell collects relatively high traffic when the surrounding macro signal is low. In the most extreme case the small cell collects even more traffic than the surrounding macro cells.

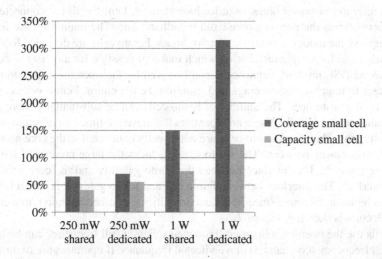

Figure 7.18 Relative traffic collected by small cells compared to adjacent macro cells

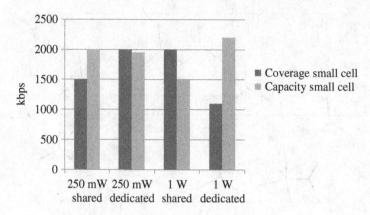

Figure 7.19   User throughputs in small cells

The user throughputs in the small cells are shown in Figure 7.19. The throughputs are between 1 and 2 Mbps in all cases. These results show that the throughputs are also satisfactory in the case of shared frequency and with different power levels. The higher power small cells collect more traffic but the user throughputs remain still at a good level. We need to note that HSPA+ features were not enabled during the trial and the backhaul was limited to 8 Mbps in Digital Subscriber Line (DSL), which reduced the throughputs.

The small cell deployment showed that the small cells transfer a significant amount of data traffic and fulfill their offloading potential. The small cell can be deployed on the shared carrier with the macro cell. High power of 1 W, or even more, is beneficial to collect more traffic to the small cell.

7.8   Femto Cells

Femto cells refer to low power home or office base stations. Femto cells are connected to the operator's network via the person's own fixed broadband line. The main use case for femto cells is to improve the indoor coverage for smartphones. Femto cells are designed for low cost implementation and for simple installation which makes it possible for any user to connect a femto cell to his DSL modem. Femto cells can be given by the operator to the user as part of the contract to improve the coverage and to minimize the churn. Femto cells could also be paid for by the customers. The number of femto cells can be substantially more than the number of outdoor macro cells. There are operators that have ten times more femto cells than macro base stations. Therefore, femto cells are not directly connected to the core network but there is a femto gateway between. The femto gateway hides the large number of femto cells from the core network. The interface between the femto gateway and the core network is a normal Iu interface. The interface between femto cell and femto gateway is called Iuh. There is no RNC in femto architecture since RNC functionalities are embedded into femto cells. The femto architecture is shown in Figure 7.20.

Femto cells use the operator's licensed frequencies. Femto cell frequency can be the same as macro cell frequency (co-channel) or a dedicated frequency. If operators are running out of capacity, then there is likely no dedicated frequency available for femto cells. Femto cells can

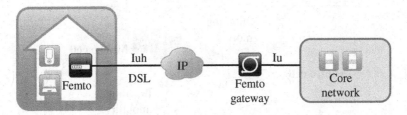

Figure 7.20 Femto architecture

be deployed on the same frequency as macro cells if femto cells are mainly targeted to improve indoor coverage in those areas with a weak macro signal. If the macro signal is strong, it will be beneficial to use a dedicated frequency for femto cells to avoid interference between macro and femto cells and to offer higher data rates with femto cells. The interference management between macro and co-channel femto cells can be obtained by a power control solution where the femto cell first measures the macro cell signal level and then sets femto cell transmission power accordingly. The interference between femto cells may also impact the user data rates in the case of dense femto deployment in apartment buildings.

Femto cells can be defined for Closed Subscriber Groups (CSGs) or for open access. The owner of a home femto cell can define the list of users who are able to access the femto cell. That feature allows the home owners to control who is able to use femto cell and DSL connection. The CSG control for pre-Release 8 UEs is located in the femto gateway. The CSG control for Release 8 UEs is located in the core network based on the information provided by the UE.

Femto cells support a listening mode which refers to the downlink receiver in the femto cell. The listening mode allows identification of those macro cells and femto cells that can be received in the area. The downlink receiver also allows measurement of the macro cell signal level which can be used when setting femto cell transmission power. Femto cells are identified by dedicated scrambling codes. Typically, a few scrambling codes are dedicated to femto cells. These scrambling codes can be indicated by all the macro cells in the area.

The mobility between macro and femto cells needs to be considered in order to provide a seamless end-user experience. The outgoing handovers from femto to macro were supported already in the first femto cells and by legacy UEs. The outgoing handover is important to maintain the ongoing call when leaving the femto cell coverage area. The incoming handover is supported in Release 9, while reselection from macro to femto is used in pre-Release 9 UEs. If there is an ongoing call in pre-Release 9 UE when entering femto cell coverage area, the call remains in the macro cell. When the call is over, the UE moves to femto cells by itself with reselection. The further mobility improvements between macro and femto, and between two femto cells, are included in Releases 10 and 11 which eventually allow soft handover for the mobility cases. It is also possible to support handover from macro to femto, and between two femtos, for the legacy Release 8 UEs in the case of coordinated femto deployment. The coordinated case refers to open access hot spots and enterprise femtos where the deployment is planned in terms of scrambling codes and the femto cells are listed in the neighborlist in the macro cell. The handover can be intra-frequency or inter-frequency. The femto mobility features for uncoordinated deployment are shown in Figure 7.21.

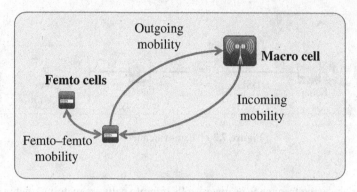

	Femto–macro (outgoing)	Macro–femto (incoming)	Femto–femto (uncoordinated)
Release 8	Hard handover	Idle reselection	Idle reselection
Release 9	Hard handover	Hard handover	Idle reselection
Release 10	Hard handover	Hard handover	Soft handover
Release 11	Soft handover[1]	Soft handover[1]	Soft handover

[1]Applies to open and hybrid access femtos,
not to closed access femtos

Figure 7.21   Femto mobility solutions for uncoordinated deployment

Most of the femto units in commercial networks are used for residential indoor coverage. Residential femtos need to support just a small number of users and there is no need for handovers between two femtos. The femto solution is also used for enterprise coverage. The enterprise solution sets higher requirements for the femto access points: higher user capacity and inter-femto mobility features are required.

Commercial femto deployments began during 2009 mainly as a fill-in solution for improving home coverage. In the long term, femto cells could also be used to off-load high traffic from the macro network. Femto cells also change the operator's business assumptions because the access point is purchased and installed by the end user and the transmission and the electricity is paid by the end user.

Practical deployments have shown that the following areas need to be considered in femto deployments:

– The supported features may be different in femto and macro cells. Typically, femto cells did not support all the same features as macro cells because different hardware and software platforms were used in femto and macro cells.
– The data rates in femto cells may be limited by the DSL backhaul. Many DSL lines are limited to 8–10 Mbps data rates that are lower than the HSDPA peak rates of 21 Mbps and 42 Mbps. It may be that the macro cells can provide even higher data rates than femto cells.

– Soft handovers are typically not available between co-channel femto and macro cells. The lack of soft handovers impacts the reliability of the mobility.
– A dedicated frequency for femto cells eats part of macro capacity and it is increasingly difficult to allocate a dedicated femto frequency when the macro cell traffic increases.
– If a dedicated femto frequency is utilized, then CSFB with LTE phones gets more complex. The CSFB target need to be decided based on UE measurements between macro and femto frequency.
– Different location areas are used in macro and femto cells. The UE performs a location area update whenever moving between the layers. Attention is required to make sure that paging messages are not lost during the change of location area.
– The maximum number of users is typically limited in femto cells, which may cause problems if using femto cells in large enterprises or in public premises.
– The femto and macro cells are typically provided by different vendors and the corresponding operability tools are also different.
– The femto cell adds co-channel interference which may impact those UEs that are in weak macro signal and are not able to connect to CSG femto cells.

Field experiences have shown that the femto solution works fine for enhancing indoor coverage at homes or in small enterprises, while it is not practical to deploy femto solutions outdoors and in public premises. The outdoor solution needs a Iub/RNC connection for smooth mobility and interference control, more transmission power, higher capacity, and feature parity with macro cells.

7.9   WLAN Interworking

WLAN (Wi-Fi) can complement mobile networks in carrying part of the data traffic and reducing the mobile network loading. In practice, all medium and high category mobile devices have Wi-Fi radio, which enables efficient offloading of data traffic to Wi-Fi. Most users can configure the mobile device to use Wi-Fi at home and in the office but it tends to be more complex to utilize public Wi-Fi networks. Mobile operators may also want to control which Wi-Fi networks are utilized in order to guarantee adequate performance and costs related especially to roaming hotspots. Two mainstream solutions are Access Network Discovery and Selection Function (ANDSF) and Hotspot 2.0. ANDSF is standardized by 3GPP while Hotspot 2.0 is defined by Wi-Fi Alliance with IEEE specifications. Both solutions are described in this section.

7.9.1   Access Network Discovery and Selection Function (ANDSF)

ANDSF based selection between Wi-Fi and mobile networks follows the same logic as Visited PLMN (VPLMN) selection in the case of mobile roaming. Most users are happy with automatic roaming network selection based on the operator priorities in the (U)SIM card. The UE uses the highest priority network out of the currently available networks. But the user is still able to override the automatic selection and select another mobile network.

3GPP has standardized ANDSF interfaces. The first version of ANDSF was included into 3GPP Release 8 and was enhanced in later releases. The interfaces allow an operator to give

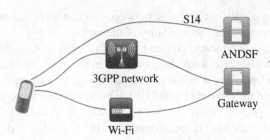

Figure 7.22   ANDSF architecture

Wi-Fi network selection policies to the terminals and define when and where to use 3GPP and Wi-Fi networks. ANDSF architecture is shown in Figure 7.22. The ANDSF server provides the priority information to the UE over the S14 interface. The S14 interface is defined based on the Open Mobile Alliance (OMA) Device Management (DM) framework. The messages are transferred on top of IP. In order to contact the ANDSF server, the UE needs an active data connection to the network. The most relevant 3GPP references are [3–5].

ANDSF can provide three kinds of information:

1. Inter-System Mobility Policy (ISMP) is the original network selection policy defined in Release 8. The same network policy information is used for all the applications in the UE. It is possible to prioritize WLAN networks based on the Service Set Identifier (SSID). An example ISMP policy:
   (a) Priority 1: use WLAN with SSID "Operator own network."
   (b) Priority 2: use WLAN with SSID "Roaming partner network,"
   (c) Priority 3: use 3GPP radio access.
2. Access Network Discovery Information: the main use is to facilitate the terminal's network discovery process to allow the terminal to learn approved Wi-Fi networks and to avoid unnecessary scanning. By using validity conditions a UE can utilize, for example, associated location information to optimize Wi-Fi scanning process and thus save battery.
3. Inter-System Routing Policy (ISRP): ISRP policies were added into ANDSF in Release 10. The idea is that it is possible to identify different applications and define separate network policies for each application. For example, QoS aware applications may use only 3GPP or proven quality Wi-Fi networks.

ANDSF also offers the following flexibility:

- The ANDSF priority information can depend on the time of the day and on the location: offloading to Wi-Fi can be used in high traffic areas like stadiums or during the mobile network's busy hours.
- The ANDSF policy can also be given to the roaming users in addition to the operator's own subscribers. The roaming case covers both own subscribers roaming in another network and other network subscribers roaming in this network.
- The initiation of the ANDSF session establishment is done by the UE. The session initiation trigger can be based on expired ANDSF policy information, UE configuration, or some other implementation dependant trigger. Also, the ANDSF server may trigger UE to initiate the session establishment. These two cases are called pull mode and push mode.

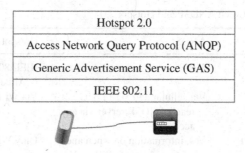

Figure 7.23 Hotspot 2.0 protocols

7.9.2   Hotspot 2.0

Hotspot 2.0 is defined by the Wi-Fi alliance and leverages the IEEE 802.11u standard which enables cellular-like roaming. Hotspot 2.0 is focused on enabling a mobile device to automatically discover those access points that have a roaming arrangement with the user's home network and then securely connect. Hotspot 2.0 is also known as Wi-Fi certified passpoint. The device can retrieve information on the access points before associating with an access point and then select the most suitable access point based on this info. The Hotspot 2.0 Release 1 specification has been completed and certified in 2013, see [6], while Release 2 is expected during 2014 [7]. A general description of Hotspot 2.0 is shown in [8].

The key protocols within 802.11u are Generic Advertisement Service (GAS) and Access Network Query Protocol (ANQP). ANQP is a query and response protocol used between the Wi-Fi network and the mobile device to discover information such as the network authentication types supported, venue name, roaming agreements in place, load, throughput of the backhaul link, and other information useful in the network selection process. ANQP messages are transferred over GAS. The Hotspot 2.0 protocols are shown in Figure 7.23. When a user with a Hotspot 2.0-capable mobile device comes within a range of a Hotspot 2.0-capable access point, it will automatically open up a dialog with that access point to determine its capabilities. This is done using ANQP packets, which allow the device to learn the capabilities of the access point. At this point, the device has not yet attached to the access point and does not yet have an IP address. The selection process can be influenced both by user preference and operator policy. Automating this manual configuration and decision-making process makes Wi-Fi usage simpler, and increases the use of Wi-Fi services. Another benefit is that Hotspot 2.0 recommends use of Wi-Fi Protected Access (WPA2) enterprise security, which harmonizes the technology landscape and leads to wider UE support of secure communication.

In short, Hotspot 2.0 works in practice as follows:

1. 802.11u-capable access point indicates Hotspot 2.0 support.
2. Device selects access point.
3. Device runs Access Network Query Protocol (ANQP) request.
4. Access point responds with requested information.
5. Device compares profile information against Hotspot 2.0 data from APs and selects the most suitable network.

Table 7.2 Differences between ANDSF and Hotspot 2.0

	ANDSF	Hotspot 2.0
Standardization	3GPP	IEEE 802.11 and Wi-Fi Alliance
Information available	Via cellular and Wi-Fi. UE can reach ANDSF server via any access	Via Wi-Fi
Prioritization between 3GPP and Wi-Fi	Yes, information on when and where to select 3GPP or Wi-Fi	Only Wi-Fi selection
Prioritization of Wi-Fi networks	Yes, different Wi-Fi networks can be prioritized	Own Wi-Fi can be prioritized but other Wi-Fi networks have the same priority
UE location information	Yes	No
Roaming support	Yes	Yes
Type of access network	No	Yes: public, private, with guest access, chargeable public, free public, etc.
Venue	No	Yes, group (residential, business), type, name etc.
Wi-Fi access point performance information	No	Yes. UE can avoid congested access points
Network authentication type	No	Yes

7.9.3 Differences between ANDSF and Hotspot 2.0

The main differences between ANDSF and Hotspot 2.0 are listed in Table 7.2. ANDSF is defined by 3GPP and the Wi-Fi information is provided via any network, while Hotspot 2.0 gives the information via the Wi-Fi network. ANDSF allows prioritization between 3GPP and the Wi-Fi network which is not supported in Hotspot 2.0. On the other hand, Hotspot 2.0 gives information about the quality of Wi-Fi networks which is not available in ANDSF. In short: ANDSF and Hotspot 2.0 are complementary solutions. ANDSF provides mobile network operator tools to tell UE when and where to use 3GPP or Wi-Fi networks according to operator business policies, for example, use roaming partner hotspot adding operator costs only during football games or in subway stations. Hotspot 2.0 enables UE to select the best Wi-Fi access point among multiple candidates. Hotspot 2.0 is also applicable for tablets without a SIM card.

3GPP brings enhanced HSPA/LTE and Wi-Fi interworking in Release 12. See Chapter 16 for a summary.

7.10 Summary

Small cell deployments are driven by the need for enhanced capacity and improved coverage. Current technology enables implementation of compact small cell products, which makes the practical small cell deployment easier than ever before. Since operators tend to use all their frequencies in macro cells, there will be no dedicated frequencies available for small cells but instead small cells need to share the frequency with macro cells. Co-channel deployment creates some challenges, including the dominance areas and uplink–downlink balancing. The simulations show that small cell deployments can boost data rates considerably. Optimized small cell rollout with four micro cells per macro sector can increase user data rates by a factor of 3.5x. Field measurements have also confirmed small cell benefits. Femto cells have been successfully used for enhancing indoor coverage at homes and in small enterprises but femto cells are not suitable for outdoor deployment. Mobile operators can complement their cellular networks by offloading data traffic to Wi-Fi. ANDSF and Hotspot 2.0 solutions provide the means for operators to control traffic offloading.

References

[1] 3GPP Technical Report 25.800 "UMTS Heterogeneous Networks", v.12.0.0, 2013.
[2] 3GPP Technical Specification 25.308 "High Speed Downlink Packet Access; Overall Description; Stage 2 (Rel 11)", v.11.6.0, 2013.
[3] 3GPP Technical Specification 23.402 "Architecture enhancements for non-3GPP accesses", v.11.7.0, 2013.
[4] 3GPP Technical Specification 24.312 "Access Network Discovery and Selection Function (ANDSF) Management Object (MO)", v.11.6.0, 2013.
[5] 3GPP Technical Specification 24.302 "Access to the 3GPP Evolved Packet Core (EPC) via non-3GPP access networks; Stage 3", v.11.7.0, 2013.
[6] Wi-Fi Alliance Hotspot 2.0 Certified Passpoint Release 1, 2013.
[7] Wi-Fi Alliance Hotspot 2.0 Release 2, 2014.
[8] "Hotspot 2.0 – Making Wi-Fi as easy to use and secure as cellular", White paper, Ruckus Wireless, 2013.

8

Advanced UE and BTS Algorithms

Antti Toskala and Hisashi Onozawa

8.1 Introduction

This chapter presents an overview of recent developments of the advanced UE as well as of advanced BTS algorithms. First the different UE receiver versions are introduced based on the work done in different 3GPP Releases for defining advanced UE performance requirements. This chapter then continues with the introduction of the different scheduling solutions in BTS and then continues to cover BTS interference cancellation. This chapter is concluded with the outlook for looking other aspects of the UE and BTS algorithms, including the latest development of Network Assisted Interference Cancellation (NAIC) being investigated in 3GPP for Release 12.

8.2 Advanced UE Receivers

The first UE implementations on Release 99 were single receiver solutions based on the traditional Rake receiver, as introduced in [1]. Work then followed to cover the following types of receivers:

- Type 1 receiver in Release 6, which introduced two-antenna receiver diversity in the UE receiver, which became a typical solution especially in data centric products such as USB dongles and ExpressCards. The practical challenges related to design include achieving a low enough correlation between the antennas, especially with the lower frequency bands below 1 GHz. The small size of a device such as a USB dongle adds to the design challenge. Figure 8.1 shows an example performance impact of the antenna correction effect as introduced in [2], illustrating that with higher frequency bands the performance benefit from RX diversity is higher than with the lower frequency bands.

HSPA+ Evolution to Release 12: Performance and Optimization, First Edition.
Edited by Harri Holma, Antti Toskala, and Pablo Tapia.
© 2014 John Wiley & Sons, Ltd. Published 2014 by John Wiley & Sons, Ltd.

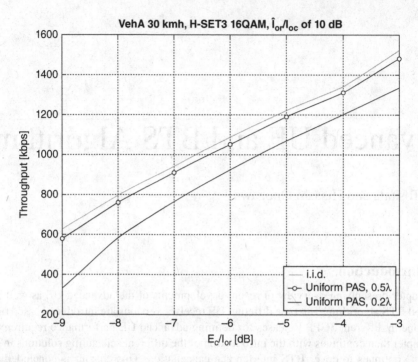

Figure 8.1 Impact to system performance for an example case with antenna correlation in the ideal case (i.i.d.) and with 2 GHz (0.5 λ) and 800 MHz cases (0.2 λ). (Holma and Toskala 2006 [2]. Reproduced with permission of Wiley)

- Type 2 with equalizer, also part of Release 6, introduced an equalizer to tackle the intra-cell interference caused by the multipath channel. With better tolerance of effects of the multipath channel, the usability of higher order modulation, in the case of HSDPA 16QAM and 64QAM, is also improved. Compared to a single antenna Rake receiver the capacity improvement is in the order of 30% [1]. The Type 2 receiver was the typical receiver for smartphone platforms that mainly had only a single antenna. The 3GPP reference equalizer was a Minimum Mean Square Error (MMSE) type equalizer, but such an equalizer was only used to generate the performance requirements, while a UE vendor has the freedom to choose another type of equalizer (for better performance or lower complexity) such as a frequency domain equalizer. In HSDPA signal reception often one needs more power for the channel estimation than available just from the CPICH; thus, using the actual data on HSDPA codes to assist the estimation process was a possible approach too.
- Release 7 with a Type 3 receiver provided both receiver diversity and equalizer. The combined use of receiver diversity and equalizer results in further improved performance, as shown in Figure 8.2 comparing the different receiver types. This soon became very common in dongles, which can also have receiver diversity. With the multimode UEs also supporting LTE, receiver diversity has become necessary in smartphones as it is a mandatory feature in LTE devices. This is paving the way for the use of the same diversity antennas for HSDPA reception in smartphones too.

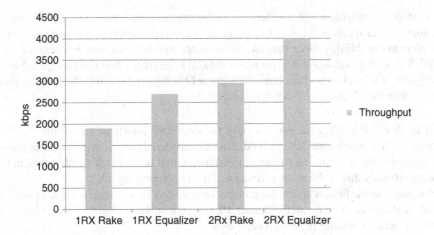

Figure 8.2   Average downlink throughput with different type of UE receivers

The use of an advanced receiver is not necessarily applied to channels other than HS-PDSCH as 3GPP does not specify the advanced receiver performance. It is implementation-dependent if any type of receiver is applied to the Dedicated Channel (DCH). Thus, for a given service running on HSDPA instead of on DCH, the performance would be boosted both by the better receiver with diversity and due to the HSDPA features – including the link adaptation and physical layer retransmissions as covered in Chapter 2.

- With the Type 3i receiver, the further element addressed is inter-cell interference. The Type 3i receiver has a combination of receiver diversity, equalizer, and inter-cell interference cancellation. This improves the performance, especially at the cell edge area when there is a dominant interfering cell. This is illustrated in Figure 8.3 with two different geometries

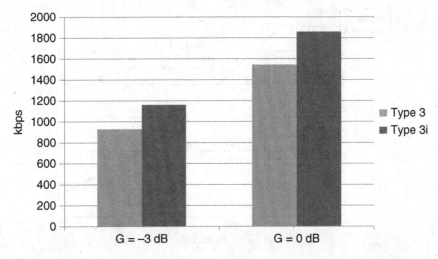

Figure 8.3   Cell edge performance with and without UE interference cancellation

(relation of the neighbor cell interference to the interference from own cell). When talk-ing about the 2i receiver it refers to a single antenna equalizer receiver with interference cancellation capabilities. Note that the performance numbers are now lower compared to Figure 8.2, as in this case the devices are in the cell edge area, while the average achievable throughput in the cell is higher. The relative benefit is higher the lower the geometry value G is, as then the neighbor cell interference is highest.

A 3i or 2i capable UE may address only the dominant interferer (as in the 3GPP test case defined) or multiple interferers could also be considered if the environment has several strong signals to cancel. Aiming to cancel interference that is too weak would result in noise enhancements only due to the noisy estimates of the interfering signals.

Performance in the field with the 3i receiver compared to Type 3 receiver is also studied in the field, as shown in more detail in Chapter 10, showing similar results with the biggest gain from the 3i receiver coming in the cell edge area.

In implementations in the field, there are different types of algorithms being used, which vary from the performance point of view depending on the solution and also depending on the environment and test cases for which the solution has been optimized. An example study in Figure 8.4 shows a test case result with different Category 13 or 14 devices (supporting 15 codes with HSDPA reception and 64QAM) and Figure 8.4 illustrates the differences between different generations of devices even from the same manufacturer, proving the importance of the advanced algorithms that are chosen to be used in a particular baseband modem. Thanks to the progress of silicon technology, advanced algorithm implementation is possible even if it requires more computational capacity. Today, the advanced receiver can be implemented with a reasonable chip size, production cost, and power consumption. The impact for end

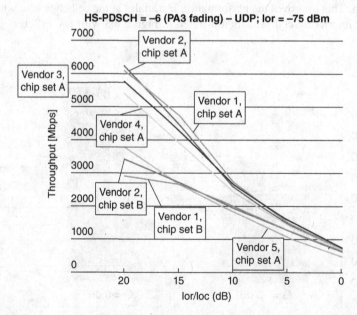

Figure 8.4   Pedestrian A 3 km/h test environment. Data from [3] courtesy of Signals Research Group

user performance is significant as a modern device can provide double the data rate in good conditions compared to the older generation device with the older chip set.

In low geometry regions, the performance is limited by Ioc rather than the multipath interference. On the other hand, the equalizer algorithm and RF circuit linearity affect the throughput in high geometry regions, where more performance difference is observed between chip sets.

From a smartphone point of view, most HSPA capable devices do not have receiver diversity, even though it is mandated for the LTE side as discussed earlier. When supporting the same band for both LTE and HSPA the necessary antennas and RF components are also already in place for HSPA receiver diversity, which is likely to make Type 3 or 3i more common in smartphone platforms too.

A further step in the receiver design is the MIMO receiver that is able to receive data from two different receiver branches, as discussed in Chapter 5. MIMO with HSDPA has not become commonly used in the devices, however, due to the problems caused for legacy devices before the introduction of the mitigation solutions discussed in Chapter 3. In the case of MIMO, one may further consider more advanced algorithms such as the Successive Interference Canceller (SIC) type of receiver. The simplest SIC receiver for two-stream MIMO is a single stage inter-stream canceller. The SIC receiver decodes the primary stream and then cancels it prior to decoding the secondary stream. The cancellation can be done using hard decisions or soft decisions. The soft decision has better reliability when the decoding of the primary stream fails but the implementation complexity is slightly high. This simple version of SIC gives most of the cancellation gain without increasing the number of Turbo decoder executions.

Also with MIMO, one can consider using Maximum Likelihood Detection (MLD). Since MLD complexity is extremely high for higher order modulation with multiple streams, simplified algorithms such as Sphere Decoding (SD) and QR Decomposition Based M Algorithm (QRM-MLD) have been proposed. There are different possibilities for the inter-stream interference cancellation with HSPA MIMO, such as those studied in [4] and references there in.

8.3 BTS Scheduling Alternatives

From a 3GPP point of view, the BTS scheduling implementation in terms of the kind of algorithm to use has been left for BTS vendor implementation; only the necessary signaling is specified to provide the scheduler with the necessary information both from the UE side, like the CQI feedback covered in Chapter 2, as well as information over the Iub interface, for example on the UE capability and possible QoS parameters.

With the initial phase of HSDPA introduction in the field with the first Release 5 devices, the type of BTS scheduler was of no major importance. When the HSDPA penetration was low, there were in many instances only a single UE in the cell, thus there was not much the scheduler could do than serve the single UE receiving data. In this case, the round robin type of scheduling was able to reach reasonable performance. The basic idea with the round robin scheduler is to serve different users one by one (assuming there is data to be transmitted) in a cyclic manner. In many cases a carrier was also shared in the power domain with legacy traffic using DCH and there were also limitations in the code space available for HSDPA use. Figure 8.5 shows an example of round robin type scheduler functionality.

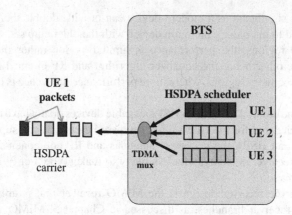

Figure 8.5 Round robin scheduler with HSDPA

The use of round robin type scheduling will not benefit from the variations in channel conditions between users as it will not try to schedule a user when the channel conditions are favorable. While a proportional fair scheduler considers basically the UE with the best channel conditions (as determined based on the Channel Quality Information (CQI) received) as the one to schedule. In order to ensure fairness it also takes into consideration that over the averaging period each UE should get the same number of scheduling instances. The most extreme form of scheduling, and also best from the system capacity, is the max C/I type scheduler, which always schedules only the user in the best channel conditions. While the total system capacity is maximized, the QoS distribution between users is poor as some users will not get a service at all while cell center UEs will have very high data rates. It is quite obvious that in a commercial system one has to use something other than a max C/I type scheduler. The proportional fair scheduler principle is illustrated in Figure 8.6.

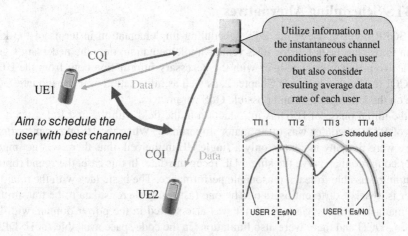

Figure 8.6 Proportional fair scheduler

The consideration of sufficient quality for each user with proportional fair scheduler can be considered with Equation 8.1, where the selection metric M_k is calculated based on the expected data rate R_k achievable as derived from CQI and the average throughput T_k experienced over the observation window.

$$M_K = \frac{R_k}{T_k}$$

(8.1)

This principle also ensures that cell edge users get schedulers and reach reasonable through-put while the data rate of the users in better conditions is also better. A scheduler could also force all users to reach an equal data rate over the averaging window, but this would lead to a clear drop in the overall system capacity.

Once the basic scheduler framework is such that suitable capacity is ensured, the next step is to ensure sufficient quality by considering the type of service. For example, services such as CS or PS voice (which both can be mapped on top of HSPA as well), or the Signaling Radio Bearer (SRB) can be given higher priority in the scheduling queue compared to the regular background PS data.

With the larger number of HSDPA users, and varying data rates, it is also important to be able to schedule users in parallel in code domains. With code multiplexing a small packet does not end up reserving the full 2-ms TTI on the carrier. Many of the earlier phase devices were also limited to using a maximum of 5 or 10 parallel codes on HSDPA, which also requires the use of code multiplexing for maximum efficiency when all or most of the carrier capacity can be used with HSDPA. Lately, most of new UEs in the market can support up to 15 codes, and thus in order to minimize the signaling overhead it makes sense to try to schedule just one UE at a time.

The operation with HSUPA also needs scheduling operation to control the uplink inter-ference level, as addressed in Chapter 2. While the code resource is not the limiting factor now due to the UE specific code tree under the UE specific scrambling code, the key is to manage the power levels and resulting total noise rise. The MAC header info on the buffer status and power headroom enable decisions in the HSUPA scheduler on whether to increase or decrease the data rate of a particular user with relative or absolute grants, but the key dif-ference from downlink is that normally some minimum data rate is transmitted by each user, while in downlink with HSDPA a UE might not get allocation at all during a particular TTI as maximum of four HSDPA users can be normally allocated simultaneously within a single TTI. The HSUPA scheduler principle is illustrated in Figure 8.7, with the BTS HSUPA scheduler determining first, based on the noise rise, whether load should be increased or decreased and then determining which user should be impacted by the actions in increasing or decreasing the uplink data rate (in principle the power ratio allocation, effectively converting to a particular data rate).

8.4   BTS Interference Cancellation

With all users sharing the spectrum, it is well known that a WCDMA system is interference limited. This is even more the case in the uplink where users do not have orthogonality like the channelization codes in the downlink, though depending on the channel the orthogonality is in many cases largely reduced.

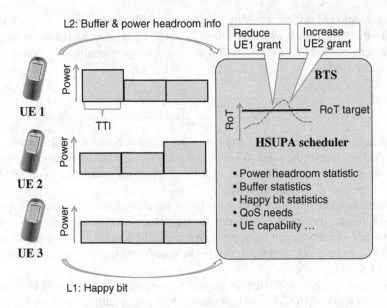

Figure 8.7 HSUPA scheduler

Similar to the earlier introduced 3i receiver for UE, in the BTS side it is also possible to cancel part of the interference caused by other users in the system. This is an item that has been investigated a lot since the 1990s. The interference cancellation or multiuser detection in a broader sense can be done in many ways. The linear equalizer type detector, such as the decorrelating detector, is a linear filter trying to suppress the interference. The Interference Cancellation (IC) receiver aims to actually estimate the interference signal and then to subtract that from the desired signal to gain better detection and decoding outcomes.

The type of IC receiver can be further classified as Serial (or Successive) Interference Cancellation (SIC) or as Parallel Interference Cancellation (PIC). The use of a SIC receiver aims to detect one user at a time, cancel that, then proceed to detecting the next user, while the PIC receiver performs the operation in parallel to all users. The use of SIC receivers easily creates some issues with the latency, especially when the number of users is large.

Often, such a receiver aims to have multiple iterations to allow better reliability, thus operating as a multistage receiver. An example principle of a PIC receiver for two users is shown in Figure 8.8, with the first phase being a traditional receiver like a Rake receiver, and

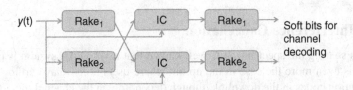

Figure 8.8 PIC receiver for two users

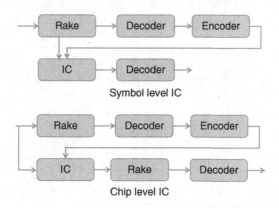

Figure 8.9 Symbol and chip level IC using output from the channel decoder

then, based on the decisions in the first stage, the interference would be cancelled from both users and further stages would then aim to improve the reliability.

Different approaches also exist based on which signal to regenerate the interfering signal to be cancelled. One may try to estimate the signal before turbo decoding, as studied for example in [5], or shown in Figure 8.8.

Another approach is to aim to decode the block of symbols for better reliability and take the signal for regeneration and cancellation operation on after convolutional or turbo decoding has been carried out, as shown in Figure 8.9. Such an approach results in better reliability of the signal to be cancelled. Counting on symbols that are too unreliable, with the interfering signal regeneration and cancellation before channel decoding, could in some situations lead to noise enhancement rather than actually improving detection. The cancellation could be done at chip level (by regenerating the spread signal as part of the IC operation) or at the symbol level (by considering, for example, the correlation properties between the codes and then removing the impact at the symbol level after the Rake receiver) as shown in Figure 8.9.

What was, in the 1990s, only of academic interest has become a reality with the increased computational powers in HSPA-capable base stations, first in narrowband CDMA, as smaller bandwidth is easier to handle due to the small chip rate, but lately as enabled in the WCDMA commercial products as well.

The use of HSUPA with turbo encoding and physical layer retransmission creates some additional challenges for the interference cancellation operation, especially if using the solution as in Figure 8.5. With efficient channel coding and HARQ the resulting signal energy at the receiver is very low and BER for channel symbols/soft bits is very high. Thus making decisions on individual symbols is not fully reliable in all situations. There are also benefits from the use of HSUPA compared to Release 99. With the higher peak data (and thus also smaller spreading factors) rates there are now individual users that momentarily typically dominate the interference situation. Addressing such a high data rate user with interference cancellation allows a clearly improving uplink data rate, especially when the solution is such that it also enables channel equalizer functionality.

The capacity benefit of an interference cancellation solution depends on both the cancellation efficiency and also on the ratio of the own cell interference and neighbor cell interference in

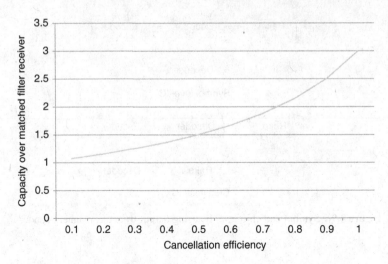

Figure 8.10   Example BTS interference cancellation performance

the particular environment. For the case with identical noise rise for both the case with and without interference cancellation, the capacity can be given as:

$$Capacity\ Increase = \frac{1+i}{1+i-\beta} \qquad\qquad (8.2)$$

With $\beta =$ to the interference cancellation efficiency (studies to be in the order of 25–40% in [1]) and i is the ratio of other-cell interference to own-cell interference. Thus, for an equal share of interference from other cells and own cell, and 40% efficiency, the resulting capacity gain would be in the order of 35%.

An example performance result is shown in Figure 8.10 for different cancellation efficiencies when the own cell and neighboring cells have an equal contribution to the interference level. Obviously the close to 100% interference cancellation is a rather theoretical case as there will be always some errors in the estimation, finite word length effects in implementation, as well as missing propagation paths in the cancellation process and so on; imperfections that degrade the practical performance.

The performance with an example interference cancellation solution is shown in Figure 8.11, illustrating how the system capacity would normally degrade due to the extra overhead with a larger number of HSUPA users, while with the use of BTS IC the capacity level is mostly retained even with the increased user numbers. The performance in a single user case would not see too much impact from the IC operation, since own transmission does not degrade the performance that much but, as the total power used by other users increases, then the overall performance goes down unless IC is used. As shown in the Figure 8.11, the performance improvement varies depending on the scenario from roughly 15 up to 80% for the larger number of simultaneous users case. From the end user data rate (average) point of view, the impact is quite significant especially in heavily loaded cases, as with seven HSUPA users (and voice traffic) one ends up sharing only approximately 1.8 Mbps total capacity while with BTS IC one gets to share approximately 3.2 Mbps capacity among active users.

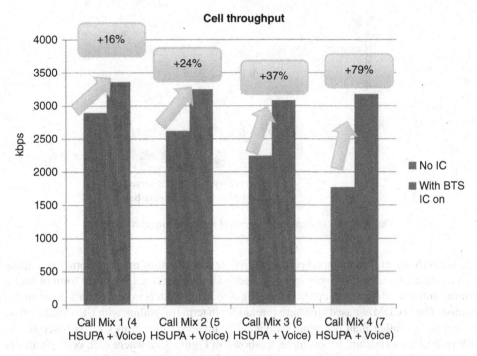

Figure 8.11 Example BTS IC measurement result of the achievable performance improvement

8.5 Further Advanced UE and BTS Algorithms

From the UE side, the link performance is further impacted by, for example, the solutions used for turbo decoding, deciding which CQI value to transmit to NodeB, or creation of the downlink power control commands (in case of DCH used for voice). Such algorithms mostly have the minimum performance criteria coming from the 3GPP performance specifications, ensuring sufficient performance for those algorithms. The mobility related algorithms, such as those used for cell search and neighbor cell measurements, are also essential to ensure good link maintenance and avoid unnecessary radio link failures.

One of the areas under investigation in 3GPP is Network Assisted IC (NAIC), where the UE would aim to cancel the strongest interferer and would obtain information from the network to make the cancellation work better. In the simplest form the UE would be provided with information on the modulation and channelization codes used, thus basing the cancellation on the soft channel bits estimates. If the solution as described in Figure 8.6 were to be used then more information, such as transport block size, would be needed to enable the actual turbo decoding operation to regenerate and subtract the interfering signal. More information on the NAIC for HSDPA UE can be found from [6]. Another approach is to aim to perform a similar operation blindly, by simply aiming to detect the interfering signal blindly. Obviously then the actual turbo decoding of the strongest (if aiming only to cancel the dominant interferer) interfering signal is not feasible as it is missing information such as the block size used with encoding and the redundancy version in use.

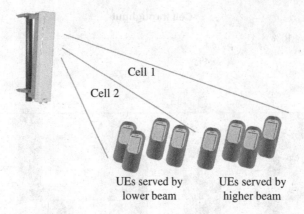

Figure 8.12   Operation with vertical sectorization with HSDPA

In the BTS algorithms the good power control algorithm was of major importance, starting from Release 99. Further, the solutions in advanced antenna technology in the form of beam-forming antennas also require sufficient intelligence in the NodeB to benefit most from such a solution. The WCDMA-based UEs have limitations in terms of dealing with UE-specific phase reference, thus rather than operating individual beams one rather uses CPICH-based beams with vertical sectorization, for example, as shown in Figure 8.12 where each cell operates as a cell on its own and thus uses its own CPICH. 3GPP is carrying on further work in this area to enable 3D-channel models. While the work formally is valid for LTE, typically a channel model is not going to be technology specific, thus enabling the use of channel models for HSPA related studies as well. The latest versions of 3GPP 3D-channel models can be found from [7].

8.6   Conclusions

In this chapter we have looked at some of the developments and possibilities of advanced UE and NodeB algorithms. Since the early days of WCDMA introduction, the technology evolution on both the UE and NodeB sides has enabled several technological improvements to also be introduced commercially in the field. One example area is interference cancellation, which has been made commercial both on the UE and on the NodeB sides.

Some of the improvements are based on specification development, like the advanced UE receiver, while most of the algorithm ones, such as BTS IC, are dependent on the manufacturer of the particular equipment. Even for the case of a 3GPP feature such as the advanced UE receiver, where performance requirements exists, the 3GPP requirements are rather representing a reference implementation while actual state-of-the-art implementations are likely to be able to perform clearly better as the studies performed on different HSPA chip sets were able to show too. The type of HSPA chip set used may have a large impact on the resulting performance, especially with higher geometry values (closer to the base station).

3GPP is working with studies in the area of performance improvements more driven by LTE, but clearly where applicability for HSPA can be seen too. Examples of such items are NAIC, with work items for LTE, termed Network Assisted IC and Suppression (NAICS) or

3D-channel modeling for enabling 3D-beamforming studies. These pave the way for new potential enhancements that could be introduced in the networks, improving state-of-the-art HSPA network performance even further. More Release 12 items are covered in Chapter 15.

References

[1] Holma, H. and Toskala, A. (2010) *WCDMA for UMTS*, 5th edn, John Wiley & Sons, Ltd, Chichester.
[2] Holma, H. and Toskala, A. (2006) *HSDPA/HSUPA for UMTS*, John Wiley & Sons, Ltd, Chichester.
[3] Signals Ahead, Volume 7, No. 5 April 26, 2011, www.signalsresearch.com.
[4] Yu, H., Shim, B. and Oh, T.W. (2012) Iterative interstream interference cancellation for MIMO HSPA+ system. *Journal of Communications and Networks*, **14**(3), 273–279.
[5] Toskala, A., Hämäläinen, S. and Holma, H. (1998) Link and system level performance of multiuser detection CDMA uplink. *Wireless Personal Communications: An International Journal Archive*, **8**(3), 301–320.
[6] 3GPP Technical Report, TR 25.800, "Study on UMTS heterogeneous networks" v 12.0.0 September 2013.
[7] 3GPP Technical Report, TR 36.873, "3D channel model for LTE", v 2.0.0, March 2014.

9

IMT-Advanced Performance Evaluation

Karri Ranta-aho and Antti Toskala

9.1 Introduction

This chapter presents a performance evaluation of HSPA Release 11 compared to the requirements for IMT-Advanced as defined by the ITU-R. First the IMT-Advanced requirements are presented and then the different features included in Release 11 3GPP specifications to address these requirements are briefly introduced to highlight how they help in achieving the improved performance. The achievable performance with different features is covered. This chapter is concluded by benchmarking with the LTE and LTE-Advanced.

9.2 ITU-R Requirements for IMT-Advanced

The International Telegraphic Union Radiocommunications sector (ITU-R) [1] started the process for the new system, the International Mobile Telecommunications Advanced (IMT-Advanced), with a circular letter distributed in 2008. The circular letter was a call for radio technology proposals, in a similar fashion as had been done earlier for the IMT-2000 process in connection with the definition of 3G radio access solutions [2]. Now the ITU-R was calling for proposals for radio technologies that could meet the requirements set for IMT-Advanced technology, widely termed the 4th generation (4G) mobile communication systems.

The ITU-R defined the requirements for the IMT-Advanced (4G) such that the system should be able to achieve the following [3]:

HSPA+ Evolution to Release 12: Performance and Optimization, First Edition.
Edited by Harri Holma, Antti Toskala, and Pablo Tapia.
© 2014 John Wiley & Sons, Ltd. Published 2014 by John Wiley & Sons, Ltd.

- Enable 100 Mbps peak data rate support for high mobility and up to 1 Gbps peak data rate for the low mobility case.
- Allow inter-working with other radio access systems.
- Enable high quality mobile services.
- Capable of worldwide roaming.
- Flexibility to allow cost efficient support of a wide range of services and applications.
- Bandwidth scalability up to and including 40 MHz, with considerations up to 100 MHz.

Furthermore, there were more detailed performance requirements in the following areas:

- Peak spectral efficiency in downlink and uplink.
- Cell spectral efficiency, ranging from 3 bits/s/Hz/cell in the indoor downlink scenario, to 0.7 bits/s/Hz/cell in the 120 km/h rural uplink scenario.
- Cell edge user spectral efficiency, ranging from 0.1 bits/s/Hz/user in the indoor downlink scenario, to 0.015 bits/s/Hz/user in the 120 km/h rural uplink scenario.
- Mobility support with up to 350 km/h (smaller data rate allowed compared to the stationary use case).
- Latency requirements for the control plane to achieve 100 ms transition time between idle and active state, and respectively to enable 10 ms user plane latency (in unloaded conditions).
- Handover interruption of 27.5 ms for the intra-frequency case and 40 and 60 ms for the inter-frequency within the band and between the bands respectively.
- VoIP capacity, with the numbers of users ranging from 30 to 50 users per sector/MHz.

The requirements related to items such as peak spectral efficiency requirements or functional requirements such as achievable latency or handover interruption time can be investigated relatively easy, while the requirements related to the average and cell edge spectral efficiency, with examples shown in Table 9.1, typically need detailed system-level studies with simulations.

It was rather obvious that new features were needed on top of Release 10 HSPA in order to meet some of these requirements with HSPA technology. As shown in the following chapters, the requirements for dealing with items such as handover delay or control plane latency were

Table 9.1 Summary of key ITU-R IMT-A requirements

System performance requirements	ITU-R requirement
Downlink peak spectrum efficiency	15 bits/s/Hz (max 4 antennas)
Uplink peak spectrum efficiency	6.75 bits/s/Hz (max 2 TX antennas)
Downlink average cell spectral efficiency	≥1.1…3.0 bits/s/Hz
Uplink average cell user spectral efficiency	≥0.7…2.25 bits/s/Hz
Downlink cell edge user spectral efficiency	≥0.04…0.1 bits/s/Hz
Uplink cell edge user spectral efficiency	≥0.015…0.07 bits/s/Hz
User plane latency	≤10 ms
Control plane latency	≤100 ms
Handover interruption	≤27.5…60 ms

already at such a level with HSPA that fundamental changes to the control plane operation in order to meet IMT-Advanced requirements could be avoided.

9.3   3GPP Features to Consider in Meeting the IMT-Advanced Requirements

While LTE-Advanced was developed during Release 10 and submitted to the IMT-Advanced process, no such activity took place for HSPA in Release 10. It was only in the next 3GPP Release, Release 11, that the following key items were studied and specified for HSPA in order to meet the ITU-R requirements defined for IMT-Advanced:

- Eight-carrier aggregation, aggregating up to eight 5-MHz carriers together, thus giving up to eight times faster data speeds compared to single-carrier operation, as covered in Chapter 3.
- Four-transmit and four-receiver antenna (4 × 4) Multiple Input Multiple Output (MIMO) in the downlink direction, roughly doubling peak data rates and system capacity for devices with four receiver antennas compared to devices with two receiver antennas, as covered in Chapter 3, enabling to meet the requirements in Table 9.2.
- Uplink beamforming, 2 × 2 MIMO and 64QAM. Uplink improvements which increase the peak data rate, user performance, and cell capacity, as discussed in Chapter 3.
- HSDPA Multiflow – increases downlink user throughputs at the cell edge, as covered in Chapter 5.

These features contribute to the HSPA peak data rate evolution from 3GPP Release 5 onwards, as shown in Figures 9.1 and 9.2. The features contributing to peak data rates are shown in Table 9.3. In the downlink direction it is possible to exceed 300 Mbps with the use of either 20 MHz of spectrum and four-transmit and four-receive antenna MIMO operation or using 40 MHz of spectrum (with eight-carrier aggregation) and two-antenna MIMO operation. In the uplink, respectively, the use of 64QAM and uplink two-antenna MIMO transmission could increase the peak data rate for 70 Mbps when aggregating two uplink carriers (dual-carrier HSUPA) but are not supported simultaneously in Release 11.

Table 9.2   Latency and handover interruption requirements

		IMT-A minimum requirement	HSPA Release 10 and prior
Downlink peak spectral efficiency		15.0 bits/s/Hz	8.6 bits/s/Hz
Uplink peak spectral efficiency		6.75 bits/s/Hz	2.3 bits/s/Hz
Bandwidth scalability		Scalable, up to 40 MHz	Scalable, up to 20 MHz
Control plane latency		≤ 100 ms	75 ms
User plane latency		≤ 10 ms	8 ms
Handover interruption	Intra-frequency	≤ 27.5 ms	0 ms
	Inter-frequency	≤ 40 ms	13 ms
	Inter-band	≤ 60 ms	13 ms

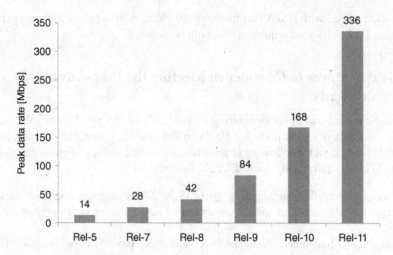

Figure 9.1 Downlink (HSDPA) peak rate evolution

When comparing the achievable peak spectrum efficiency values with IMT-Advanced requirements and improvements achieved in Release 11 HSPA, it can be noted that the IMT-Advanced requirements can be reached, as shown in Table 9.4, with the introduction of

- 4×4 MIMO for peak downlink spectral efficiency;
- 2×2 MIMO and 64QAM for peak uplink spectral efficiency;
- eight-carrier HSDPA for bandwidth.

Note that Release 10 refers to the features included up to and including Release 10. The peak spectral efficiency values in Table 9.4 are based on simply calculating the maximum achievable peak data rate and dividing by the bandwidth required.

The Release 11 work was finalized in the first half of 2013.

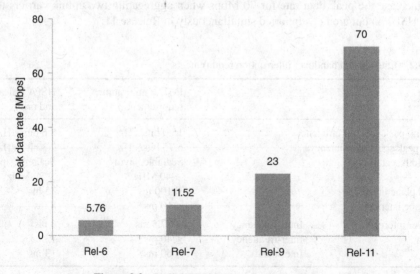

Figure 9.2 Uplink (HSUPA) peak rate evolution

Table 9.3   HSPA peak rate feature evolution from 3GPP Release 5 to Release 11

	Downlink (HSDPA) features	Uplink (HSUPA) features
Release 5	HSDPA (single carrier, no MIMO, 16QAM)	
Release 6		HSUPA (single carrier, no MIMO, QPSK)
Release 7	2 × 2 MIMO (28 Mbps), 64QAM (21 Mbps)	16QAM
Release 8	2 × 2 MIMO + 64QAM and dual carrier + 64QAM	
Release 9	Dual carrier + 2 × 2 MIMO + 64QAM	Dual carrier + 16QAM
Release 10	Four-carrier + 2 × 2 MIMO + 64QAM	
Release 11	Eight-carrier + 2 × 2 MIMO + 64QAM and four-carrier + 4 × 4 MIMO + 64QAM	Dual carrier + 2 × 2 MIMO + 64QAM

Table 9.4   Peak spectral efficiency and bandwidth scalability requirements

	IMT-Advanced minimum requirement	HSPA Release 10	HSPA Release 11
Peak spectral efficiency, downlink	15.0 bits/s/Hz	8.6 bits/s/Hz (2 × 2 MIMO + 64QAM)	17.2 bits/s/Hz (4 × 4 MIMO)
Peak spectral efficiency, uplink	6.75 bits/s/Hz	2.3 bits/s/Hz (16QAM)	6.9 bits/s/Hz (2 × 2 MIMO + 64QAM)
Spectrum flexibility	Scalable bandwidth, up to 40 MHz	Scalable up to 20 MHz (four-carrier HSDPA)	Scalable up to 40 MHz (eight-carrier HSDPA)

How each of these features individually contribute to meeting the IMT-Advanced requirements on average and cell edge spectrum efficiency based on the simulations is studied in the following sections. The peak spectrum efficiency and other resulting improvements are also addressed.

9.4   Performance Evaluation

9.4.1   Eight-Carrier HSDPA

Release 8 Dual-Carrier HSDPA (DC-HSDPA) has been widely taken in to use in many HSDPA networks, and it provides data rates of up to 42 Mbps, as introduced in Chapter 3. In Release 9 the solution was extended to cover dual-carrier operation on two different frequency bands, while further carriers were added with up to four carriers in Release 10 and up to eight downlink carriers in Release 11.

The resulting peak data rate (with the use of a 40 MHz spectrum for eight carriers) reaches 336 Mbps with 2 × 2 MIMO. The same peak rate can also be reached with four carriers and 4 × 4 MIMO, but notably the specification does not support simultaneous operation of 4 × 4 MIMO and more than four carriers. Release 11 also enables aggregating non-adjacent carriers on the same frequency band.

Aggregating multiple carriers brings substantial benefits for the end user because any free resources across all carriers can be used flexibly. This provides a dynamic pool of resources, with the multicarrier capable devices able to use any of the carriers with 2 ms resolution, thus allowing an even load with the bursty traffic too, leading to high trunking gain with the dynamic packet sharing over a large number of carriers.

Unlike the single carrier device, the multicarrier capable device may be re-allocated to the carrier or carriers that are experiencing the best propagation and interference conditions every 2 ms, thus maximizing the resulting system throughput.

The gains can be seen in Figure 9.3, which shows the cumulative distribution of the average user throughput and the mean packet call delay for macro cells with an average cell load of 1 Mbps.

The gains depend significantly on the load in the system. If the load is very high and there are a large number of users, then there will be fewer free resources on the other carriers and there is a high likelihood that there will be several users in a good propagation condition on each carrier, thus resulting in lower gains.

For meeting the requirements for the ITU-R IMT-Advanced system, the eight-carrier HSDPA allows aggregation of 40 MHz of spectrum, meeting the requirement for flexible system bandwidth support up to 40 MHz.

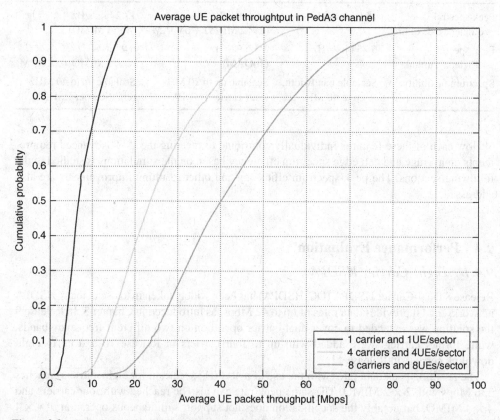

Figure 9.3 Cumulative distribution probability of the average device packet throughput for one, four, and eight carriers at low offered system load

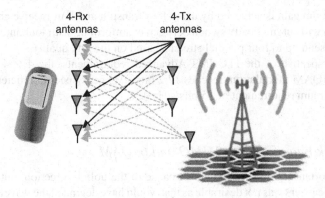

Figure 9.4 4 × 4 MIMO operation in downlink

9.4.2 Four-Antenna MIMO for HSDPA

The MIMO solution with two transmit and two receiver (2 × 2) antennas was introduced in Release 7, as covered in Chapter 3. To further push the multiantenna transmission capability, MIMO operation with four transmit and four receiver antennas (4 × 4) was introduced. In 4 × 4 MIMO the NodeB uses four transmit antennas and the UE uses four receiving antennas forming four different reception images for each four transmitted signals between the transmitter and receiver, as shown in Figure 9.4. This enables transmission of up to four parallel streams, which doubles the peak data rate compared to 2 × 2 MIMO and also improves the typical cell capacity and user data rates directly benefiting from additional receive antennas, even if four-stream transmission is not possible due to less favorable radio conditions.

This can be seen in Figure 9.5, showing the average macro cell throughput assuming an ideal channel. It can be seen that adding Rx antennas gives more benefits than adding Tx antennas,

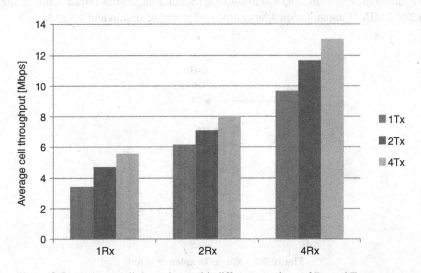

Figure 9.5 Average cell throughput with different numbers of Rx and Tx antennas

while the maximum gain is achieved by using four transmit and four receive antennas. In that case, the system will automatically switch between beamforming with four antennas and using the antennas to send up to four parallel streams based on the link quality.

From the perspective of the ITU IMT-Advanced requirements, the 4 × 4 MIMO, when coupled with 64QAM modulation, reaches 17.2 bits/s/Hz peak spectral efficiency, exceeding the IMT-A minimum requirement of 15 bits/s/Hz.

9.4.3   Uplink Beamforming, MIMO and 64QAM

It was also important to improve the performance in the uplink direction, but adding a very large number of carriers was not desirable as that would have degraded the waveform properties and resulted in very difficult power amplifier requirements. Instead, improvements were made for the single carrier transmission (dual-carrier uplink had also been defined earlier) to reach the bits/s/Hz requirement. The following solutions were introduced:

- Uplink beamforming allows uplink dual-antenna transmission to provide better data rate coverage and lower interference from neighboring cells.
- It will also double the peak uplink rate using dual-stream MIMO transmission and triple it when coupled with 64QAM modulation. Analogous to the downlink MIMO, the mobile device uses two transmit paths and antennas to form a complex radio wave pattern in the multiple base station receive antennas, as shown in Figures 9.6 and 9.7. In favorable radio conditions, this yields over 2 dB in the link budget, which translates into up to 30% higher average uplink data rates throughout the cell and up to 40% higher data rates at the cell edge. In another analogy to downlink MIMO, in very good channel conditions and when a high received signal-to-noise ratio is possible, the user equipment may use dual-stream MIMO transmission with two orthogonal beam patterns. This effectively doubles the raw bitrate on the physical layer.
- When further moving from 16QAM to 64QAM modulation, the raw bitrate is tripled together with 2 × 2 MIMO transmission when compared to single stream and 16QAM.

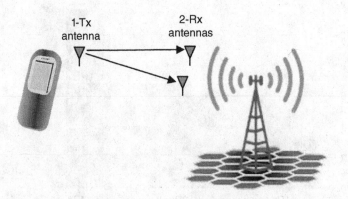

Figure 9.6   Single Tx antenna uplink

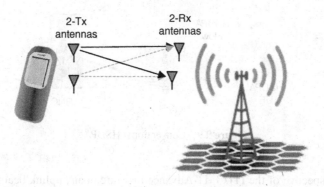

Figure 9.7 Dual Tx antenna uplink enabling beamforming transmit diversity and 2 × 2 MIMO

To achieve received signal-to-noise ratios that are high enough to make dual-stream transmission with 64QAM possible over a significantly large area, four or even eight receiver antennas, or a combination of both, may need to be deployed. Yet again this is analogous to what happens in the downlink.

Dual-antenna transmission in the uplink should be viewed as two separate features. First, there is uplink beamforming, which is possible and beneficial in most environments and which helps in the full cell area by providing better uplink data rate coverage. Second is the uplink dual stream MIMO, which is possible only in more limited scenarios and doubles the uplink peak rate.

Introducing 64QAM modulation does not require two transmit antennas, but when aiming for the highest peak rates, it needs to be coupled with uplink MIMO.

Figure 9.8 shows the gains of beamforming on the average and cell edge throughput. It can be seen that gains of up to 20 and 80% respectively can be achieved.

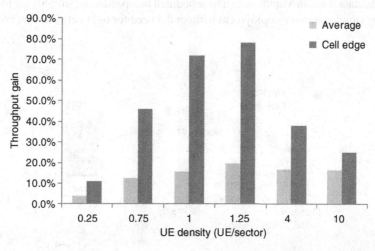

Figure 9.8 Average and cell edge (10%) gain from uplink beamforming in 2.8 km ISD network

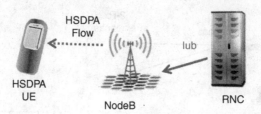

Figure 9.9   Conventional HSDPA

From the perspective of the ITU IMT-Advanced requirements, uplink beamforming helps achieve the average and cell edge performance requirements. The uplink 2×2 MIMO, together with 64QAM modulation, achieves 6.9 bits/s/Hz peak spectral efficiency, exceeding the IMT-A minimum requirement of 6.75 bits/s/Hz.

9.4.4   HSPA+ Multiflow

Another feature enabling a better use of resources in cellular systems is HSPA+ Multiflow, which is designed to improve cell edge data rates. Multiflow, as covered in more detail in Chapter 5, enables the transmission of data from multiple cells to a user device at the common cell edge, instead of transmitting the data via a single cell as in HSDPA today. This is illustrated in Figures 9.9 and 9.10 for dual-cell Multiflow operation. With inter-BTS Multiflow the RNC splits the incoming data stream to two independent flows that are transmitted to the UE by two independent NodeBs on the same carrier frequency. With intra-site Multiflow the NodeB MAC layer splits the single data flow coming from the RNC just like in multicarrier HSDPA operation, but the transmission takes place over neighboring sectors of the NodeB on the same carrier frequency, as opposed to multicarrier operation where the data is transmitted on different carrier frequencies, but on the same sector.

Each of the data flows in Multiflow can be scheduled independently, simplifying the concept and enabling simple inter-site deployment without the need for tight network synchronization.

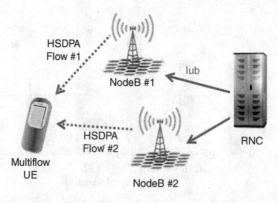

Figure 9.10   HSPA+ Multiflow

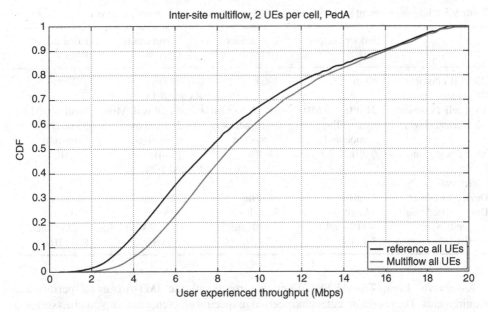

Figure 9.11 Cumulative probability distribution of user throughputs with and without Multiflow

The presence of two HSDPA flows from two cells leads to a doubling of the power available for the desired signal at the device, which is used to increase the overall user throughput. For 3GPP Release 11, Multiflow is considered for up to four different flows over two different frequencies, enabling the radio network to send data from two different base stations and up to four different cells to a user device.

Figure 9.11 shows the cumulative distribution of the throughput experienced by the user with and without Multiflow (including both intra-site and inter-site Multiflow devices). At the low values of the cumulative distribution, users at the cell edge gain particular benefit from Multiflow, since they are the most likely to receive transmissions from multiple cells with adequate signal quality.

From the perspective of the ITU IMT-Advanced requirements, HSPA+ Multiflow helps achieve the required performance at the cell edge.

9.4.5 Performance in Different ITU-R Scenarios

The key parameters of the ITU-R deployment environments are described in Table 9.5.

The system performance requirements for a candidate IMT-Advanced radio technology are shown in Table 9.6. The evaluation criteria state that it is sufficient to meet each requirement type in three out of four deployment scenarios to qualify as an IMT-Advanced radio.

The evaluation of Long Term HSPA Evolution performance against these requirements was both a major simulator development effort and a long-lasting simulation campaign. The following figures (Figures 9.12, 9.13, 9.14, and 9.15) show the simulation results against the ITU IMT-Advanced set requirements.

Table 9.5 Key deployment scenario parameters for the system performance evaluation

	Indoor hotspot InH	Urban micro UMi	Urban macro Uma	Rural macro RMa
BS-to-BS distance	60 m	200 m	500 m	1732 m
BS antenna elements		Up to 8 Rx and 8 Tx		
Total cell Tx power	21 dBm/20 MHz	41 dBm/10 MHz	46 dBm/10 MHz	46 dBm/10 MHz
BS antenna height	6 m, ceiling mounted	10 m, below rooftop	25 m, above rooftop	35 m, above rooftop
BS antenna gain	0 dBi	17 dBi	17 dBi	17 dBi
Device antenna elements		Up to 2 Rx and 2 Tx		
Device Tx power	21 dBm	24 dBm	24 dBm	24 dBm
Device speed (high speed)	3 km/h (10 km/h)	3 km/h (30 km/h)	30 km/h (120 km/h)	120 km/h (350 km/h)
Carrier frequency	3.4 GHz	2.5 GHz	2.0 GHz	800 MHz

As shown, Long Term HSPA Evolution meets all the IMT-Advanced performance requirements. The results do not include cell edge spectral efficiency features, such as scenario-optimized scheduler or HSPA+ Multiflow, but assume a fairly advanced interference suppressing receiver in the device.

9.4.6 Latency and Handover Interruption Analysis

Table 9.2 listed the ITU-R IMT-Advanced requirements on latency and handover interruption as well as what the UMTS system was already able to achieve. The requirements should be understood as limits that the technology can fulfill, with state-of-the-art implementation and best case deployment.

Table 9.6 IMT-Advanced system performance requirements

	Indoor hotspot InH	Urban micro UMi	Urban macro UMa	Rural macro RMa
Downlink average spectral efficiency	3.0 bits/s/Hz	2.6 bits/s/Hz	2.2 bits/s/Hz	1.1 bits/s/Hz
Downlink cell edge spectral efficiency	0.1 bits/s/Hz	0.075 bits/s/Hz	0.06 bits/s/Hz	0.04 bits/s/Hz
Uplink average spectral efficiency	2.25 bits/s/Hz	1.8 bits/s/Hz	1.4 bits/s/Hz	0.7 bits/s/Hz
Uplink cell edge spectral efficiency	0.07 bits/s/Hz	0.05 bits/s/Hz	0.03 bits/s/Hz	0.015 bits/s/Hz
High speed spectral efficiency	1.0 bits/s/Hz	0.75 bits/s/Hz	0.55 bits/s/Hz	0.25 bits/s/Hz
VoIP capacity	50 users/MHz	40 users/MHz	40 users/MHz	30 users/MHz

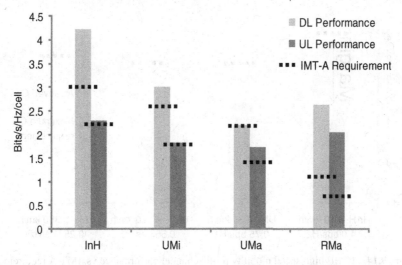

Figure 9.12   LTHE cell average spectral efficiency vs. IMT-A requirement

The user plane latency was analyzed in the same way as was done for LTE-Advanced, looking at the lowest possible air interface delay, and coupling it with HARQ round-trip time averaged with the HARQ retransmission probability. These components are illustrated in Figure 9.16 and Table 9.7.

Control plane latency analysis also followed the practice used in performance evaluation of LTE-Advanced, consisting of the best case latencies from the UE initiating a connection request to the dedicated connection being set up. The latency analysis is detailed in Table 9.8.

Due to the make-before-break nature of soft handover there is no interruption in the intra-frequency handover of WCDMA. HSDPA intra-frequency handover with Multiflow also

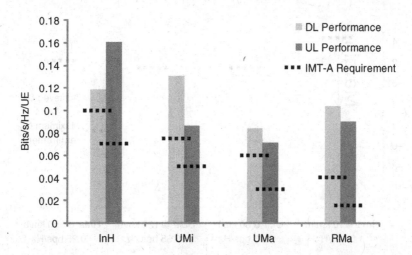

Figure 9.13   LTHE cell edge spectral efficiency vs. IMT-A requirement

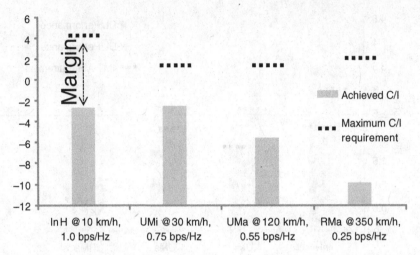

Figure 9.14   LTHE high speed mobility traffic channel performance vs. IMT-A requirement

experiences zero break as first the UE connects to the target cell, and only after that does the UE disconnect from the source cell. Furthermore, with HSPA technology it is possible, with RNC based bicasting and using enhanced HSDPA serving cell change, to achieve zero-break even without Multiflow. In uplink, HSUPA benefits from soft handover and thus by definition there is no break in uplink intra-frequency handovers.

Inter-frequency and inter-band handovers are effectively the same and when executing intra-BTS timing maintained hard handover, even though the handover is break-before-make, the reacquisition of the synchronization is extremely fast. Thus the very basic Release 99 WCDMA standard was already able to meet the handover interruption requirements set much later for IMT-Advanced.

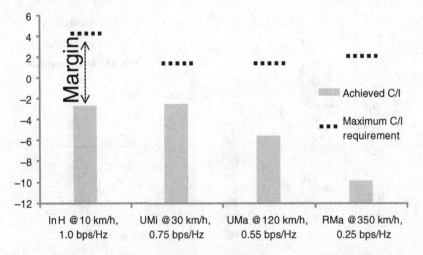

Figure 9.15   LTHE voice outage performance vs. IMT-A requirement

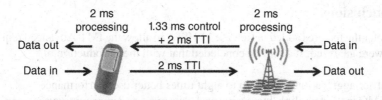

Figure 9.16 HSPA user plane latency components

Table 9.7 HSPA user plane latency analysis

	Downlink	Uplink
NodeB processing	2 ms	2 ms
Air interface delay	3.33 ms	2 ms
UE processing	2 ms	2 ms
Total with zero BLER	**7.33 ms**	**6 ms**
HARQ RTT	12 ms	16 ms
Total with 10% BLER	**8.53 ms**	**7.6 ms**

Table 9.8 HSPA control plane latency analysis

Delay component	Delay [ms]
Average delay due to RACH scheduling period	0.66
Time from the beginning of PRACH preamble to the beginning of acquisition indication and channel assignment	2
Average time from the beginning of AICH to the beginning of UE uplink transmission start	3.33
Time from start of UE uplink transmission to start of UL data transmission (UL synch)	10
Transmission of RRC and NAS request	2
NodeB and RNC processing delay (L2 and RRC)	10
Transmission of RRC connection setup	3.33
Processing delay in the UE (L2 and RRC)	12
Transmission of RRC connection setup complete	2
Processing delay in collapsed architecture NB/RNC (Uu → Iu)	
Iu transfer delay	
CN processing delay (including UE context retrieval of 10 ms)	
Iu transfer delay	
Processing delay NodeB amd RNC (Iu → Uu)	10
Transmission of RRC security mode command and connection reconfiguration (+TTI alignment)	3.67
Processing delay in UE (L2 and RRC)	16
Total delay (Requirement 100 ms)	**75**

9.5   Conclusions

In this chapter the features added in Release 11 HSPA to allow the IMT-Advanced requirements to be met were analyzed. It could be concluded that with the introduction of

- eight-carrier aggregation, giving up to eight times better user performance;
- 4×4 MIMO in the downlink, increasing peak data rates, and improving system performance;
- uplink beamforming, 2×2 MIMO and 64QAM; uplink improvements increase the peak data rate, user performance, and cell capacity;
- Multiflow – increases downlink user throughputs at the cell edge;

Release 11 HSPA, sometimes referred too as Long Term HSPA Evolution, can also meet the criteria as IMT-Advanced technology as defined by the ITU-R. To meet the increased network capacity demand caused by growing mobile data traffic, HSPA has continued to be improved in parallel with LTE and LTE-Advanced.

HSPA is currently the radio access technology serving the most wireless broadband users worldwide and the load on the system is expected to increase further in the coming years, thus creating demand for performance improvements in the future, facilitating introduction of new performance improvements features to the commercial HSPA networks expected to be serving more than 2 billion users worldwide by end of 2014.

References

[1] Holma, H. and Toskala, A. (2012) *LTE Advanced: 3GPP Solution for IMT-Advanced*, John Wiley & Sons, Ltd, Chichester.
[2] Holma, H. and Toskala, A. (2010) *WCDMA for UMTS*, 5th edn, John Wiley & Sons, Ltd, Chichester.
[3] ITU-R report, M.2134, "Requirements related to technical performance for IMT-Advanced radio interface(s)".

10

HSPA+ Performance

Pablo Tapia and Brian Olsen

10.1 Introduction

HSPA+ is nowadays a mature network technology that has been deployed globally for a number of years. Technologies such as 64QAM, Advanced Receivers, and Release 8 Dual Cell are well adopted today in a large number of network and devices.

This chapter analyzes the performance of various HSPA+ features in live network environments with real traffic loads. The maturity of the tested networks helped achieve performance levels that are close to the theoretical expectations from the technology itself, unlike previous hardware and software versions that often presented problems and design limitations. Even so, wide variations in performance amongst different infrastructure vendors can still be noticed, as will be discussed in detail in Sections 10.3.7 and 10.4.3. In some cases, the scenarios analyzed present challenges due to coverage, traffic load, or interference, as can be expected from real network deployments.

In the various sections, several test scenarios are analyzed, including different RF environments in stationary and mobile conditions. A detailed performance analysis is offered for each scenario, focusing on special interest areas relevant to each particular case; special care is taken to analyze the behavior of the link adaptation mechanisms, which play an important role in optimizing the performance of the data transmission and thus the spectral efficiency offered by the technology.

In general, the results show that the HSPA+ technology can provide an excellent data experience both in terms of throughput – with typical downlink throughputs over 6 Mbps with a single carrier configuration – and latency (typically under 60 ms). Networks with Release 8 Dual Cell effectively double the downlink speeds, offering throughputs between 5 and 30 Mbps, and latency as low as 30 ms. These performance levels are sufficient to offer any data service available today, including real-time services which have quite stringent delay requirements.

HSPA+ Evolution to Release 12: Performance and Optimization, First Edition.
Edited by Harri Holma, Antti Toskala, and Pablo Tapia.
© 2014 John Wiley & Sons, Ltd. Published 2014 by John Wiley & Sons, Ltd.

This chapter also highlights some of the challenges that are present in real HSPA+ deployments, including throughput fluctuations both in uplink and downlink, which makes it hard to offer a consistent service experience across the coverage area. As will be discussed, some of these challenges can be mitigated by careful RF planning and optimization, while others can be tackled with proper parameter tuning. In some cases, the technology presents limitations that cannot be easily overcome with current networks and devices, as is the case of uplink noise instability due to the inherent nature of the HSUPA control mechanisms.

Another challenge that HSPA+ networks present is the delay associated with state transitions; while the connected mode latency can be very low, the transition between various Radio Resource Control (RRC) states can hurt the user experience in the case of interactive applications such as web browsing. Chapter 12 (Optimization) provides some guidelines around various RF and parameter optimizations, including tuning of the state transition parameters.

Finally, Section 10.6 provides a direct performance comparison between HSPA+ and LTE under a similar network configuration, and shows that dual-carrier HSPA+ remains a serious competitor of early versions of LTE, except in the case of uplink transmission, where the LTE technology is vastly superior.

10.2   Test Tools and Methodology

The majority of the performance measurements that are discussed in this chapter are extracted from drive tests. While network performance indicators are important to keep track of the overall network health from a macroscopic point of view, these metrics are not very useful to capture what end users really perceive. To illustrate what the HSPA+ technology can offer from a user point of view, the best course of action is to collect measurements at the terminal device.

There are many different choices of tools that collect terminal measurements, and within these tools, the amount of metrics that can be analyzed can sometimes be daunting; in this section we provide some guidelines around test methodologies that can help simplify the analysis of HSPA+ data performance.

The first step towards measuring performance is to select the right test tools. It is recommended that at least two sets of tools are used for this:

- A radio tracing tool that provides Layer 1 to Layer 3 measurements and signaling information, such as TEMS, QXDM, Nemo, or XCal. These tools can typically be seamlessly connected to devices that use Qualcomm chip sets, although the device may need a special engineering code to enable proper USB debugging ports.
- An automatic test generation tool that is used to emulate different applications automatically, such as File Transfer Protocol (FTP) or User Datagram Protocol (UDP) transfers, web browsing, pings, and so on. These tools are a good complement to the radio tracing tool and typically provide application level Key Performance Indicators (KPIs), some Layer 3 level information, and often offer options to capture IP packet traces. Example tools in this space are Spirent's Datum, Swiss Mobility's NxDroid, and AT4's Testing App. A low cost alternative to these tools is to use free application level tools, such as iPerf (for UDP) or FTP scripts, combined with Wireshark to capture Internet Protocol (IP) traces. Capturing IP packet traces is not recommended by default unless required for specific application performance analysis or troubleshooting purposes.

Once the test tools are set up it is recommended to perform a series of quick tests to verify basic connectivity and performance. These could be done using an online tool such as SpeedTest or similar, or using a specific application test tool. Test values from Section 10.3.6 can be used as a reference for expected throughput and latency values:

- UDP down and up transfers, to check that the terminal is using the HSDPA and HSUPA channels and that the speeds that are obtained are aligned with the given radio conditions.
- 32-byte pings to verify low latency. These measurements should be performed with the terminal in CELL_DCH state; the state change can be triggered by the transmission of a large chunk of data, and verified in the radio tracing tool.

Depending on the objective of the tests, it may be a good idea to lock the phone in 3G mode to avoid Inter Radio Access (IRAT) handovers and be able to measure the performance in the poorer radio conditions. During the collection phase the amount of information that can be analyzed is limited due to the real-time nature of the task, but it is nonetheless recommended to monitor specific metrics such as:

Received Signal Code Power (RSCP) and Ec/No
Areas with poor radio signal (RSCP below -105 dBm or Ec/No below -12 dB) can be expected to present various challenges such as low throughput, interruptions, radio bearer reconfigurations, or IRAT handovers.

Number of servers in the active set
Areas with many neighbors may indicate pilot pollution. In these areas it is common to observe performance degradation, sometimes caused by a large amount of HS cell changes in the area due to ping-pong effects.

Used channel type
If the used channel type is "Release 99" instead of HSDPA and HSUPA, there could be a problem with the network configuration: channels not activated or conservative channel reconfiguration settings.

Instantaneous Radio Link Control (RLC) Level Throughput
Low throughputs in areas with good radio conditions need to be investigated, since these could indicate problems with the test tools, such as server overload. If the throughputs never reach the peaks it could indicate a problem with the network configuration, for example 64QAM modulation not activated or improper configuration of the number of HS codes.

HS Discontinuous Transmission (DTX) or Scheduling Rate
The scheduling rate indicates the number of Transmission Time Intervals (TTIs) that are assigned to the test device. A high DTX can be indicative of excessive network load, backhaul restrictions, or problems with the test tool.

Figure 10.1 illustrates some of the most relevant metrics in an example test tool:

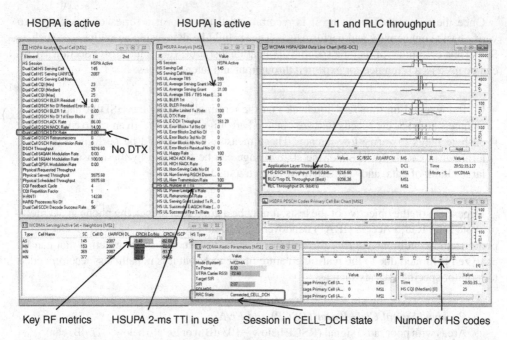

Figure 10.1   Key metrics to inspect during HSPA tests. Example of drive test tool snapshot

During the test care should be taken to make sure that there are no major impairments at the time of the data collection, including issues with transport, core, or even test servers. Once the data is collected, a deeper analysis should be performed to inspect items such as:

- Channel Quality Indicator (CQI) distribution.
- Throughput vs. CQI.
- Number of HS-PDSCH codes assigned.
- Downlink Modulation assignment.
- Transport block size assignment.
- HS cell change events.
- HSUPA format used (2 ms vs. 10 ms).

In addition to RSCP and Ec/No, the CQI measurement indicates the radio conditions perceived by the HSDPA device, permitting the operator to estimate the achievable throughputs over the air interface. Typical reference CQI values are 16, beyond which the 16QAM modulation is in use, and over 22 in the case of 64QAM. However, although indicative, absolute CQI values can be misleading, as will be discussed later on Section 10.3.

The throughput can be analyzed at the physical or RLC layer. The physical layer throughput provides information about the behavior of the scheduler and link adaptation modules, while the RLC throughput indicates what the final user experience is, discounting overheads, retransmissions, and so on.

The following sections analyze these metrics in detail and showcase some of the effects that can be present when testing in a real network environment. Sections 10.3.6, 10.3.7 (for single carrier), and 10.4.1 and 10.4.2 (for dual cell), present performance summaries, including reference throughput curves based on realistic expectations.

10.3 Single-Carrier HSPA+

This section discusses the performance of HSPA+ Single Carrier with 64QAM and Type 3i receivers. This configuration represents a common baseline for typical HSPA+ deployments; therefore a deep analysis of its performance can be used as a basis to understand more advanced features, such as dual cell or 2 × 2 MIMO.

Downlink 64QAM was one of the first features to be available in HSPA+ networks and devices since its definition in 3GPP Release 7 standards. It enables up to 50% faster downlink data transmissions in areas with a good Signal to Interference Ratio (SINR) thanks to a higher modulation and the utilization of bigger transport block sizes. HSPA+ mobile devices with 64QAM are capable of theoretical speeds of 21 Mbps for single carrier devices, which in practice provide around 16 Mbps of average IP level throughput in excellent radio conditions. To take full advantage of this feature it is recommended to combine it with an advanced receiver (Type 3i), that will increase the probability of utilizing the higher modulation. Section 10.5 analyzes in detail the performance gains from 64QAM and Type 3i as compared to Release 6 HSDPA.

Both these features are relatively simple to implement and are found today in all high end smartphones, as well as in many low–medium end devices.

10.3.1 Test Scenarios

The performance of HSPA+ single carrier will be analyzed over multiple radio conditions, including stationary and mobility scenarios, based on measurements from a live network.

The network was configured with one transmit and up to four receive antennas, with advanced uplink receiver features in use, including 4-Way Receive Diversity and Uplink Interference Cancellation. In the downlink direction, the HSDPA service was configured with a maximum of 14 PDSCH codes. The test device was a HSPA+ Cat 14 HSDPA (21 Mbps) and Cat 6 HSUPA (5.76 Mbps), featuring a Type 3i receiver.

The stationary results discussed in Sections 10.3.2 to 10.3.5 were obtained under various scenarios with good, mid, and poor signal strength. The most relevant metrics are discussed to illustrate different characteristics of the HSPA+ service, such as the dependency on RSCP and CQI, the assignment of a certain modulation scheme, and others.

The different RF conditions discussed correspond to typical signal strength ranges that can often be found in live networks, spanning from −80 dBm to −102 dBm of Received Pilot Signal Power (RSCP). The test results were collected during a period of low traffic to be able to show the performance of the technology independently from load effects.

Section 10.3.7 also analyzes the HSPA+ performance under mobility conditions. As will be noted, the performance in stationary conditions is usually better than in mobile conditions, where effects from fading and HS cell changes can degrade the quality of the signal. While it is important to analyze both cases, it should be noted that the majority of data customers access their services while in stationary conditions, therefore stationary results are closer to what customers can expect from the technology in practical terms.

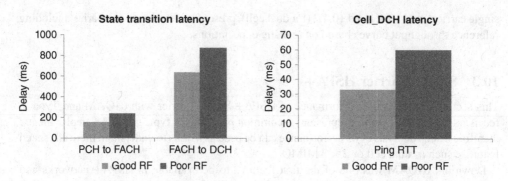

Figure 10.2 HSPA+ latency values: state transition delay (left) and connected mode latency (right)

10.3.2 Latency Measurements

There are several types of delay that can be experienced in a HSPA+ network, depending on the radio state on which the terminal initiates the data transfer: Paging Channel State (Cell_PCH), Forward Access Channel State (Cell_FACH), or Dedicated Channel State (Cell_DCH). In addition to these, there is also the delay from terminals that are in IDLE mode, however this mode is rarely utilized by modern devices which will generally be in PCH state when dormant.

The PCH or FACH delay will be experienced at the beginning of data transfers, and can be quite large: as Figure 10.2 (left) shows, the transition from FACH to DCH alone can take between 600 and 800 ms, depending on the radio conditions. On the other hand, once the HS connection is assigned, the latency can be very small, between 30 and 60 ms in networks with an optimized core network architecture. HSPA+ latency can be further reduced with newer devices such as dual-carrier capable smartphones, as will be discussed in Section 10.4.1.

Connection latency plays an important role in the end user experience, especially with interactive applications such as web browsing, and should be carefully optimized. One option to improve this latency is to enable direct transitions from PCH to DCH, which is shown in Figure 10.3. Section 12.2.6 in Chapter 12 discusses the optimization of RAN state transitions in more detail.

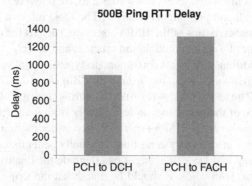

Figure 10.3 RTT improvement with direct PCH to DCH transition (left) vs. conventional PCH to FACH to DCH (right)

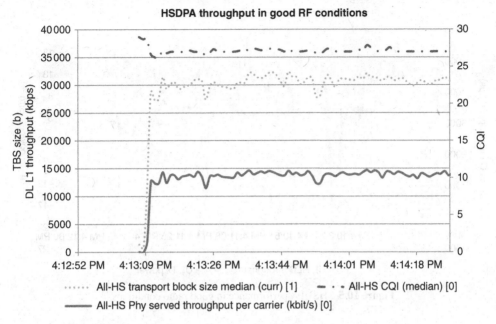

Figure 10.4 HSDPA Single-carrier performance (good radio conditions)

10.3.3 Good Signal Strength Scenario

Figure 10.4 illustrates the performance of HSDPA under good radio conditions. The pilot signal in this location was around −80 dBm, with an idle mode Ec/No of −4 dB. To simplify the presentation of the results, the measurements have been filtered to show one data sample per second.

In these test conditions, which are good but not perfect from an RF point of view (the average CQI was around 27), the average RLC throughput obtained was 13.5 Mbps. The use of 64QAM modulation was close to 100% during this test, as expected based on the CQI.

Note that based on the reported average CQI (27), the 3GPP mapping tables indicate the use of a maximum of 12 HS-DSCH codes, and a Transport Block Size (TBS) of 21,768 bits [1]. In contrast, the UE was assigned 14 codes, with an allocated transport block size around 30,000 bits – which is close to the maximum achievable with CQI 29 (32,264 bits). In this case, the infrastructure vendor is compensating the CQI reported by the device making use of the NAK rate: the link adaptation mechanism is smart enough to realize that the device is able to properly decode higher block sizes, and is thus able to maximize the speed and spectral efficiency of the network.

The CQI measurements are defined by 3GPP to specify the reporting at the UE level, but the RAN algorithms are left open for specific vendor implementation. Furthermore, the UE reported CQI can be modified with an offset that is operator configurable, and therefore the absolute values may vary between different network operators. On the other hand, CQI can be a useful metric if all these caveats are considered, since it can provide a good indication of the Signal to Interference and Noise Ratio (SINR) perceived by the mobile at the time of the measurement.

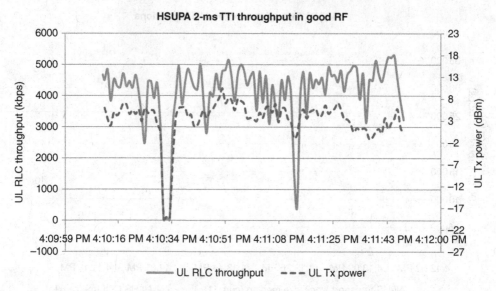

Figure 10.5   HSUPA performance in good radio conditions

Figure 10.5 shows a snapshot of an UDP uplink transfer in the same test location. As can be observed, the HSUPA throughput presents higher fluctuation than the downlink case, considering that the location is stationary.

The average throughput achieved in these conditions was 4.1 Mbps, with peak throughputs of up to 5.2 Mbps. The UE transmit power fluctuated between -1 and $+10$ dBm, with an average value of 4 dBm. During the same period, the measured RSCP fluctuated between -78 and -81 dBm, which indicates that the UL fluctuations were significantly more pronounced than any possible fluctuation due to fading.

The explanation for the HSUPA power and throughput fluctuations can be found in the nature of the HSUPA 2-ms TTI service. The power in HSUPA is controlled by two mechanisms: an uplink closed-loop power control that adjusts the power on the Dedicated Physical Control Channel (DPCCH) based on quality (UL BLER); and an UL Power assignment (grant) determined by the NodeB scheduler. The scheduling grant provides an offset for the Enhanced Dedicated Physical Data Channel (E-DPDCH) power relative to the DPCCH, which in the case of Category 6 Release 7 terminals can range between 0 (-2.7 dB) and 32 ($+28.8$ dB) [2]. Figure 10.6 illustrates the most relevant UL transmission KPIs for this case.

As can be observed, there was a period of UL inactivity (around 4:10:34) in which the UE indicated that it had no data to transmit (happy rate of 100%) – this was probably due to a problem with the test server or the tool itself. Another interesting event occurs around 4:11:08, when there is another sudden dip in throughput. At that time, the UE data buffer was full, as reflected by the happy bitrate of 0%, but the NodeB reduced the UL scheduling grant from 28 to 18, a drop of 10 dB in UE transmit power. This may have been caused by instantaneous interference, or the appearance of some other traffic in the cell. At this time, the UE throughput is limited by the allocated max transmit power.

Another source of power fluctuation is the Uplink Block Error Rate (UL BLER). In the case presented, the UE was transmitting with 2-ms TTI frames during all this time, which makes use of both Spreading Factor 2 (SF2) and SF4 channels during transmission. The data

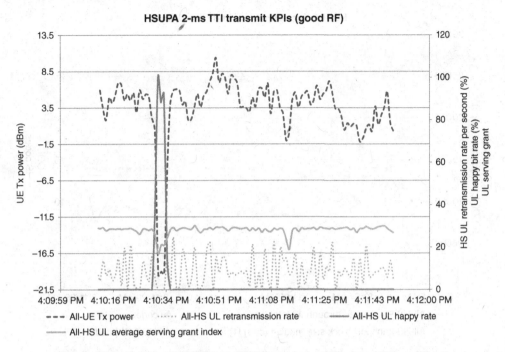

Figure 10.6   Relevant HSUPA 2-ms TTI Transmission KPIs (good RF conditions)

transmitted over the SF2 channels has very low error resiliency and results in a higher number of retransmissions, as can be observed in Figure 10.6: even though the overall retransmission rate is 10% across the whole test period, there are long periods in which the average retransmission rate is higher than 20%. These spikes cause the UE to transmit at a higher power, which brings the retransmission rate to a lower level and thus in turn leads to a reduction in the UE power; this cycle repeats itself multiple times during the transmission, resulting in the observed power fluctuation.

10.3.4   Mid Signal Strength Scenario

The mid signal strength scenario corresponds to the typical experience for the majority of customers. The tests in this section have been performed with a pilot strength of −92 dBm, and an idle mode Ec/No of −5 dB. As in the previous case, the tests were performed at a time when there was no significant load in the network. Under these conditions, the average downlink RLC throughput recorded was 9 Mbps. In this case, the network assigned 14 codes most of the time; as can be observed in Figure 10.7 there is a period of time when the network only assigns 13 codes, possibly due to the presence of other user traffic transmitting at the time.

The analysis of mid-RF level conditions is quite interesting because it permits observation of the modification of different modulation schemes (in the case of HSDPA) and transport formats (for HSUPA). In the downlink case, the RLC throughput fluctuated between 5 Mbps and 11.5 Mbps, with typical values around 8.5 Mbps. These throughput fluctuations can be explained by the commuting between 16QAM and 64QAM modulation schemes, as explained in Figure 10.8.

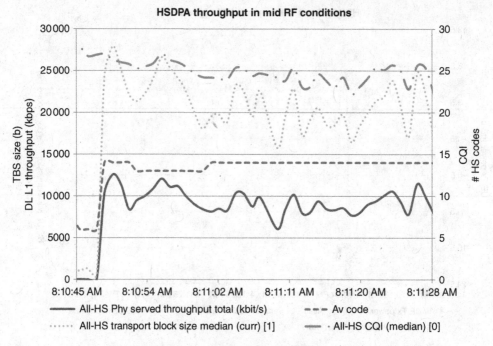

Figure 10.7   HSDPA throughput in mid RF conditions

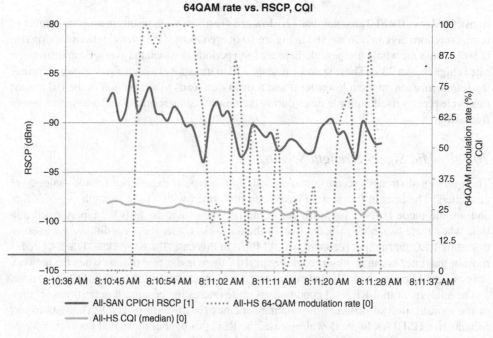

Figure 10.8   64QAM modulation assignment with changing CQI values

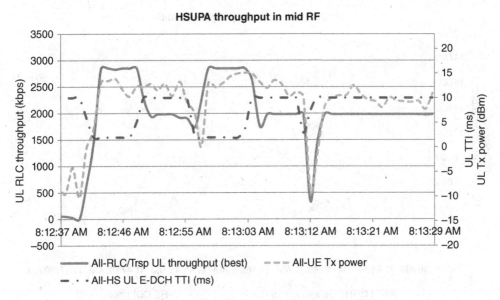

Figure 10.9 HSUPA transmission in mid RF conditions

In this test, the fluctuations in signal strength resulted in the reported CQI swinging around 25. The 64QAM modulation was used intermittently, coinciding with CQI values of 25 or higher, with average transport block size ranges between 16,000 and 28,000 bits. As discussed earlier, the network link adaptation mechanism is not strictly following the 3GPP mapping tables, which suggests that 64QAM would only be used with CQI 26 or higher.

The HSUPA transmission in mid RF conditions also experiences throughput fluctuations due to the adaptation of the transport format block, which at this point switches between 10-ms and 2-ms TTI to compensate for the more challenging radio conditions. This effect can be observed in Figure 10.9.

The typical uplink RLC throughput varies between 1.8 Mbps and 2.8 Mbps, with an average throughput during the test of 2.1 Mbps. The typical UL transmit power varies between 0 and 15 dBm, with an average value of 10 dBm. Unlike in the case of HSDPA, there is no gradual degradation between transport formats. The throughput profile presents a binary shape with abrupt transitions between 10-ms TTI (2 Mbps) and 2-ms TTI (2.8 Mbps). The UE transmitted power in both cases is quite similar with, of course, more power required in the 2-ms TTI case.

10.3.5 Poor Signal Strength Scenario

This section analyzes the performance of the service under coverage limited scenarios, in which the signal strength is weak, but there's no heavy effect from interference. The RSCP in this case varied between the HSDPA case (−102 dBm) and the HSUPA case (−112 dBm).

In the case of HSDPA, the pilot strength measured during the test was around −102 dBm, with an idle mode Ec/No of −9 dB. There was some interference from neighboring cells: at times during the test up to three neighboring cells could be seen within 3 dB of the serving cell; however this didn't impact the performance significantly.

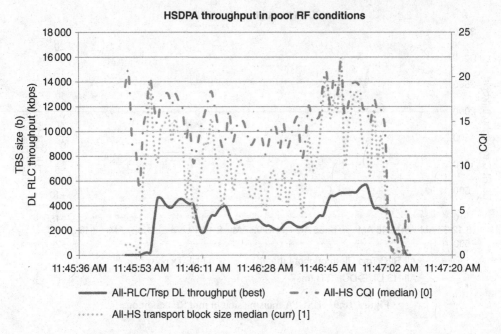

Figure 10.10 HSDPA performance in weak signal strength conditions

There was a significant fluctuation of signal strength at the test location, with RSCP values ranging from −107 dBm to −95 dBm. In these conditions, the HSDPA throughput experienced large variations, ranging between 2 and 6 Mbps. The changes in throughput can be explained by the fluctuations in signal strength, which can indirectly be observed in the CQI changes in Figure 10.10.

The performance measurements under poor RF conditions are extracted from the period with the weakest radio conditions (average RSCP of −102 dBm). In this case, the average CQI was 13, and the corresponding average throughput was 2.5 Mbps.

As in previous cases, the transport block sizes assigned were significantly higher than expected: in the worse radio conditions (CQI 13), the network assigns around 5000 bits instead of the 2288 bits from 3GPP mapping tables. Also, all 14 codes are used instead of the 4 suggested by the standards. This results in an overall improved user experience at the cell edge, in addition to significant capacity gains.

The variation in signal strength during this test permits showcasing the behavior of link adaptation when switching between QPSK and 16QAM. As Figure 10.11 illustrates, during the tests there were a few occasions when the 16QAM modulation was used; these correspond to periods when the signal strength was stronger (typically over −98 dBm), with reported CQI values over 15. The QPSK modulation was used when the CQI value remained under 15.

In the case of HSUPA, the test conditions were quite challenging, with RSCP of −112 dBm and an idle mode Ec/No of −13 dB. Under these conditions, the UE was using the 10-ms TTI transport format and transmitting at full power all the time, as shown in Figure 10.12.

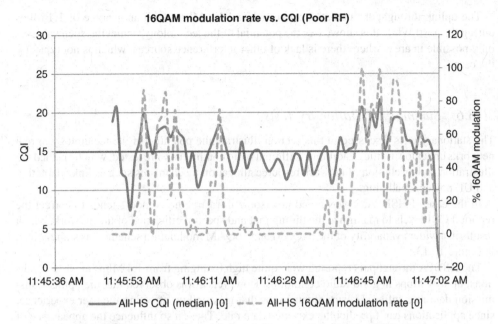

Figure 10.11 16QAM modulation assignment with changing CQI values

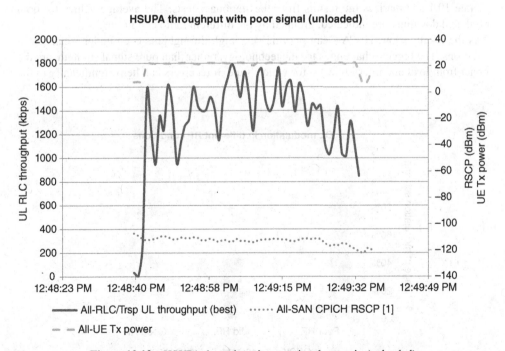

Figure 10.12 HSUPA throughput in poor signal scenario (unloaded)

The uplink throughput fluctuated between 1 and 1.8 Mbps, with an average of 1.4 Mbps during this test. While this showcases the potential of the technology, achieving such values is only possible in areas where there is lack of other interference sources, which is not typically the case.

10.3.6   Summary of Stationary Tests

The stationary tests discussed in this section illustrate the potential of the technology in real networks under low traffic conditions. Different radio conditions were tested, which enabled the illustration and discussion of network resource allocation mechanisms such as link adaptation and UL power scheduling.

In the case of HSDPA, it was noted how some infrastructure vendors tends to correct the reported CQI levels to maximize the throughput and spectral efficiency of the network, which results in a wider availability of the 64QAM and 16QAM modulation schemes, as summarized in Figure 10.13.

The HSUPA throughputs reported were quite high – ranging from 1.4 Mbps in weak conditions, up to 4 Mbps in good radio conditions. However, it was observed that the uplink transmission data rate suffered severe fluctuations that may end up affecting the user experience, since applications can't predict the expected data rate. These also influence the appearance of uplink noise spikes that may lead to network congestion – this will be discussed in more detail in Section 12.2.7 of Chapter 12 (Optimization).

Table 10.1 summarizes the results from the stationary tests. The average values for both uplink and downlink are indicated, grouped by the different radio conditions.

As the results show, HSPA+ can provide very high throughputs even under weak signal conditions. The biggest challenges for the technology, rather than pure signal strength, are the impact from load and interference – for which Type 3i receivers can help significantly in the

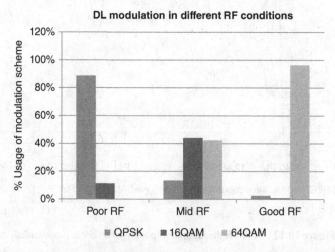

Figure 10.13   Distribution of HSDPA modulation schemes depending on radio conditions

Table 10.1 HSPA+ throughputs in stationary conditions

	Good signal strength CQI = 27.2	Typical signal strength CQI = 25	Poor signal strength CQI = 13.6
HSDPA	13.5 Mbps	9 Mbps	2.5 Mbps
HSUPA	4.1 Mbps	2.1 Mbps	1.4 Mbps

case of downlink transmission – and mobility related aspects. These topics will be discussed later on this chapter as well as in Chapter 12 (Optimization).

10.3.7 Drive Test Performance of Single-Carrier HSPA+

This section analyzes the performance under mobile conditions of a typical HSPA+ single-carrier network with live traffic. The tests were performed in an urban environment that wasn't heavily loaded. As in the previous sections, the tests were executed using a UDP stream to avoid side effects from TCP that may result in lower throughputs due to DTX.

Figure 10.14 illustrates the drive route and the RSCP values at different points: the darker colors correspond to higher values of RSCP, and the smaller dots indicate areas where the interference levels are high.

Figure 10.15 shows the RLC throughput profile along the drive. The RLC level throughput is a realistic measurement of what the user will experience, since it does not include MAC retransmissions. As expected from a wireless network, the data speed fluctuates depending on the RF conditions, especially around handover (cell change) areas. The areas in the chart with low RLC throughput typically correspond to areas with poor coverage or high interference, and the cases where it gets close to zero are often related to problems during the HS cell change, as will be discussed later on this section.

The overall RLC level throughput distribution is illustrated in Figure 10.16. In this route, the throughputs are generally high, typically over 2 Mbps, with peaks of over 14 Mbps. The average throughput in this particular case was 6.7 Mbps, which is a typical value to be expected from real networks in which radio conditions are never perfect. In the same figure, a chart indicates the different modulation schemes used in this drive. In this case, 64QAM was used over 20% of the time, but more important is the fact that QPSK was only applied in 25% of the route, which corresponds to a network with good radio conditions. As discussed in Section 10.3.3, the use of the different modulation schemes greatly depends on the vendor implementation of the link adaptation algorithm. The general trend is to use as many codes as possible first, and higher order modulation after that.

Figure 10.17 summarizes the experienced RLC throughput and usage of 64QAM grouped by different RSCP ranges, to illustrate the dependency of these on the underlying radio conditions. This relationship is quite linear in typical RSCP ranges. On the same figure, the results from the stationary tests presented before are overlaid, to permit comparison between the two sets of tests.

Note the wide differences between throughputs and modulation share at any given radio condition between stationary and mobility tests. This indicates that the results from a drive test in an area will always provide a conservative view of what the network can really provide.

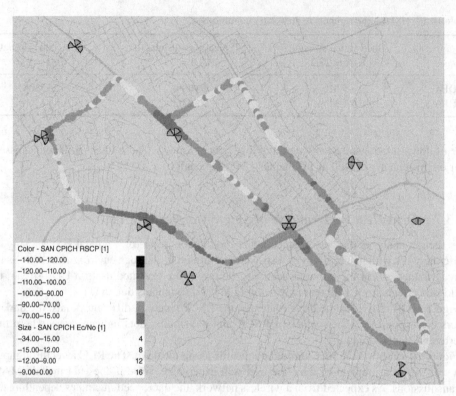

Figure 10.14 HSPA+ single-carrier drive (urban). The color of the dot represents RSCP, the size indicates Ec/No

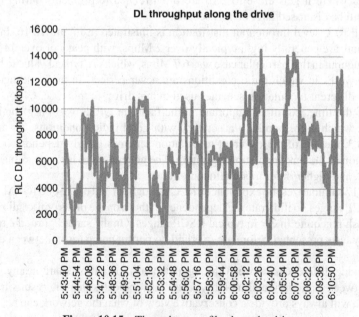

Figure 10.15 Throughput profile along the drive

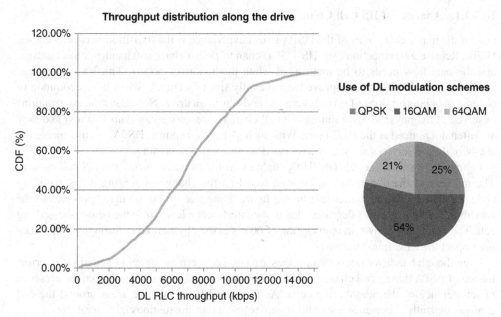

Figure 10.16   DL throughput distribution and DL modulation scheme used along the drive

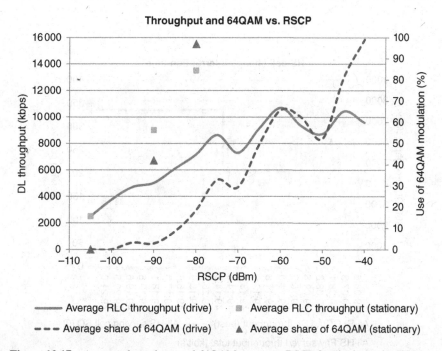

Figure 10.17   Average throughput and 64QAM usage vs. RSCP for single carrier HSPA+

10.3.7.1 Analysis of HS Cell Changes

One of the major challenges of the HSPA+ data experience is the transition between sectors. Unlike Release 99 connections, the HS-DSCH channel doesn't have soft handover mechanisms and the data flow needs to be interrupted when the handover occurs. The HS cell change performance in HSPA+ has improved significantly since its launch, when it was common to observe data interruptions of up to several seconds during a drive. Nowadays the interruptions observed at the physical layer during the cell changes are very small (sub 100 ms) and they are often not noticed at the RLC layer. With such short interruptions HSPA+ could enable the use of real-time services such as voice over IP or live video streaming.

Figure 10.18 shows the HS-DSCH throughput profile of a call during the HS cell change. The moment of the transition can be seen based on the change of serving cell scrambling code, indicated with the darker line on the figure. Note that in the moments previous to the transition the throughput is degraded due to the interference level from the new-to-be-serving cell. The figure also shows an interruption of 60 ms in the physical layer throughput that does not impact the RLC transmission.

Even though handover interruption times are not a concern anymore, in general the performance of HSPA during cell changes presents significant challenges derived from the presence of interference of the neighboring cells. As Figure 10.19 shows, the areas around the cell changes typically experience lower throughputs and larger fluctuations in general.

The presence of ping-pong effects can cause additional degradation in throughput, due to the continuous switching back and forth between sectors. Figure 10.20 shows an example in which the terminal performs multiple serving cell changes, between two sectors with

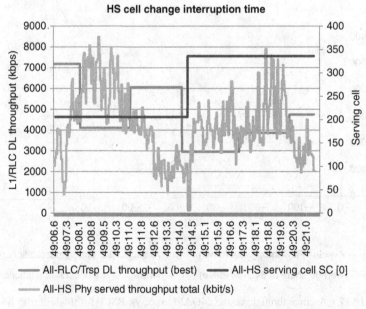

Figure 10.18 Example of HS cell change interruption time

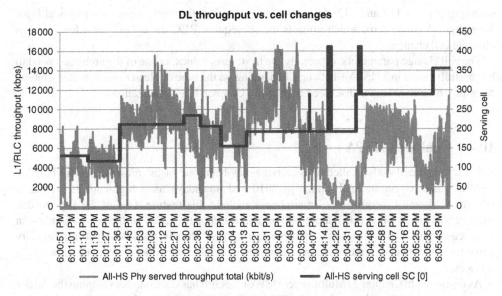

Figure 10.19 Throughput profile vs. HS cell changes along a drive

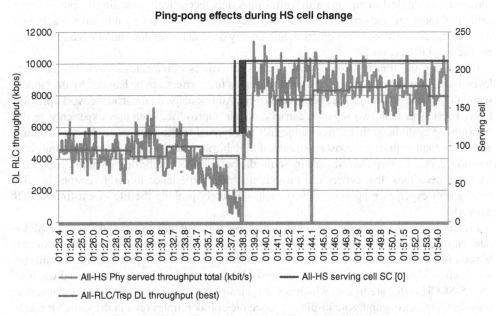

Figure 10.20 Ping-pong effects during a HS cell change

scrambling codes 117 and 212. As can be observed, this ultimately brings the physical layer throughput down to zero, which impacts the subsequent RLC throughput for a few seconds after the cell change.

The cell change parameters, especially the hysteresis values, is one of the areas that need to be carefully tuned in a HSPA+ network, depending on the type of cluster and operator strategy regarding mobility. This is discussed in detail in Chapter 12 (Optimization).

10.4   Dual-Cell HSPA+

Release 8 dual cell – also called dual carrier – has been one of the most successful HSPA+ features since its commercial launch in 2010 and nowadays most high-end, as well as a significant share of medium-tier, smartphones include this feature. One of the keys to this success was the fact that this feature only required minor upgrades on the network side – in some cases purely software since multiple carriers had already been deployed for capacity purposes – and it enables the operator to compete with other networks offering Release 8 LTE services.

As discussed in Chapter 3 (Multicarrier HSPA+), 3GPP has continued developing the carrier aggregation concept, with new items in Release 9 and Release 10 such as uplink dual carrier, dual-band dual-carrier, and multicarrier aggregation; however at the time of writing this book such features were not commercially available.

The Release 8 downlink dual cell feature enables the aggregation of two adjacent UMTS carriers into a single data pipe in the downlink direction, theoretically doubling the bitrate in the entire cell footprint, and providing some efficiency gains due to trunking efficiency – although these gains are questionable in practice, and greatly depend on vendor implementation as will be discussed in Section 10.4.3.

Combining the transmission of two carriers introduces efficiencies in mobility or inter-layer management, since now load balancing across the carriers can be handled by the NodeB scheduler instead of through load based handovers. Additionally, dual-carrier networks provide larger burst sizes than two separate carriers, which improve the data user experience of the smartphone customers; this is due to the fact that typical traffic profiles are bursty in nature and rather than requiring a constant stream of high bitrate, they can benefit from having short bursts of data at high speed followed by semi-idle periods. As will be discussed in Chapter 13, on many occasions dual carrier can provide a similar experience to early versions of LTE from a user experience point of view, which will likely prolong the life of existing UMTS networks.

Figure 10.21 illustrates the maximum and typical throughputs expected from a HSPA+ dual-carrier network, as compared to a single-carrier one. The difference between maximum bitrates in the lab as compared to field measurements has to do with practical load conditions and configuration of the service, since the absolute max speeds can only be achieved when all 15 HS-SCCH codes are in use, which is not typically possible in the live network. Note that only downlink throughputs are displayed, since the uplink bitrates remain the same for single and dual carriers.

This section presents results from stationary and drive tests in multiple live dual-carrier networks, including different vendors and traffic loads. As will be discussed, there are many

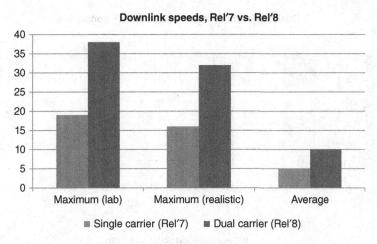

Figure 10.21 Typical downlink bitrates, dual vs. single carrier

factors that affect the throughput provided by the feature, which results in wide variations between different networks. Some of these factors include:

- Baseband configuration. In some networks, the lack of baseband resources can limit the number of HS codes that can be used, thus limiting the maximum throughput achievable by dual carrier.
- Vendor implementation. Some vendors treat dual-carrier users differently than single carriers, leading to a de-prioritization of scheduling resources for dual carrier, which results in lower achievable throughputs.
- Load and interference.
- Terminal performance.

10.4.1 Stationary Performance

The following results show a summary of tests performed around the network in Figure 10.25. This network was configured with two UMTS carriers in a high frequency band and presented low traffic in the sectors at the time of the tests. Several stationary tests were conducted in multiple locations, representing various RF conditions, which are reflected in the RSCP and CQI noted in Figure 10.22. These tests were conducted with a USB data stick with Type 3i receiver capabilities.

As the results show, in very good RF conditions (CQI 29) the average downlink speed reported at the RLC layer was 33.4 Mbps, which is far from the theoretical maximum that the dual-carrier technology can deliver (42 Mbps). This is due to the practical limitations on configuration and usage of HS-PDSCH codes, which in this case was set to 14: with this number of codes, the maximum physical layer throughput is 32.2 Mbps according to the 3GPP mapping tables.

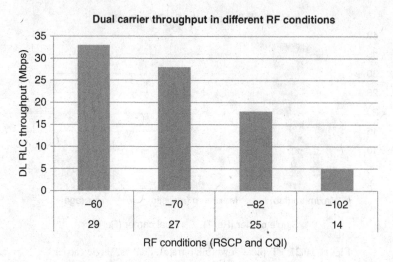

Figure 10.22   Dual carrier downlink throughput in different RF conditions (stationary)

As discussed in Section 10.3 the infrastructure vendor adjusts the assigned transport block size and modulation, based on internal algorithms, to try and maximize the offered throughput. This will not be discussed in detail in this section, since it is the same effect illustrated in the case of single-carrier devices.

While it is possible to achieve over 30 Mbps in perfect radio conditions, in more realistic conditions, the expected downlink throughput would be lower: the value recorded in our tests was 28 Mbps for good radio conditions, while in more typical conditions it was around 18 Mbps; in areas with poorer signal strength the throughput reduces to around 5 Mbps. These results are in line with the expectation that dual-carrier devices would double the downlink throughput, which is summarized in Table 10.2 below. Note that the results are perfectly aligned, even though these were tested in different networks at different times using different devices – however the infrastructure vendor was the same.

One additional benefit that can be noticed when using dual-carrier devices is a reduction of latency of up to 10 ms as compared to Cat 14 devices. There is really not a fundamental reason for this, and the improved latency is likely due to the fact that dual-carrier devices have better UE receivers and processing power. The ping tests in Figure 10.23 show that dual-carrier devices can achieve connected mode latency values as low as 26 ms in good radio conditions. In poor conditions, latency degraded down to 53 ms. A test performed in the same location with a single-carrier data stick showed average latency values of 65 ms.

Table 10.2   Comparison of single vs. dual carrier stationary performance

	Good RF (CQI 27) (Mbps)	Mid RF (CQI 25) (Mbps)	Poor RF (CQI 14) (Mbps)
Single carrier	13.5	9	2.5
Dual carrier	28	18	5

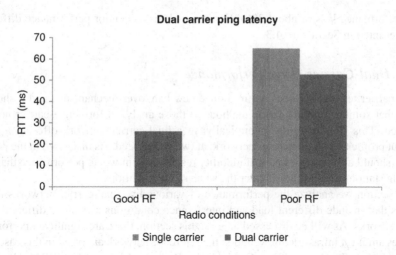

Figure 10.23 Dual carrier ping latency in good (left) and poor (right) radio conditions

Figure 10.24 compares the downlink performance of two different infrastructure vendors. One of the vendors had a limitation on the baseband units that limited the amount of codes used by the dual-carrier to a maximum of 26 (13 per carrier), while the other vendor could use up to 28 codes in this particular configuration.

As Figure 10.24 shows, in the case of the vendor with baseband limitations (Vendor 1), the maximum achievable throughput would be around 26 Mbps in perfect RF conditions, and around 15–18 Mbps in more typical conditions. Another interesting fact to note is that the link adaptation method of this vendor is more conservative than Vendor 2, thus resulting in an

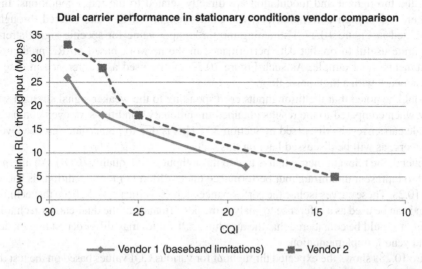

Figure 10.24 Dual carrier performance comparison between vendors (Vendor 1 with baseband limitations)

overall performance loss of about 20%. Further analysis of vendor performance differences will be presented in Section 10.4.3.

10.4.2 Dual-Carrier Drive Performance

The dual-carrier technology does not introduce new handover mechanisms, and it should be expected that similar link adaptation methods to those analyzed for single-carrier networks are utilized. This should result, theoretically, in a dual-carrier network offering a similar throughput profile as a single-carrier network, at twice the speed. As discussed in the previous section, it should also be expected that mobility results present worse performance than those obtained in stationary conditions, under the same radio conditions.

In this section we analyze the performance of various live dual-carrier networks through drive tests that include different load and interference conditions as well as different infrastructure networks. As will be discussed later on this section, there are significant performance differences amongst infrastructure vendors; the reference network analyzed in this case shows the vendor with the better tuned Dual Carrier Radio Resource Management (RRM) algorithms.

Figure 10.25 shows a drive route on a network in an urban environment configured with two UMTS carriers in a high frequency band, in which the traffic load was evenly distributed in both carriers. This network presented some challenges, including high traffic load, and some high interference areas as can be observed on the left chart (darker spots). The chart on the right illustrates the downlink throughput along the drive route.

The average throughput along this drive was 10.5 Mbps, with typical throughputs ranging from 2 to 20 Mbps, depending on the radio conditions. Due to the high traffic in the area, the number of HS codes per carrier available during the drive fluctuated between 11 and 14, depending on the area, with an overall average of 12 codes along the drive. Figure 10.26a shows the throughput distribution, which presents a bimodal profile that follows the RF conditions in the drive. The 64QAM modulation was used 32% of the time, as indicated in Figure 10.26b.

Both the throughput and modulation are directly related to the radio conditions. In this case, there was a high correlation between pilot strength (RSCP) and offered throughput, as illustrated in Figure 10.27. Estimating the throughput value for specific signal strengths can be quite useful to predict data performance in the network based on RF prediction or measurements, for example. As such Figure 10.27 can be used as a reference in case of a loaded network with a high noise floor.

It should be noted that the throughputs corresponding to the weaker signal strength values are low when compared to the results obtained in stationary conditions, or even compared to the single carrier results discussed in Section 3.7. This is due to higher interference levels in this network, as will be discussed later on this section.

In general, the relevant metric that is tied to throughput is link quality (CQI). An alternative method to represent the throughput performance under different radio conditions is shown in Figure 10.28. These curves isolate the performance of the specific network from the technology itself and can be used as a reference to analyze the performance of the dual-carrier technology; however, it should be considered that the absolute CQI values may differ depending on device type and vendor implementation.

Figure 10.28a shows the expected throughput for various CQI values based on the test drive. The stationary measurements discussed earlier are also displayed for comparison purposes: mobility measurements tend to be lower for each particular radio condition, especially in

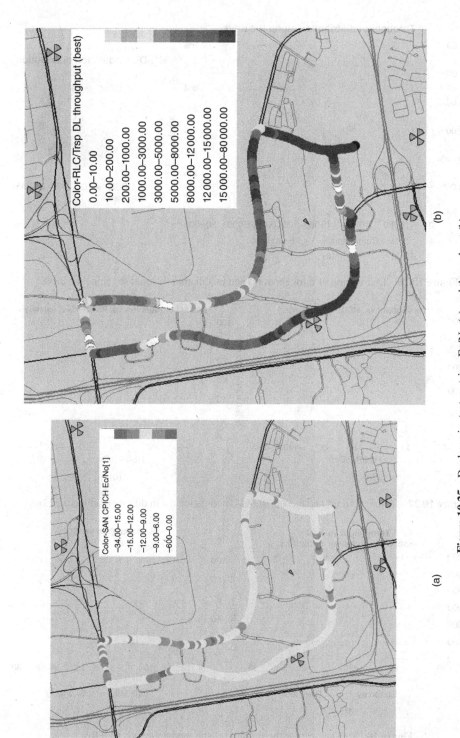

Color-RLC/Trsp DL throughput (best)

0.00–10.00
10.00–200.00
200.00–1000.00
1000.00–3000.00
3000.00–5000.00
5000.00–8000.00
8000.00–12000.00
12000.00–15000.00
15000.00–80000.00

(b)

Color-SAN CPICH Ec/No[1]

–34.00––15.00
–15.00––12.00
–12.00––9.00
–9.00––6.00
–6.00–0.00

(a)

Figure 10.25 Dual carrier test metrics: Ec/No (a) and throughput (b)

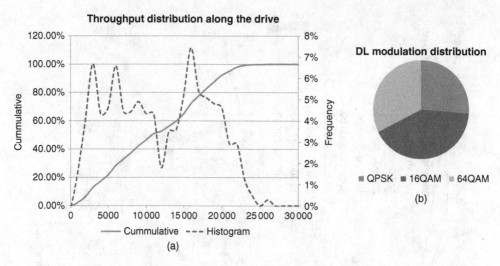

Figure 10.26 Distribution of throughput (a) and modulation schemes (b) along the drive

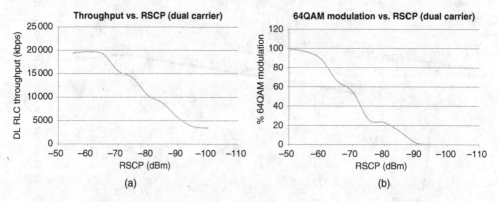

Figure 10.27 Throughput (a) and use of 64QAM modulation (b) in different radio conditions

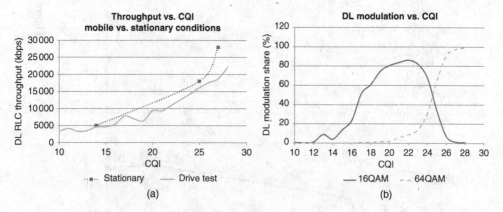

Figure 10.28 Dual carrier throughput (a) and modulation vs. CQI (b)

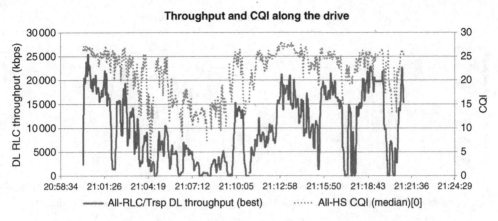

Figure 10.29 Throughput and CQI profile along the drive

cases of good CQI levels. Figure 10.28b illustrates the assignment of the different downlink modulation schemes, which shows that 64QAM is only utilized in the higher CQI ranges (23–30), as expected.

Figure 10.29 illustrates the throughput profile along the drive to better analyze the correlation between interference and throughput. The chart shows how the dual-carrier performance suffers in areas with high interference: while the typical throughputs along the drive were around 10 to 15 Mbps, the areas with low CQI (below 15) experienced throughputs below 5 Mbps.

The poor throughput was heavily linked to areas with weaker signal strength (RSCP) values, however the underlying problem was one of interference due to pilot pollution. Pilot pollution is a very typical situation in real deployments, especially in urban areas, and is associated with the reception of many pilots with similar signal strengths. As discussed in Section 10.3.5, HSPA+ can still provide very good performance levels even at low signal strength conditions; however, it cannot deal so well with interference, in particular when it comes from multiple cells simultaneously.

The profile also presents dips of CQI coinciding with cell change transitions, sometimes creating periods of data interruption. A detailed analysis of this situation showed that pilot signal experienced sudden power drops over a short distance, which is usually indicative of heavily downtilted antennas. In any case, these problems are not related to the network technology itself, rather to the RF planning of the network, but are a good example to highlight the importance of a proper RF planning exercise. Chapter 11 (Planning) will discuss this topic in more detail and provide guidelines and tips to properly plan and optimize the HSPA+ network.

Figure 10.30 provides a breakdown of the throughput allocated on each of the two carriers assigned to the dual-cell device during the transmission.

As the chart shows, the throughputs in both carriers are balanced, with minor differences in throughput (5.8 Mbps average in carrier 1 vs. 5.1 Mbps average in carrier 2), which can be explained by slightly different load and interference conditions in each carrier, as is typically the case in live networks. Note that this behavior may vary depending on vendor implementation and configuration of the traffic balancing parameters in the network.

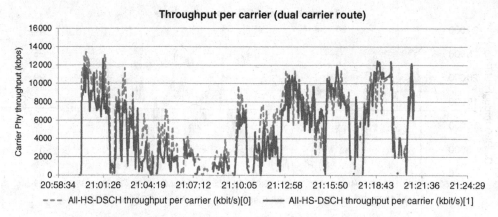

Figure 10.30   Throughput per carrier (dual-carrier transmission)

10.4.3   Impact of Vendor Implementation

During the stationary performance discussion it was noted how there were significant performance differences with different vendor suppliers, especially in the case of hardware restrictions (e.g., baseband limitations). This section analyzes further performance differences between vendors that are rather related to the design of the radio resource management methodologies, in particular the NodeB scheduling function and link adaptation.

When analyzing the throughput profiles for two different infrastructure vendors, it was found that one vendor was outperforming the other by 20–30%, especially in the areas with mid to good radio conditions. Such a difference has a significant impact on user experience and network capacity.

The difference between both vendors can be mostly explained by the strategies around link adaptation, and in particular, the assignment of the downlink modulation schemes. As Figure 10.31 shows, the vendor with the best performance assigns both the 16QAM and 64QAM modulation schemes well before the other one.

Vendor 2 starts assigning the 16QAM modulation around CQI 16, while Vendor 1 doesn't reach the same threshold until CQI 20: that is, the first vendor starts transmitting at a faster

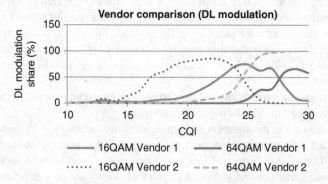

Figure 10.31   Downlink modulation assignment with different vendors

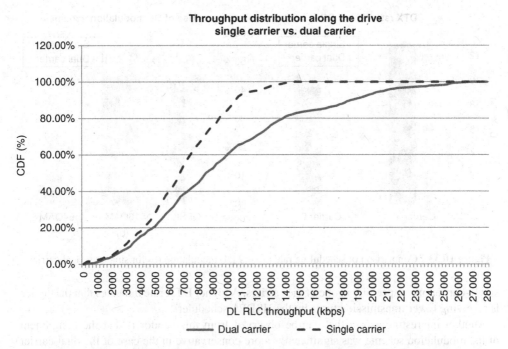

Figure 10.32 Single carrier vs. dual carrier throughput distribution

rate with approximately 4 dB worse quality than the second. While an aggressive setting of the modulation could arguably result in higher error rate, such an effect is controlled by utilizing the NAK rate in the decision-making process of the algorithm. In a similar fashion, the 64QAM modulation is available with 2 dB of worse signal quality, and more importantly the modulation is used at 100% rate when radio conditions are sufficiently good, which is not the case for the other vendor.

Furthermore, when analyzing the behavior of Vendor 1 it was found that there are additional effects that degrade the performance of the dual carrier service. The following example presents a comparison between a single carrier and dual carrier test drive in Vendor 1's network. The same device was used for the tests, changing the terminal category from 24 to 14, and the same drive route was followed (drive route from Figure 10.14). Figure 10.32 shows the throughput distribution for both the single-carrier and dual-carrier device.

The results from this exercise show that the dual-carrier device only achieved 45% higher throughput on average as compared to the single-carrier device; and even more important, in the lower throughput ranges (poor radio conditions), there was no noticeable difference between both devices.

When analyzing the reasons behind this behavior it was noted that this vendor prioritizes the traffic from single-carrier devices over dual carrier to try and achieve the same level of QoS independently of the terminal capabilities (in other vendor implementations it is possible to select the priority of dual-carrier devices in the scheduler, rather than being hardcoded). That explains part of the problem, since the network was loaded at the time of the tests; as Figure 10.33 shows, the dual-carrier device experienced a higher DTX rate than the

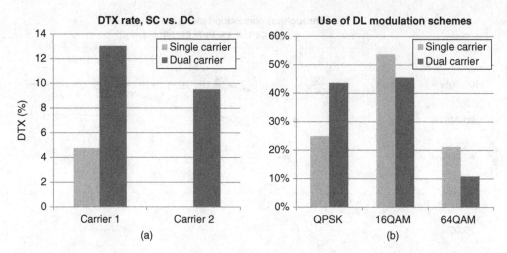

Figure 10.33 Comparison of scheduling rate (a) and DL modulation (b) for single vs. dual carrier

single-carrier one, especially in one of the carriers. A higher DTX rate indicates that the device is receiving fewer transmission turns by the HSDPA scheduler.

Another interesting aspect that can be observed from this vendor is that the assignment of the modulation scheme was significantly more conservative in the case of the dual-carrier device, as can be appreciated in Figure 10.33b: the single-carrier device makes more use of 64QAM and 16QAM than the dual-carrier device, even though the radio conditions were the same for both carriers.

In the case of the other vendor tested, the dual carrier was found to perform in line with the single-carrier performance, effectively doubling the user throughput at any point of the cell footprint. These are important performance differences that should be considered when selecting the infrastructure vendor.

10.5   Analysis of Other HSPA Features

This section analyzes some of the HSPA+ features that are commercially available in networks today, apart from Release 8 dual cell, which was reviewed in the previous section: 64QAM, advanced receivers, single-carrier MIMO, and quality of service.

At the time of writing this book it wasn't possible to test commercially available versions of other interesting features, such as dual-cell HSUPA, CS over HS, or dual-carrier MIMO. While some of these features may become commercially available in the future, it is likely that they will not be widely adopted due to the required modification and penetration of devices incorporating such features.

Other relevant HSPA+ features that are not discussed in this section include CPC, and enhanced FACH and RACH, which have already been discussed in Chapter 4.

10.5.1   64 QAM Gains

Downlink 64QAM was one of the first HSPA+ features to be commercially launched, and one that will typically be found in any HSPA+ phone. As such, all the results discussed in this chapter have assumed that this feature was available by default; however it is interesting to analyze how the feature works and what benefits it brings as compared to previous Release 6 HDSPA terminals.

The use of 64QAM can increase by 50% the peak speeds of previous HSDPA Release 6 in areas with very good signal strength and low interference. This section presents an "apples to apples" comparison of a device with and without 64QAM capability, along the same drive route. In order to make the results comparable in both cases we used exactly the same device, in which we could turn on and off the availability of 64QAM by selecting the desired UE category (Cat 14 vs. 10). The advanced receivers feature (Type 3i) was active for both device categories.

Figure 10.34 shows the signal strength along the drive route; as can be observed, there were many areas along the drive that were limited in coverage, therefore the results from this drive would provide a conservative estimate of the potential gains from the technology.

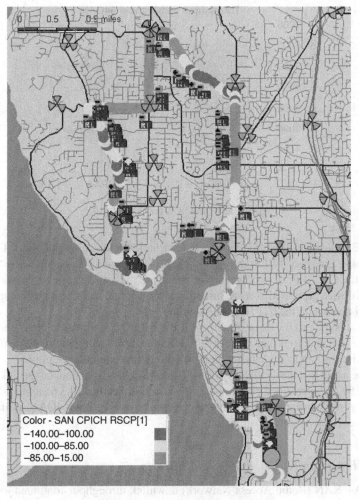

Figure 10.34   64QAM drive area

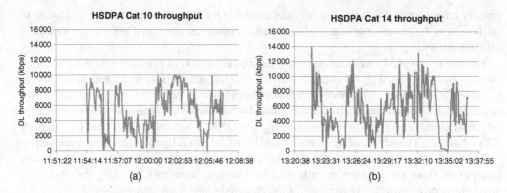

Figure 10.35   Throughput profile of Category 10 (a) vs. Category 14 terminal (b)

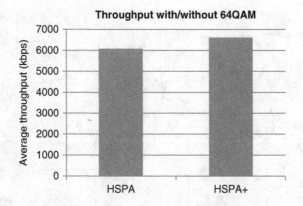

Figure 10.36   64QAM throughput gain

Figure 10.35 illustrates the throughput profile of both terminal configurations: Category 10 (on the left) and Category 14 (on the right). The Cat 14 device follows the same throughput envelope as the Cat 10 device; however the peaks are higher in the good throughput areas. The 64QAM device presented average throughput peaks of up to 14 Mbps as compared to the 10 Mbps registered with the Cat 10 configuration.

Figure 10.36 summarizes the availability of the 64QAM modulation along the route: the Cat 14 device was able to utilize 64QAM in 25% of samples along the route, which resulted in an overall increase in average throughput of 10% in this particular example.

It should be noted that the availability of 64QAM is heavily dependent on the link adaptation method used by the infrastructure vendor, as discussed in Section 10.3.6; some infrastructure vendors apply very conservative settings which results in very low 64QAM utilization, and therefore lower capacity gains overall.

10.5.2   UE Advanced Receiver Field Results

As the HSPA+ RAN (Radio Access Network) downlink throughput continued to evolve via high order modulation and MIMO, which fundamentally only improve speeds in good SNR

Figure 10.37 Location of stationary tests for advanced receivers

areas, there was a need to improve the user throughputs in poor SNR areas near the inter-cell boundary areas. These areas need to combat multi-path fading and inter-cell interference created from neighboring cells. Implementing advanced receivers in the UE (User Equipment) as discussed in more detail in Chapter 8, offers an attractive way to improve the downlink user experience in poor SNR areas and provide a more ubiquitous network experience over more of the geographical footprint with no network investment or upgrades to the RAN. Advanced receivers were first introduced into HSPA consumer devices starting around 2010. In only a few years, they have become a mainstream technology that is in nearly every new high-tier and medium-tier HSPA+ smartphone in the USA[1].

One of the main challenges for operators is to provide solid in-building coverage to areas where it is historically difficult to build sites, such as in residential areas. Another practical challenge with HSPA+ is cell border areas, especially those locations where three or more pilots are present at relatively the same signal strength. In these locations, in addition to the degradation due to high interference levels, HS cell changes have a negative effect on the throughput because of the constant ping-ponging – sometimes causing interruptions in the data transmission. One of the benefits of Type 3i receivers is that the session can cancel those interference sources, and thus remain connected to the same serving cell.

Figure 10.37 shows various stationary test locations where the performance of advanced receivers was evaluated. The locations were chosen to try to represent typical coverage challenges that are experienced in residential areas and measure the range of throughput improvements that might be expected. The results from these tests using a cat 14 UE are shown in Table 10.3. The Qualcomm QXDM (Qualcomm eXtensible Diagnostic Monitor) tool was used to change the device chip set configuration from single receiver (Type 2) to receiver diversity (Type 3) and to receiver diversity with interference cancellation (Type 3i).

---

[1] By end of 2012, nearly 70% of HSPA+ phones in T-Mobile USA had advanced receivers.

Table 10.3 Stationary test results with different UE receiver types

Stationary location	RSCP (dBm)	Type 2 Rx (Mbps)	Type 3 Rx (Mbps)	Type 3i Rx (Mbps)	Type 3 over Type 2 gain (%)	Type 3i over Type 2 gain (%)
1	−60	10.7	12.0	13.0	+12	+21
2	−80	8.3	8.9	9.2	+7	+11
3	−100	2.0	3.4	4.0	+70	+100

As Table 10.3 shows, the largest gain is achieved on location #3 (cell edge) and most of it is derived from implementing receiver diversity (Type 3). There is an additional improvement by implementing interference cancellation (Type 3i) too, with the largest gains also being realized near the cell edge. Location #3 in the table below had three pilots in the active set with roughly the same RSCP; however, not surprisingly, the performance improvement of interference cancellation is a function of the number of interfering pilots and the load on those neighboring cells.

The same three receiver configurations were also tested under mobile conditions, following the test route illustrated in Figure 10.38.

The mobility testing drive route was chosen to represent both the good RSCP and Ec/No areas near the NodeB and also areas where there were multiple pilots, which may cause a UE to ping-pong during HS cell reselection. The area at the top of the map in particular was quite challenging because there were as many as three pilots covering this area all with an RSCP of around −100 dBm at the street level. Figure 10.39 below shows the throughput distribution for the three different test cases.

The mobility results show again that the largest gains in throughput performance are achieved at the cell edge, which can be as high as 100%. The average gains are more modest in the 20–40% range between Type 2 and Type 3i receivers. On the higher throughput range, Type 3i provides a significant boost and enables the receiver to achieve higher peaks than with Type 3 or Type 2 configurations.

As these results show, encouraging advanced receiver adoption in UEs is very important to improving HSPA+ network spectral efficiency and improving service in terms of increased speeds throughout the network.

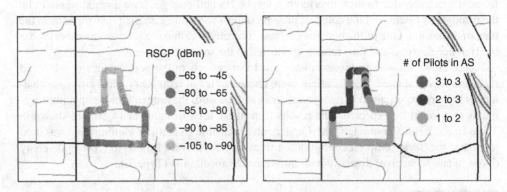

Figure 10.38   UE Advanced receiver drive route

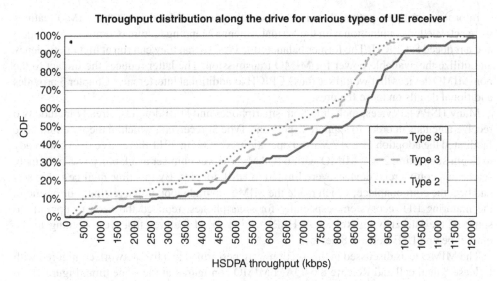

Figure 10.39   Throughput distribution for different types of UE receivers

10.5.3   2 × 2 MIMO

As introduced in Chapter 3, 3GPP Release 8 provides the framework for support of HSPA+ 2 × 2 MIMO (Multiple Input–Multiple Output) with 64QAM in the downlink. This feature set provides up to 42 Mbps in the downlink with a single HSPA+ 5 MHz carrier. Most of the performance gains associated with 2 × 2 MIMO come from spatial multiplexing where two data streams are sent in parallel to the UE on the downlink. By exploiting multipath on the downlink, the throughput can effectively be doubled when the SNR (Signal to Noise Ratio) is high enough and the two signal paths are uncorrelated.

HSPA+ 2 × 2 MIMO allows operators to leverage their existing HSPA+ network infrastructure and RAN equipment to further evolve the spectrum efficiency of HSPA+ utilizing many of the same underlying technologies and building blocks that are being implemented in Release 8 LTE.

Release 8 HSPA+ 2 × 2 MIMO is a relatively simple network upgrade. Generally, HSPA networks are deployed with two antenna ports with two coax cable runs per sector to be used for receiver diversity, so from a technology enablement perspective there are already enough antenna ports and coax cable installed to support 2 × 2 MIMO operation. It is often the case that these two antenna ports are implemented with a single cross-polarized antenna (XPOL). For single-carrier operation, typically only one power amplifier is deployed per sector, so an additional power amplifier will need to be deployed on the second antenna port to support 2 × 2 MIMO. In many networks the second power amplifier may have already been deployed to support the second carrier and/or Release 8 HSPA+ dual cell deployments. Once the second power amplifier is deployed with Release 8 compliant baseband, only a software upgrade is needed to activate the 2 × 2 MIMO with 64QAM feature set. If dual cell is already deployed, the network will be backward compatible with Release 8 dual cell 42 Mbps, Release 7 64QAM 21 Mbps, and pre-Release 7 HSPA devices; however, the devices will not be able to use MIMO and dual cell at the same time.

In order to mitigate potential issues with existing non-MIMO devices the MIMO features were activated in conjunction with the Virtual Antenna Mapping feature (Common Pre-Coder) and a reduced S-CPICH. The former balances the PAs, so that they can run at higher efficiency and utilize the available power for MIMO transmission. The latter reduces the impact to the non-MIMO devices that would see the S-CPICH as additional interference. Chapter 3 provides additional details on these features.

Many HSPA+ devices, including both smartphones and USB dongles, already include two receive antennas in order to support Type 3 and Type 3i receivers, which in theory could have facilitated the adoption of 2 × 2 MIMO operation. Already in 2011 there were networks ready to begin supporting 2 × 2 MIMO with 64QAM, however chip set availability was extremely scarce and commercial devices were limited to USB dongles. By this time dual-cell chip sets had become commonplace, which made the MIMO upgrade look uninteresting. Furthermore, the maturing LTE ecosystem, especially for smartphones, reduced the need to upgrade to some of the late arriving Release 8 HSPA+ high-speed features that were coming to the market behind LTE Release 8 devices.

The MIMO tests discussed in this section were performed in a live network configured with Release 8 dual cell and Release 8 64QAM MIMO configured at the same time. Figure 10.40 shows the location of the stationary tests and the drive route used to analyze the technology.

Table 10.4 summarizes the stationary field results when MIMO was activated. The Virtual Antenna Mapping and MIMO with 64QAM feature set was activated with the S-CPICH set to 3 dB below the P-CPICH. The stationary throughput results show that when the SIR (Signal-to-interference ratio) dropped below 10 dB there is very little measurable gain from MIMO. The same device was used to perform the MIMO and dual-cell testing. Once the MIMO features were activated, no changes were made to the dual-carrier network configurations. The results show the relative comparison between HSPA+ Release 8 dual cell vs. Release 8 MIMO

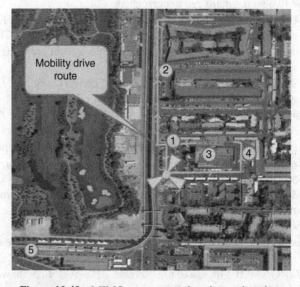

Figure 10.40   MIMO test route and stationary locations

Table 10.4   MIMO vs. DC performance measurements

Stationary location	RSCP (dBm)	SIR (dB)	HSPA+ MIMO-42 (Mbps)	HSPA+ DC-42 (Mbps)	MIMO gain over single carrier (%)	Location relative to NodeB
1	−40	24	28	32	75	LOS
2	−60	17	14	19	47	LOS
3	−60	13	15	18	67	NLOS
4	−70	10	11	15	47	NLOS
5	−80	9	7	15	−7	NLOS

with 64QAM. An additional column is introduced to estimate the performance gain of MIMO as compared to a single carrier mode configuration.

HSPA+ MIMO operates in dual-stream and single-stream operation. The figure below shows the high correlation between device throughput and dual-stream operation along the drive route. As Figure 10.41 shows, the higher throughputs are achieved in areas where the device is requesting a high percentage of two transport blocks (dual stream). In other areas the UE only requests one transport block size (single stream) and the throughput rates drop accordingly.

MIMO is configured on the cell level, however it is recommend to be deployed on the market level because once a device is in an active data session and a P-CPICH from a non-MIMO cell falls into the active set the session will not be able to use MIMO again until the session is restarted in a MIMO configured cell. This can be observed toward the end of the drive on Figure 10.41. Note that mobility was maintained when there was a cell reselection to a non-MIMO cell, but in this particular vendor implementation, the UE will remain in single-carrier non-MIMO mode during the data session even if it re-enters MIMO coverage.

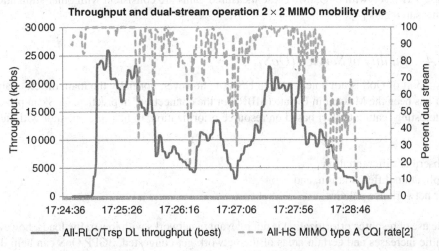

Figure 10.41   Effect of MIMO dual stream on downlink throughput

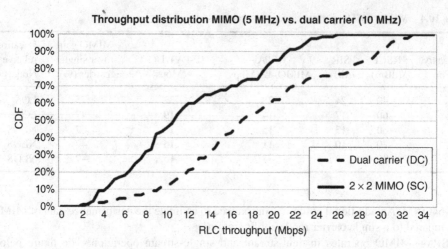

Figure 10.42   Throughput comparison, MIMO vs. dual carrier

Figure 10.42 compares the performance of HSPA+ MIMO (single carrier) with Release 8 dual carrier, utilizing the same test device that was reconfigured from single-carrier 2×2 MIMO mode to dual-carrier mode. On average dual carrier outperforms MIMO by 60–70%, with relatively small performance differences in areas with good SIR. In this case, most of the drive was in fairly good radio conditions, so the throughput results may be slightly optimistic compared to more challenging mobility conditions. As expected, at the lower throughputs levels there was little to no gain in MIMO performance and the test device performance was similar to what you would expect from a device not configured with MIMO operating on a single HSPA+ 5 MHz carrier.

By estimating the single carrier performance based on the dual cell performance that was collected in the field, MIMO shows an overall 15–25% improvement in spectral efficiency over Release 7 HSPA+ with 64QAM. These measured gains are consistent with other simulations and measurements [3].

10.5.4   Quality of Service (QoS)

The support of QoS is not a new item in HSPA+; however, normally the main differentiation factor has been the Maximum Bitrate (MBR) for the connection. In practice, QoS can provide an interesting feature toolset based on resource prioritization, to help the operator with items such as:

- subscriber differentiation
- application differentiation, and
- fair network use policies.

The use of resource prioritization is not relevant in networks with low traffic loads; however, as traffic increases and certain areas of the network get congested, 3GPP QoS can help deal with congestion in a cost efficient manner.

Table 10.5 Summary of QoS test configuration and conditions

Sector conditions	Medium-light load; tests during busy hour; RSCP = −93 dBm
Priority level of regular users	Priority 3 (2x)
Priority level of test users	Priority 1 (4x), Priority 2 (3x), and Priority 4 (1x)
Traffic profile	FTP DL, FTP UL, web browsing

This section illustrates the behavior of utilizing quality of service prioritization in HSPA+ data networks, in particular the impact of assigning different data traffic priority levels to multiple users, under different loading conditions and traffic profiles.

The tests were performed in a live network configured with four different priority levels, which are configured following the guidelines summarized in Table 10.5: the users with the lowest priority (Priority 4) were set to receive approximately half of the throughput of users with Priority 3, one third of the throughput of users with Priority 2, and one fourth of the Priority 1 users.

Two different sets of tests were executed, one with a single device sharing the sector with existing regular network users, and one in which three devices with a mix data profile were running simultaneously. The data profile consisted of multiple cycles of FTP upload and downloads of small and large files; and web download of multiple web pages (CNN, Expedia, Pandora, YouTube, Facebook, and others). Note that with this profile the devices often experience gaps in data transmission and reception due to the bursty nature of the web profile. The charts in Figure 10.43 illustrate the performance of the file downloads (a) and web browsing (b) for different priorities and load conditions.

The results show the impact of the different priority levels, which are only remarkable when the sector becomes loaded with the presence of other test devices. As can be noticed, the experienced throughputs do not necessarily follow the weights that were configured in the scheduler: even in the conditions with artificial load, there is not a significant differentiation between the Priority 2 and Priority 4 cases.

The following sets of tests illustrate the performance with more aggressive priority settings: Priority 1 (40x), Priority 2 (20x), Priority 3 (10x), and Priority 4 (1x). The objective of these

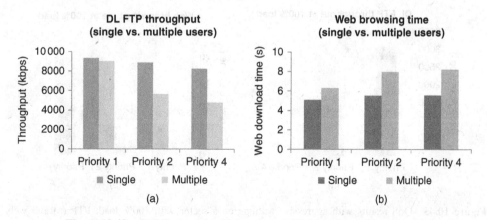

Figure 10.43 QoS Tests with initial settings: FTP (a) and web browsing (b)

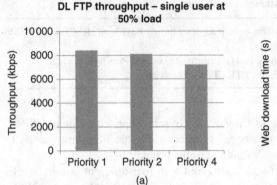
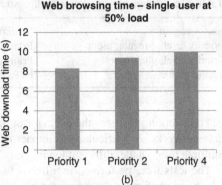

Figure 10.44 QoS tests with a single user in a loaded sector (a); evaluation of effect of different priorities (b)

settings was to heavily deprioritize the data transmission of Priority 4 users. Figure 10.44 shows the result of a single user test with different priority levels in a sector with medium–high load (50% TTI utilization), under good radio conditions. The default priority in the network was set to Priority 3.

As the tests results show, even with these aggressive settings the differentiation at the user experience level is barely noticeable; for example it takes only 20% more time for the Priority 4 user to download a web page as compared to the Priority 1 user. The differentiation only becomes clear in the case of overload, as Figure 10.45 illustrates.

The overload situation was created by having multiple users transmitting in the sector, achieving 100% TTI utilization. Only under these conditions is the differentiation clear, and it follows the guidelines set in the scheduler configuration. As such, this feature can be useful to manage extreme congestion situations when there is a clearly defined type of traffic that the operator needs to control. However it cannot be expected that it will provide a clear differentiation of the subscriber data experience across the network, since this greatly depends on the location of said subscriber.

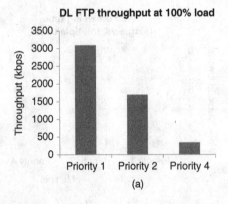
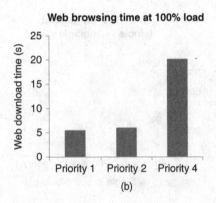

Figure 10.45 QoS results with aggressive settings on a sector with 100% load: FTP (a) and web performance (b)

10.6   Comparison of HSPA+ with LTE

The HSPA+ technology has continued to evolve over the years to provide potentially similar downlink air-interface speeds as LTE Release 8. In the USA, most LTE Release 8 deployments have initially been configured with 10 MHz of spectrum in the downlink. HSPA+ Release 8 dual cell uses the same bandwidth configuration of 10 MHz in the downlink. Both technologies also utilize 64QAM in the downlink, so if 2×2 MIMO was for some reason disabled on LTE, they would in theory provide very similar peak speed performance in excellent RF conditions. So the real difference in performance between the two systems should consist of the gains from 2×2 MIMO and any frequency domain scheduling gains that can be measured.

This section analyzes the performance of an experimental LTE system that is co-located with an existing HSPA+ network configured with dual cell. In order to perform the testing and compare the drive results a single drive was performed while collecting measurements on both HSPA+ dual cell and LTE with 2×2 MIMO activated. The HSPA+ network is a commercial network that has been in operation for a number of years, in which regular traffic load was present, while the LTE system was pre-commercial with no active LTE devices. HSPA+ dual cell and LTE Release 8 commercial dongles operating in the 2100 MHz band were utilized to perform this test. There is a one-to-one overlay in a relatively small geographic five-site cluster area. Both HSPA+ and LTE share the same antennas and utilize the same frequency band, so the coverage of the two systems should be very similar from a construction and deployment perspective. The drive route was chosen so that the measurements would be taken inside the overlay area. In an attempt to try and make the comparison between the two technologies as fair as possible, care was taken to only collect measurement data on the interior of the LTE coverage area: driving areas that were on the fringe of the LTE coverage where no inter-site interference was present would yield an overly optimistic LTE performance compared to an HSPA+ network with more neighbors and hence more inter-site interference.

Figure 10.46 shows the results of a mobility drive that compares HSPA+ dual cell to LTE that is using a 10 MHz downlink channel. The 2×2 LTE network achieved downlink peak

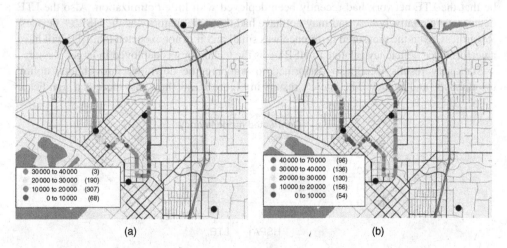

| (a) | (b) |

Figure 10.46   Mobility throughput (kbps): HSPA+ dual carrier (a); LTE 10 MHz (b)

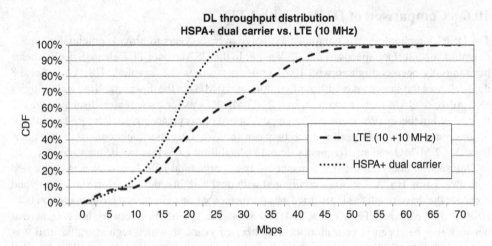

Figure 10.47 Comparison of HSPA+ vs. LTE (downlink). HSPA+ throughputs have been adjusted to account for non-scheduling periods (DTX) due to existing network traffic

speeds of nearly 68 Mbps, while during the same drive we measured a HSPA+ dual cell max speed of 32 Mbps.

Figure 10.47 illustrates the throughput distributions along the drives. The HSPA+ throughputs have been adjusted to account for non-scheduling periods (DTX rate) due to existing traffic, to facilitate the comparison between both technologies. It should be noted that the HSPA+ network also suffered from additional inter-cell interference due to the existing commercial traffic.

On average, the LTE network provided around a 50% increase in downlink throughputs relative to HSPA+ dual carrier during the mobility drive while performing multiple FTPs in parallel. One interesting observation from the drive is that the performance of LTE was slightly worse than HSPA+ in very poor RF conditions, which was not expected. One reason could be that the LTE network had recently been deployed with little optimization. Also the LTE devices were relatively new and may not have had the same maturity as the HSPA+ devices. Regardless, even the measured difference was small and it is not expected that LTE will have a performance deficiency relative to HSPA+ as the technology is deployed in larger clusters.

Stationary results were also performed to compare the downlink throughputs, uplink throughputs, and latency. The latency results in Figure 10.48 show that there is a significant

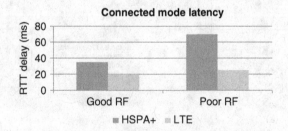

Figure 10.48 Latency comparison, HSPA+ vs. LTE

Table 10.6 Summary of stationary tests performance, DC-HSPA vs. LTE

	Good signal strength (Mbps) RSCP: −65 dBm, Ec/No: −3 dB RSRP: −80 dBm, RSRQ: −5 dB	Typical signal strength (Mbps) RSCP: −90 dBm, Ec/No: −5 dB RSRP: −94 dBm, RSRQ: −7 dB	Poor signal strength (Mbps) RSCP: −102 dBm, Ec/No: −9 dB RSRP: −116 dBm, RSRQ: −8 dB
HSPA+ DL	31	18	5
LTE DL	55	35	9
HSPA+ UL	2.7	2.1	1.4
LTE UL	13	14	5

improvement in an LTE network due to the shorter TTI length (1 ms vs. 2 ms). It is interesting to note that the LTE latency shows a significantly better performance than HSPA+ in poor RF conditions. This may be explained by the fact that HSUPA uses 10-ms TTI in poor RF, which increases delay. On the other hand, 10-ms TTI may provide a better coverage than 1-ms TTI in LTE.

Stationary measurements are important because they give an indication of what a typical stationary user might experience on a wireless system, since often wireless customers are not necessarily mobile when using data services. Stationary measurements in good, medium, and poor RF conditions give an indication of how the air interface of a technology performs under realistic RF conditions that are geographically distributed over a wireless network coverage area.

The results in Table 10.6 show that the downlink throughputs of LTE Release 8 provide nearly twice the throughputs of HSPA+ dual cell. The uplink results of LTE are even more impressive; however, it should be noted that the HSPA+ network only used 5 MHz in the uplink direction. LTE uplink throughputs are between 3.5 to 7 times faster than HSPA+ which is consistent with 2x difference in bandwidth and the 2x potential increase due to higher modulation.

Overall, from a customer's perspective, HSPA+ dual cell provides downlink throughputs that are still competitive with a Release 8 LTE system operating with 10 MHz of bandwidth. However, LTE offers vastly superior uplink speeds that facilitate the adoption of uplink-heavy applications such as video conferencing.

10.7 Summary

This chapter has provided a detailed analysis of HSPA+ performance for downlink and uplink. Multiple HSPA+ features have been discussed, including Release 8 dual cell, MIMO, and advanced receivers. During the analysis, it has been shown how different vendor implementations can have a significant impact on data performance, mostly driven by the implementation of their specific link adaptation methods.

Table 10.7 summarizes the expected data rates in commercial networks under various radio conditions; it also provides the performance of an equivalent LTE network that was unloaded at the time of the tests, for comparison purposes.

In terms of latency, the results have shown that, depending on the radio conditions, HSPA+ networks will offer typical round-trip latency values between 25 and 70 ms.

Table 10.7 Overall performance summary: stationary throughput tests

	Good signal strength (Mbps)	Typical signal strength (Mbps)	Poor signal strength (Mbps)
HSDPA single carrier	13.5	9	2.5
HSDPA dual carrier	28	18	5
HSUPA	4.1	2.1	1.4
LTE 10 MHz downlink	55	35	9
LTE 10 MHz uplink	13	14	5

Overall, the HSPA+ technology offers a compelling data performance even when compared to LTE networks – furthermore, as will be discussed in more detail in Chapter 13, the quality of experience offered to the end customer is still competitive with networks offering higher nominal speeds.

References

[1] 3GPP TS 25.214: Physical Layer Procedures (FDD), Release 8.
[2] 3GPP TS 25.213: Spreading and Modulation, Release 8.
[3] "MIMO in HSPA; the Real-World Impact", GSMA, July 2011.

11

Network Planning

Brian Olsen, Pablo Tapia, Jussi Reunanen, and Harri Holma

11.1 Introduction

The proper planning of the HSPA+ is a very important activity that will determine to a large extent the coverage, capacity, and quality of the service. No matter how much effort is invested in maintaining a network, if the original planning is flawed it will be a constant headache for the engineering teams.

The main focus area when deploying HSPA+ are the radio nodes, including NodeB site location and antenna and power configuration. This chapter provides practical guidelines to take into account when performing the RF planning, however it does not include the amount of detail that can be found in specialized RF planning books such as [1]. Although the main emphasis is put on the Radio Access Network (RAN) nodes, basic considerations around core and backhaul design are also discussed.

In addition to deployment aspects, the chapter also provides useful information to help manage capacity growth, including multilayer management and network dimensioning guidelines.

11.2 Radio Frequency Planning

The performance at the radio layer has a direct impact on the experienced data rates on the application layer and, ultimately, on the experienced user data rate at the application layer. The stronger and cleaner the signal, the higher the data rate that can be achieved; and vice versa, in challenging radio conditions the throughput will be reduced since the physical layer adapts the Modulation and Transport Format to avoid excessive packet loss. The radio signal is typically measured in terms of Received Signal Code Power (RSCP) and interference levels: Energy per chip over Interference and Noise (Ec/No) for downlink and Received Signal Strength Indicator (RSSI) for uplink.

HSPA+ Evolution to Release 12: Performance and Optimization, First Edition.
Edited by Harri Holma, Antti Toskala, and Pablo Tapia.
© 2014 John Wiley & Sons, Ltd. Published 2014 by John Wiley & Sons, Ltd.

The main challenge when planning the radio network is to do so considering both coverage and interference performance, with consideration for future traffic growth. HSPA+ systems utilize a single frequency reuse, which means that the same frequency is used in all adjacent sectors and therefore is highly susceptible to interference. Many of the problems found in today's networks can be tracked down to poor Radio Frequency (RF) planning. In many cases, the problems are masked by adding additional capacity or new sites to the network, but they will eventually need to be addressed through proper optimization.

The placement and orientation of the antenna is the best tool the operator has to ensure proper RF planning, therefore this activity should be done very carefully. To help with this task it is advisable to use network engineering tools like RF Planning and Automatic Cell Planning (ACP) tools, which will be covered in Section 11.2.3.

When configuring antenna parameters, the operator needs to strike a balance between coverage and capacity. When a technology is first deployed, the network needs to provide service in as many locations as possible, so coverage will normally be the initial driver. Ensuring a good level of coverage with HSPA+ can be quite challenging, especially when evolving the network from a previous GSM layout, since in UMTS/HSPA+ it is not possible to use overlay (boomer) sites. Having sectors inside the coverage area of other sectors is very harmful due to the single frequency reuse nature of the technology. Contrary to common wisdom, in some cases, the coverage versus interference tradeoff will result in the reduction of transmit powers, or even the deactivation of certain sectors.

A second challenge when building a coverage layer is deciding the transmit power to be used. In dense urban areas, having too much signal strength beyond a certain point can actually hurt the network performance. The reason is that signals will propagate outside of the desired coverage zone and interfere with sectors in other places – this in turn may create the sensation of a "coverage hole" when in reality the problem is that the network is transmitting with too much power.

Along the same lines, UMTS networks experience so-called "cell breathing" when the network gets loaded, which shrinks the effective coverage of the sectors due to an increase of interference at the cell borders. The "more power is better" guiding principle doesn't always apply, and often ends up causing significant performance degradation as network traffic grows. This is why it's important to try and design the network for a certain load target, for example 50%. To do this in practice the operator should use realistic traffic maps during the planning stage.

Achieving pilot dominance is very important for the long-term health of the network. Well defined cell boundaries will minimize problems related to pilot pollution and will enable a more gracious cell breathing as the network gets loaded. Figure 11.1 illustrates a network in which the original RF configuration (a) has been optimized to minimize interference in the area (b). As the maps show, the optimized network has clearer sector boundaries, although these are still far from perfect in some areas given the challenges from the underlying terrain.

Considering that every network is different, the operator needs to make the appropriate decisions taking into account the terrain, the frequency bands in use (low bands are more prone to interference, high bands are more challenged in coverage), and the antenna configuration, among other issues. Since typically UMTS networks are uplink limited – both from coverage and interference points of view – one good idea is to explore the option to have additional receive antennas and enable the 4-Way Receive Diversity feature in the NodeB. This feature can significantly improve uplink performance, as was discussed in Chapter 10.

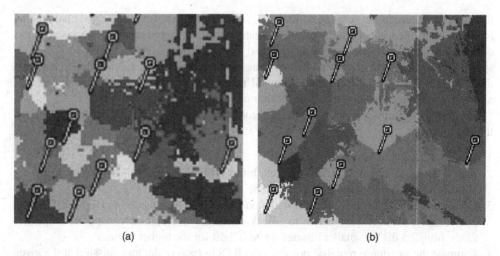

(a) (b)

Figure 11.1 Example of poorly defined boundary areas (a) and optimized boundaries (b) in the same network

In summary, below are a few guidelines to UMTS radio planning:

- Deploy based on traffic location: the closer the cell is to the traffic source, the lower the interference introduced in the system, and the more efficient the air interface will be.
- Proper definition of cell boundaries, avoiding zones with pilot pollution by creating a dominant sector in each particular area.
- Limit footprint of sectors: avoid overshooting sectors that can capture traffic outside their natural range. By the same token, do not abuse antenna downtilts, which are often used as the only method to achieve interference containment and can create sudden attenuation areas within the cell footprint.
- Do not use excess power, try to serve the cell with the minimum power required to keep the service at the desired quality level – taking into consideration service coverage inside buildings.
- In case of overlaid GSM deployment, identify and correct harmful inherited configurations such as boomer sites, sectors that are pointing at each other, and unnecessary capacity sites.

11.2.1 Link Budget

The first step toward planning the radio network is to perform an estimation of the coverage area that the technology can provide. While this is a complicated exercise that needs to consider many different factors, a first order approximation is to analyze the maximum pathloss that can be endured both in the uplink and downlink directions to achieve a certain level of quality. This is known as link budget calculation. This section will provide useful guidelines to help calculate the link budget for HSDPA and HSUPA, however more detailed explanation on computing link budget can be found in network planning books such as [1].

At a very high level, the link budget exercise will estimate the maximum pathloss permitted to transmit with a given quality; there will be many factors affecting the link budget, such as technology, network configuration, type of data service, network load, and interference conditions. The link budget calculations for uplink and downlink are quite different, given the different nature of the service: while HSUPA is transmitted over a dedicated channel that is power controlled, the HSDPA service is carried over a shared channel whose power is determined based on operator configuration and load demands.

In practice, HSPA+ networks are UL coverage limited. Therefore, the first step to compute the link budget would be to estimate the cell range based on the required UL speed, and afterwards calculate the corresponding DL bitrate for that particular pathloss.

The following steps describe the generic process used to estimate the link budget on the uplink direction:

1. Determine required bitrate at cell edge.
2. Estimate required signal quality for such bitrate (Eb/Nt). This is provided by link level simulations – sample mapping values can be found in [2].
3. Compute UE Equivalent Isotropical Radiated Power (EIRP) considering UE Max Transmit power and a HSUPA back-off factor. Back-off factors for commercial power amplifiers range from 2.5 dB for smaller bitrates, up to 0.5 dB for the higher bitrates.
4. Estimate the minimum required power at the BTS to receive the transmission at the given bitrate (receive sensitivity):
   (a) calculate the noise floor of the system;
   (b) calculate UL noise rise based on target system load;
   (c) calculate the processing gain based on the target bitrate required;
   (d) estimate the UL noise increase in the cell when the UE will be transmitting at the required bitrate;
   (e) compute the sensitivity considering all previous calculations.
5. Estimate the difference between the UE EIRP and the required sensitivity, considering adjustment factors such as:
   (a) BTS antenna gain and cable losses;
   (b) fading margin;
   (c) soft handover gains.

Table 11.1 illustrates an example calculation for a network with a target 50% load.

The maximum pathloss values provided are calculated for outdoor transmissions; in the case of indoor, an additional in-building propagation loss should be subtracted, which is typically in the 15–20 dB range. Other adjustments should also be taken into account: for example, the use of uplink advanced receivers, or receive diversity methods that were described in Chapter 8.

Once the Maximum Allowable Pathloss (MAPL) has been determined, the HSDPA link budget exercise will estimate the downlink bitrate that is possible at that particular cell location. The following steps summarize the process:

1. Determine power breakdown in the cell: overhead of control channels, power for voice, and available power for HSDPA.
2. Calculate the received signal levels for the HS-DSCH channel, own cell interference, and noise:
   (a) calculate EIRP values considering BTS antenna gains, cable and feeder losses and adjust with MAPL values;
   (b) calculate noise floor considering thermal noise density and UE noise figure.
3. Calculate the Geometry (G) factor at that particular location.
4. Estimate SINR for the given G factor, and calculate the bitrate at that particular level based on pre-defined mapping tables.

Table 11.1 Example HSUPA outdoor link budgets for 64, 256, and 512 kbps at 50% load

Parameters and calculations		Target UL data rates (kbps)		
		512	256	64
UE Tx power	UE maximum transmit power (dBm)	21.0	21.0	21.0
	UE cable and other losses (dB)	0.0	0.0	0.0
	UE transmit antenna gain (dBi)	0.0	0.0	0.0
	Power back off factor (dB)	1.1	1.7	2.3
	Mobile EIRP (dBm)	19.9	19.3	18.7
Required power	Thermal noise density (dBm/Hz)	−174.0	−174.0	−174.0
at the BTS	BTS receiver noise figure (dB)	3.0	3.0	3.0
	Thermal noise floor (dBm)	−105.2	−105.2	−105.2
	UL target loading (%)	50%	50%	50%
	UL noise rise (dB)	3.0	3.0	3.0
	Required Eb/Nt (dB)	3.9	4.0	4.8
	Processing gain (dB)	8.8	11.8	17.8
	Interference adjustment factor (dB)	1.2	0.7	0.2
	NodeB Rx sensitivity (dBm)	−105.7	−109.3	−114.9
Maximum	BTS antenna gain (dBi)	18.0	18.0	18.0
pathloss	BTS Cable connector combiner losses (dB)	4.0	4.0	4.0
	Slow fading margin (dB)	−9.0	−9.0	−9.0
	Handover gain (dB)	3.7	3.7	3.7
	BTS body loss (dB)	0.0	0.0	0.0
	Maximum allowable pathloss (dB)	*134.3*	*137.3*	*142.3*

The geometry factor indicates how susceptible that location is to receiving interference from external cells. It is defined as:

$$G = \frac{P_{own}}{P_{other} + P_{noise}}$$

Where:

- P_{own} is the power from the serving cell.
- P_{other} is the interference power from the rest of the cells.
- P_{noise} is the thermal noise power, N_o.

It is important to estimate the right geometry factor for the required scenario, as it will define whether the network is interference or coverage limited. In many cases, geometry, and not noise, will be the limiting factor in HSDPA networks. Geometry factors range from over 20 dB, for users close to the cell site, to below −5 dB, for users at the cell edge in pilot polluted areas. Typical values for cell edge in urban areas are around −3 to −4 dB.

One other important factor to consider in the HSDPA link budget exercise is the *orthogonality factor(α)* which takes into account the impact of multipath propagation on the Signal to Interference and Noise ratio (SINR) [3]. The orthogonality factor is 1 for non-multipath environments, such as rural areas, and 0.5 for typical environments.

The SINR will determine the achievable bitrate for the HSDPA service. In interference limited conditions, the SINR for HSDPA can be estimated as:

$$SINR = SF_{16} \times \frac{P_{HSDPA}}{(1 - \alpha)P_{own} + P_{other} + P_{noise}} = SF_{16} \times \frac{P_{HSDPA}}{P_{TOT\_BS}\left(1 - \alpha + \frac{1}{G}\right)}$$

Where:

- SF_{16} is the processing gain corresponding to the spreading factor 16 (12 dB).
- P_{TOT_BS} is the total power transmitted by the BTS.
- P_{HSDPA} is the power available for the HS-DSCH channel.

The power allocated to the HSDPA service depends on the specific vendor implementation (static vs. dynamic allocation) and operator configuration. Nowadays, most HSDPA networks allocate the power dynamically, with a low power margin to maximize the bitrates in the sector.

Table 11.2 provides some example values for the required SINR to achieve certain downlink bitrates, based on HSDPA Release 5 [4].

The HSDPA advanced receivers will significantly improve SINR, especially at challenging cell locations such as the ones found at cell borders in urban areas (low G factors).

Table 11.3 provides some example SINR and throughput under various network configuration and geometry factors, for a cell with a minimum HSUPA bitrate of 64 kbps (MAPL of 142 dB). The maximum BTS power is set to 20 W.

Table 11.2 SINR requirement for various DL bitrates

SINR	Throughput with 5 codes	Throughput with 10 codes	Throughput with 15 codes
0	0.2	0.2	0.2
5	0.5	0.5	0.5
10	1.0	1.2	1.2
15	1.8	2.8	3.0
20	2.8	4.2	5.5
25	3.1	5.8	7.8
30	3.3	6.7	9.2

Table 11.3 Example cell edge DL bitrates for various HSDPA configuration scenarios, for MAPL = 142 dB and $\alpha = 0.5$

Configuration	G-factor (dB)	HS power (W)	SINR (dB)	DL bitrate (Mbps)
1	14	12	12.3	2.0
2	14	7	10.0	1.2
3	0	7	5.6	0.6
4	-3	7	3.5	0.4
5	-3	12	5.8	0.6

Table 11.4 Calculated DL bitrates for various locations for Max HSDPA power allocated = 7 W, G factor at cell edge = −3 dB

Pathloss (dB)	UL bitrate (kbps)	G factor (dB)	SINR (dB)	DL bitrate (Mbps)
142	64	−3	3.5	0.4
137	256	7	9.0	0.9
134	512	13	12.2	2.0

These results highlight the importance of the geometry factor in the achievable HSDPA data rates. The scenario with high G factors (coverage limited) can achieve significantly higher throughputs than those with lower G factor (interference limited). This is one of the reasons why too much cell coverage overlap with HSDPA should be avoided. This also explains why assigning higher power to the cell will not be an effective method to extend the cell coverage in the case of interference limited networks. Table 11.3 also shows that increasing the power available for the HS-DSCH channel can help improve the bitrates at the cell edge; as expected, such improvement will be higher in coverage limited situations.

Table 11.4 calculates the downlink bitrates for various locations in the cell corresponding to the MAPL calculated for UL 64, 256, and 512 kbps. The network follows the configuration #4 from Table 11.3 for a network limited by interference.

11.2.2 Antenna and Power Planning

Selecting the right antenna for a site is just as important as selecting the right tires for a car. Like tires, they are small part of the overall site cost, but they are often overlooked even though they have a tremendous impact on achieving the desired coverage and capacity performance.

Some of the most common antenna characteristics are the peak antenna gain and horizontal 3 dB beamwidth, however they shouldn't be the only ones considered when evaluating antenna selection for a site. The antenna's vertical 3 dB beamwidth, which is inversely proportional to the peak main lobe gain, is just as important. Other important considerations are the front-to-back ratio, front-to-side ratio, and the power of the 2nd and 3rd sidelobes relative to the mainlobe. Figure 11.2 illustrates some of these parameters.

The antenna packaging and radome are also important considerations. As more physical antennas are packaged into the same radome (i.e., Dual POL and Quad POL) there is a tradeoff made in the shape of the antenna pattern. The peak antenna gain is often used as the most important figure of merit, but the intended geographical area the antenna is attempting to serve is just as important. So ideally the antenna shape should match the intended coverage in all three space dimensions. As an extreme example, an antenna on a mountaintop with a high gain and small vertical beamwidth may appear to provide the best signal strength over a large geographical area on a two-dimensional propagation prediction, but in practice there may actually be coverage holes closer to the sites because the coverage is being provided by the lower sidelobe of the vertical pattern where there may be nulls in the pattern that are not very well behaved. In this situation it is often better to use an antenna with a larger vertical

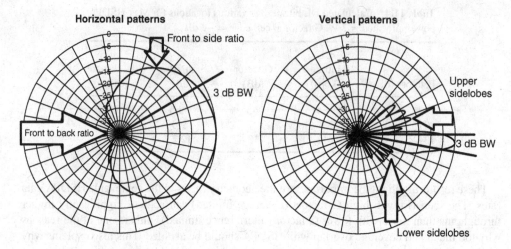

Figure 11.2   Example horizontal and vertical antenna patterns

beamwidth and less peak gain so that the antenna pattern does not contain any nulls or large changes in gain toward the intended coverage area.

There are often practical limitations regarding which antennas should be utilized as often in mature networks antennas are shared between multiple technologies as discussed in Section 11.6.4. In many cases, the operator may need to transmit in multiple bands, for which it is possible to use multiband antennas. These antennas can operate in a large range (700–2600 MHz) and present a pretty good performance, which makes them a good choice for future capacity growth as well as carrier aggregation scenarios.

Antenna sharing may limit optimization such as azimuth, tilt, or height adjustments on a per band/technology basis, so a compromise must be struck in antenna design that may not be optimal for either technology as each technology on a site may not share the exact same coverage objectives. Other practical aspects may limit the adjustment of the ACL (Antenna Center Line) height. For example, sites located on high-rise rooftops may not have the ability to lower the ACL as the network densifies and the coverage objectives for a site are reduced, so downtilt and reductions in the transmit power may need to be utilized to optimize the coverage. However, as antenna downtilts are deployed, care must be taken not to place any nulls of the antenna pattern toward the intended coverage area.

The goal of the RF design is to ensure that the "well behaved" portion of the antenna is pointed at the intended coverage area. Often, antenna adjustments such as downtilt are over-used and the top portion of the vertical antenna pattern, which can have high variations in gain, is used to serve the coverage area leading to large drops in signal strength over the sectors intended coverage footprint. In an attempt to create a "dominant" coverage area the excessive downtilt can create significant coverage issues, which can become especially evident in sectors with low antenna height. Consider Figure 11.3, which illustrates the typical antenna range of a rooftop antenna in a suburban neighborhood (around 10 m height). Figure 11.3b shows the 2D pattern of an antenna with 6° vertical beamwidth that is downtilted by 6°.

In the case of low antenna height, downtilting by 6 degrees or more pretty much points the main lobe of the antenna confines to less than 100 meters next to the sites; furthermore,

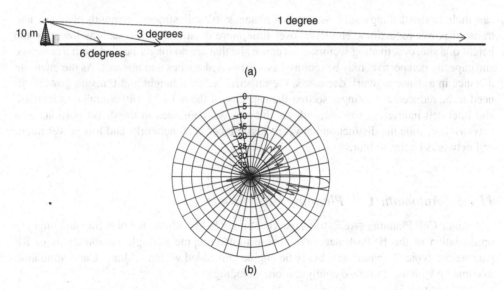

Figure 11.3   Effect of excessive antenna tilt on footprint (a) and antenna pattern (b)

considering the practical effects of real antenna patterns, the boundary area of the sector will be exposed to unpredicted attenuation that can change every few meters, due to the steep slope of the antenna pattern around the beamwidth zone. This often results in the creation of very hot areas next to the antenna, and artificially attenuated signal strength and signal quality inside the sector footprint leading to undesirable spotty coverage holes in between neighboring sites. The antenna pattern gain is reciprocal for the uplink and downlink, so the excessive downtilt impact is especially harmful to the uplink as it is the limiting link in terms of coverage. Since cell selection is controlled by the downlink coverage of the pilot, a more effective way to control the coverage footprint without negatively affecting the uplink is to adjust the pilot power.

The pilot power should be utilized to cover the intended coverage objective; however, as the pilot power is reduced the overall power of the Power Amplifier (PA) must be reduced by the same ratio in order to maintain the proper Ec/No quality. A typical rule of thumb is to ensure that the CPICH power is between 8 and 10% of the max power of a carrier and that the overhead power for control channels doesn't exceed 25% of the max carrier power. The recommend CPICH power will place the transmit Ec/No of a carrier at −10 to −11 dB when the entire power of the carrier is fully utilized at maximum load. Using a lower ratio risks eroding the Ec/No quality, which will limit the sector coverage area. Using a higher ratio will increase the sector coverage at the expense of less total power available for customer traffic, which may limit the overall sector capacity.

As additional sites are added and the network becomes more dense care must be taken to review the coverage of the neighboring sites and to make adjustments so that sites are not just integrated in terms of neighbor list optimization, which might mask interference problems, but to incorporate new coverage objectives for all the neighboring sites.

As the network densification continues and new coverage and capacity objects are introduced, the intended coverage of each site will need to be optimized holistically. These changes can include small things such as antenna changes, tilt adjustments, azimuth changes, and transmit power reductions. However, over time more drastic changes, such as lowering the height of a site, deactivating sectors or an entire site that are no longer needed from a coverage and capacity perspective, may be required as a network densifies and matures. As the inter-site distance in a dense network decreases, the effective antenna height and transmit power will need to be reduced accordingly so that the capacity of the network can continue to increase and inter-cell interference is reduced. As this process continues in the dense portions of a network over time the distinction between classic macro cell networks and low power micro cell networks begins to blur.

11.2.3   Automatic Cell Planning (ACP) Tools

Automatic Cell Planning (ACP) tools can be of great help with the RF planning and ongoing optimization of the HSPA+ networks. These tools compute multiple combinations of RF parameters, typically based on a heuristic engine embedded within a Monte Carlo simulator, to come up with an optimized configuration, including:

- optimum location of the sites based on a set of predefined locations;
- height of the antennas;
- azimuth orientations of the sector antennas;
- antenna downtilts (mechanical and electrical);
- pilot powers.

The accuracy of the results of the ACP tool depends to a large extent on the input data supplied. Given that traffic is not uniformly distributed in real networks, it is recommended to utilize geolocated traffic maps. Ideally this information will be provided from a geolocation tool such as those discussed in Chapter 12; however, this can also be achieved through configuration of typical clutter maps. Geolocation information can also be used to tune the propagation maps used in the ACP exercise; otherwise the tool can be tuned with drive test data to complement the input from the prediction engine.

The following example shows the result of an ACP optimization on a live HSPA+ network. The original network was experiencing quite challenging radio conditions, both in terms of signal strength (RSCP) and quality (Ec/Io). In the exercise, the tool was configured to only modify electrical tilts, and the optimization targets were set to RSCP = −84 dBm and Ec/Io = −7 dB.

Figure 11.4 illustrates the modification of antenna tilts in the network. The original tilts are on the left, and the optimized settings on the right. In this particular case a large number of antennas were uptilted, shifting the median tilt value from 7 to 4 degrees.

As a result of the tilt changes, the RSCP in the region improved significantly: the number of customers served with RSCP higher than −87 dBm increased from 65 to 85%. Furthermore, with the new settings, the whole area was covered at a level higher than −102 dBm, while previously only 85% of the population was at this level. The quality in the area was also significantly improved, as can be appreciated in Figure 11.5. The results of this test indicate that the network was coverage limited, and the excessive downtilt was making the problem worse.

After the implementation of the ACP proposed changes, the operator should closely monitor the performance in the cluster. If a geolocation tool is not available, it is recommended to

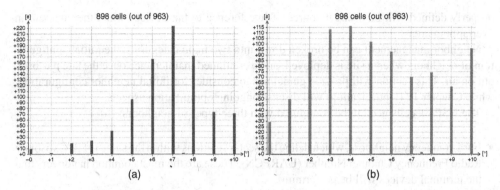

Figure 11.4 Baseline (a) and recommended antenna tilts (b)

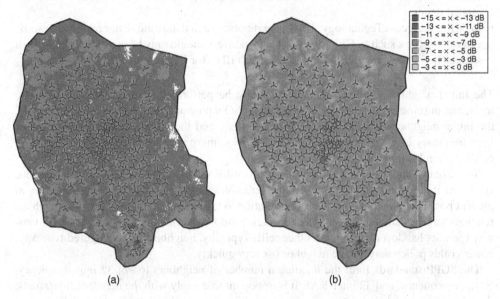

Figure 11.5 Ec/No improvement after ACP execution. Baseline (a), Optimized (b)

perform a thorough drive test in the area where new changes have been made and compare the main RF metrics against the results predicted by the ACP tool.

Once the network carries a significant amount of traffic, it will be possible to perform these optimization exercises automatically with the use of a Self-Organizing Networks (SON) tool. These tools are discussed in Chapter 12 (Network Optimization).

11.2.4 Neighbor Planning

The configuration of neighbors is a very important task in a HSPA+ system since cell footprints can be very dynamic due to traffic growth and the deployment of new sites for coverage or capacity purposes. Furthermore, due to the single frequency reuse, neighbors that are not properly defined will be a direct source of interference to the devices when they operate in overlapping areas.

Neighbor management can be broken down into two main categories, the initial configuration of neighbors when a site is deployed, and a continued maintenance once the site has been put on air. This section will provide guidelines to consider for initial neighbor configuration, while Chapter 12 (Optimization) will discuss ongoing optimization aspects.

The HSPA+ cells need to be configured with three types of neighbors:

- Intra-frequency neighbors, which define the sector relations within the same UTRA Absolute Radio Frequency Channel Number (UARFCN). These are the most common handovers that the terminal devices will be performing.
- Inter-frequency neighbors, which define the sector relations between different frequency channels or carriers. These handovers are normally performed in case of overload or coverage related issues.
- Inter Radio Access Technology (IRAT) neighbors, which define the sector relations towards external systems such as GSM or LTE. The handovers should only be used for special cases, such as loss of coverage or to transition back to LTE, for LTE enabled networks.

The initial neighbor configuration will typically be performed with the help of planning tools, and in some cases, automatically based on SON plug and play functionality. Normally, the initial neighbors will include the first two tiers, and the co-sited neighbors in case of intra-frequency layers. Section 11.3.2.2 will provide more specific information on neighbor configuration for multiple layers.

Special care should be taken with nearby sectors that use the same scrambling code; in these situations it is recommended to review the scrambling code plan and reassign the codes to prevent possible collisions. Another consideration to take into account is symmetry: neighbor relations can be defined as one-way (enables handovers from one cell to other), or two-way (permits handover to/from the other cell). Typically, neighbors are configured two-way; however this policy may fill the neighbor list very quickly.

The 3GPP standards limit the maximum number of neighbors to 96: 32 intra-frequency, 32 inter-frequency, and 32 inter-RAT. It is important to comply with the neighbor limitations, as in some vendor implementations a wrong configuration will lock the sector.

The neighbors need to be configured in two different groups, which are broadcasted in the System Information Block (SIB) 11 and SIB 11bis. The highest priority neighbors should be configured in SIB 11, which admits up to 47 neighbors, and the remaining will be allocated SIB 11bis.

Inter-frequency and IRAT handovers require the use of compressed mode, therefore these neighbor list sizes should be kept small (below 20 if possible) to speed up the cell search and reduce handover failure probability.

11.3 Multilayer Management in HSPA

The rapid growth of smartphone users and their demand for ever faster throughputs has required fast capacity expansions in HSPA+ networks. As it will be discussed in Section 11.4, capacity expansions can be done by adding cells/BTS or constructing micro cells; however, in case operators have additional frequencies, the fastest way to expand the capacity is to add carrier frequencies. This can be done simply without any site visit in the case that the RF unit is capable to handle several carriers with enough power. Therefore, currently most high traffic networks include cells with two, three, or even four carriers on the same or different bands. The typical case is multiple carriers on the high band and one carrier on the low frequency band – if available. The traffic sharing between all the carriers – also called "layering" – is a very important task of network planning and optimization. This section considers both intra-band and inter-band layering.

11.3.1   Layering Strategy within Single Band

The single band layering strategy refers to traffic sharing between multiple carriers within one band, for example, 2100 MHz carriers. In this section some typical layering strategies are discussed.

11.3.1.1   Layering for Release 99 Only

The starting point of 3G deployments was a single carrier on a high band and then gradually a second carrier was added in high traffic areas. A 3G deployment on 2100 MHz is considered in this section, and will be referred to as U2100. In the first UMTS networks, the addition of a second carrier was mainly caused by the Release 99 Dedicated Channel (DCH) traffic increase. The load was shared equally between the two layers using, for example, Radio Resource Control (RRC) connection setup redirection, where the RRC setup message informs the UE to move to another carrier and therefore the UE completes the RRC setup in this other carrier. The load in this context can refer to uplink noise rise or number of connected users, for example. The load sharing is illustrated in Figure 11.6.

Typically the second carrier deployment at this stage was not done for large contiguous areas but rather only for hotspots and, therefore, the second carrier cell coverage was much larger than that of the first carrier. This is due to the lower interference levels on the second carrier, which do not have direct adjacent cell neighbors like the first carrier cells have. In this scenario, all UEs must be directed to the first (contiguous) carrier very aggressively in idle mode. To achieve this, a negative Qoffset2 is used for first carrier neighbors on the second carrier, as shown in Figure 11.7.

Similarly, in dedicated mode the UEs must be pushed to the first carrier and the inter-frequency handover has to be triggered based on RSCP threshold rather than Ec/No. The reason for this is that good Ec/No areas in the second carrier can extend over several first

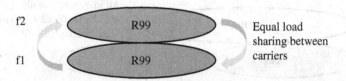

Figure 11.6 Load sharing between Release 99 carriers

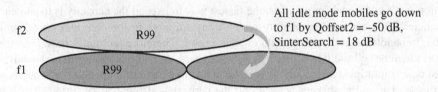

Figure 11.7   Pushing UEs to f1 in idle and dedicated mode

carrier neighboring cells and can over complicate the neighbor planning. Therefore the inter-frequency handover should be triggered on RSCP instead (typical values of around −103 dBm are recommended, although this greatly depends on the specific sector situation). Each second carrier cell should be configured with inter-frequency neighbors from the first carrier of the same sector, in addition to inter-frequency neighbors extracted from the intra-frequency neighbor list of the first carrier.

11.3.1.2   Early Layering for HSDPA

At the initial phases of HSDPA deployment, the layering strategy can be very simple, just add HSDPA on the second carrier and keep all Release 99 on the first carrier as shown in Figure 11.8. This way the additional interference caused by HSDPA activation does not affect voice traffic (note that HSDPA typically uses full base station power during the transmission).

The neighboring strategy can follow the same as that described in the Release 99 layering scenario. This type of layering strategy causes some challenges:

- In order to have all the voice calls on the first carrier the UEs should be moved to the first carrier immediately after the release of HSDPA (or Release 99 packet call) call on the second carrier. This causes extra signaling load on the first carrier due to cell updates in the case that the UE was in cell_PCH state.
- As each call is always initiated on the first carrier, the signaling load on the first carrier could cause extremely high load. The signaling traffic can lead to higher uplink noise rise due to heavy RACH usage, which can impact the coverage area of the first carrier.
- As HSDPA is only served on one single carrier, the throughput per user can degrade rapidly as the HSDPA traffic increases. In case of non-contiguous HSDPA layer, the HSDPA service would only be available in certain hotspots.

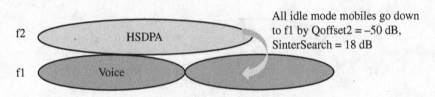

Figure 11.8   Early layering strategy with HSDPA

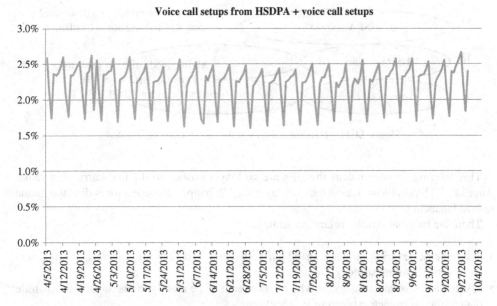

Figure 11.9 Share of voice call connection establishments

Figure 11.9 shows that most connection establishments are HSDPA packet calls. This is explained by the fact that smartphones can create very frequent data connections compared to the number of voice calls. Therefore, using the initial layering setup would lead to several carrier changes for roughly 98% of the calls. This would mean excessive signaling load with increased uplink noise rise on the first carrier.

As the HSPA service grows, the layering strategy should be adjusted as described in the following section.

11.3.1.3 Layering with Increasing HSDPA Traffic

The rapid HSDPA traffic increase, decrease of HSDPA throughput per user and the signaling traffic explosion due to the continuous layer changes, required a modification of the layering strategy, as shown in Figure 11.10.

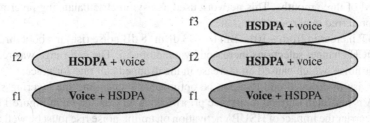

Figure 11.10 Layering with increased HSDPA traffic

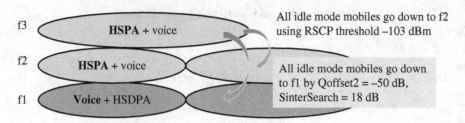

Figure 11.11   Handling of UEs at the carrier coverage border

This layering strategy means the UEs are no longer pushed to the first carrier but rather voice or HSDPA calls are handled by the layer the UE happens to select, or is directed to due to load balancing.

There are two main traffic balancing strategies:

- a symmetric layering strategy, in which both voice and data are balanced in all layers (symmetric layering strategy), or an
- asymmetric layering strategy, in which voice tends to be mainly in one layer that has little data traffic (asymmetric layering strategy).

In this second case, the load balancing can be done based on the number of HSDPA users so that carrier 2 (and 3 or 4) handle most of the HSDPA calls while carrier 1 handles most of the voice calls. In networks with this configuration, typically 70–80% of all HSDPA calls are equally shared among HSDPA carriers 2 and 3, while 70–80% of voice calls are allocated to carrier 1.

At the border of any carrier the neighbors need to be set according to Section 11.3.1.1 that is, all UEs at the coverage border of certain carrier should be directed to any of the continuous coverage carriers. Further load balancing at the coverage border of certain carrier can be done so that idle mode cell reselection is targeted to a different carrier than dedicated mode handover, as shown in Figure 11.11.

11.3.1.4   Layering with HSUPA

When HSUPA was activated, the layering strategy was typically such that HSUPA was activated together with HSDPA. This caused a rapid increase of uplink noise rise as indicated in Figure 11.12 which shows the Received Total Wideband Power (RTWP) in an example network over a period of three months. This network used an asymmetric balancing layering strategy (i.e., voice preferred in carrier 1).

The RTWP increased from −101 dBm to −93 dBm (8 dB noise rise) in about three months, and the E-DCH channel selections increased by a factor of 5. The cell range was reduced due to the higher noise, which caused an increase in the dropped call rate for voice.

This increase in dropped call rate can be optimized by changing the layering strategy so that HSUPA is activated only on non-voice prioritized layers, as shown in Figure 11.13.

For each carrier the impact of HSUPA activation on uplink noise rise must be well controlled in order to maintain the same coverage footprint. In addition to the deactivation of HSUPA,

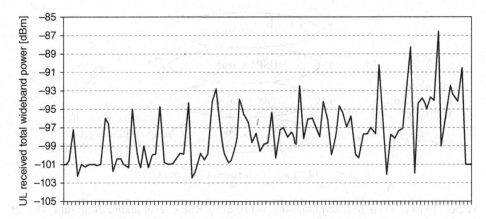

Figure 11.12   HUSPA (E-DCH) selections increase of RTWP

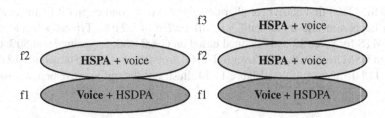

Figure 11.13   HUSPA activation on non-voice prioritized layers

which is a rather extreme measure, Chapter 12 describes other techniques that can help mitigate the increased noise rise caused by smartphone traffic.

With the introduction of dual-cell HSDPA (DC-HSDPA) at least two carriers should be used everywhere in the network where DC-HSDPA capability was required. In cases with three carrier contiguous deployments, the DC-HSDPA can be activated in the second and third carrier leaving the first carrier slightly less loaded and therefore offering a better quality (Ec/No). This higher quality can be utilized when terminals are making handovers or cell resections from GSM and also as a fall-back layer within the WCDMA network (used before handing over to GSM). In the case of four carrier deployment in WCDMA, two DC-HSDPA cell pairs can be configured, that is, one for carriers 1 and 2 and another one for carriers 3 and 4. It should be noted that the primary cell of DC-HSDPA must (typically) have HSUPA activated and therefore the carrier that has the largest HSUPA coverage should be selected as the primary cell.

11.3.1.5   Layering with GSM

The GSM layer is used as a fall-back layer where the UEs are directed in idle and connected mode due to lack of coverage. The idle mode threshold for Ec/No is typically around −16 dB Ec/No with minimum *rx level* at around −101 dBm as shown in Figure 11.14. In this particular example, the GSM layer is in the 900 MHz band and UMTS in the 2100 MHz band.

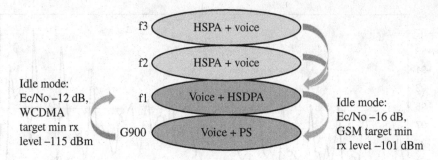

Figure 11.14 Layering between G900 and U2100

Handover to G900 is initiated at around −110 dBm RSCP level and/or at around −16 dB Ec/No. Handovers and idle mode cell reselection to G900 is only triggered from the first carrier and therefore G900 neighbors are only defined between G900 and U2100 first carrier. Return from G900 to U2100 first carrier (or alternatively also to second carrier if that is continuous) should not happen in idle mode until −12 dB Ec/No in U2100. Typically handovers from G900 to UMTS for voice are only defined based on G900 load, for example, at 80% time slot utilization in GSM triggers the handover to U2100 provided that Ec/No target of −12 dBm and RSCP of −115 dBm are fulfilled. Figure 11.14 illustrates the traffic movement across layers.

11.3.2   Layering Strategy with Multiple UMTS Bands

Layering strategy with multiple WCDMA bands greatly depends on the number of carriers on different frequency bands. The most typical deployment scenario is that the low band (at 900 MHz i.e., U900 or at 850 MHz i.e., U850) has fewer carriers than the high band U2100. This means that in order not to overload the low band carrier(s) the UEs must be pushed aggressively to U2100 and the fewer low band carriers there are, the more aggressively UEs need to be pushed to the high band. Otherwise the low band cell will be highly loaded (practically blocked) due RRC connection requests or cell update signaling. Figure 11.15 shows the typical case which is used as an example scenario in the detailed analysis throughout this chapter.

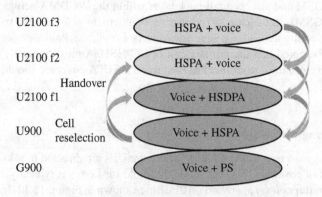

Figure 11.15 Typical multiband layering scenario for WCDMA

The multiband layering scenario in Figure 11.15 shows that there are no handovers or cell reselections to the G900 layer any more, except at the U900 coverage border where idle and dedicated mode mobility between WCDMA and GSM can be as shown in Section 11.3.1.5. The traffic sharing between U900 and U2100 layers can be done so that in idle mode all the UEs are pushed aggressively from U900 to the first carrier of U2100 and handovers from U900 are towards the second U2100 carrier to balance the load, provided that the second U2100 carrier has continuous coverage.

11.3.2.1 Multiband Layering Parameters

In this section some of the most critical design aspects of layering between U2100 and U900 are discussed in detail.

Idle Mode

Figure 11.16 shows the importance of Sintersearch parameter setting at the U2100 layer. In the case that the Sintersearch is set too high and therefore UEs start to measure U900 very early (at high Ec/No values at U2100 layer) and reselect aggressively to U900, the RRC setup success rate and RRC success rate (1-RRC drop rate) collapses. The worst values for RRC setup success rate are around 10%, indicating that the single U900 carrier cannot handle the traffic from three or four U2100 layers. Therefore the UEs should not be allowed to even measure the U900 carrier until the quality (Ec/No) at the U2100 layer falls to around 6 dB above the minimum acceptable Ec/No limit. This, combined with the Qoffset for RSCP of around 4 dB and Ec/No around 2 dB (both requirements mean U900 should be better than U2100), can protect the U900 layer from getting too much traffic from U2100.

The reselection from U900 to U2100 should be very aggressive, basically measuring all neighbors (intra- and inter-frequency) all the time and having 2–4 dB negative offsets for RSCP and Ec/No thresholds at the U2100 layer. The negative offset means that U2100 quality

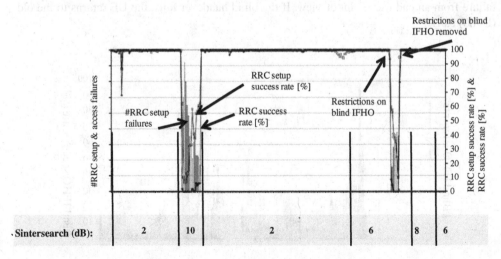

Figure 11.16 Importance of SinterSearch

can be lower than U900 for the reselection. The reselection from U900 to the first carrier of U2100 should be performed quite early, as the first carrier of U2100 has less HSDPA (and no HSUPA) traffic and therefore Ec/No performance is better than for other U2100 carriers.

Dedicated Mode

In dedicated mode, the handover target (U2100) carrier should be different from the idle mode reselection target carrier. The idle mode target is the first frequency while the handover target could be the second frequency, provided that there is more than one single continuous coverage layer in U2100. Then, similarly to the idle mode control, the UEs must be pushed to the second U2100 carrier as soon as possible. This can be done by using, for example, blind handover during the RAB setup phase where the handover from U900 to U2100 is triggered without any inter-frequency measurements. This blind handover can be also assisted by intra-frequency measurements from event 1a or 1c, or any periodic RSCP reporting that might active, or then by inter-frequency RSCP measurement report from RRC connection request and/or cell update messages. Based on these measurement results the handover is not entirely blind as some information about target cell RSCP is available. Then, the handover from U900 to U2100 can be triggered based on either real U2100 target cell RSCP or U900 source cell RSCP offset by a threshold to take the penetration loss difference between the bands. The blind handover (or blind handover assisted by measurements) performance is very sensitive to the antenna directions for different bands. If there is even a slight difference between the antenna directions of the two band cells, the blind handover performance suffers greatly unless the target cell (U2100) RSCP measurements are not available from RRC connection request or cell update messages. Figure 11.17 shows the blind handover success rate for a sector where U900 and U2100 have individual antennas around 6 degrees difference in directions. The blind handover success rate cannot reach >90% levels even if the threshold for own cell (U900) RSCP is increased from −95 dBm to −90 dBm. It should also be noted that when RSCP threshold is increased the traffic in the U900 layer increases, causing excessive load in the U900 layer. It should be noted that failure in blind handover does not cause any dropped call or setup failure from an end user point of view. If the blind handover fails, the UE returns to the old (U900) carrier.

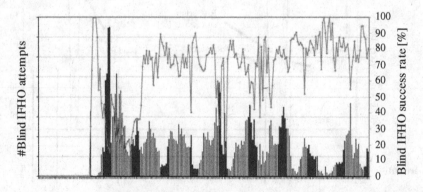

Figure 11.17   Blind handover success rate

Another possibility is to use compressed mode and actual inter-frequency measurement at the end of the packet call. This is done so that, instead of moving the UEs to Cell_PCH, URA_PCH, or idle mode, the UEs are ordered first to compressed mode and in the case that the U2100 carrier is found to have adequate RSCP and Ec/No the handover is executed instead of moving UE to cell_PCH, URA_PCH, or idle mode. This way the achievable handover success is around 98 – 99%.

Handovers from U2100 to U900 should only be done in the case of lack of coverage in the U2100 layer. And in the case of hotspot coverage for certain U2100 carriers, the handover should be first done to the continuous coverage carrier within the U2100 layer before handover to the U900 layer. The handover thresholds from U2100 to U900 are typically set to be similar to the previous chapter's U2100 to G900 handover thresholds, as discussed in Section 11.3.1.5.

11.3.2.2 Neighbor Planning

The neighbor relations are planned so that from each U2100 carrier having continuous coverage there are neighbors to the U900 carrier of the same sector (same BTS) and inter-BTS neighbors are only needed in the case that the antenna directions between the same sector bands are not the same. Intra-band neighbors within the U2100 carriers of the same sector are not defined except in the case of U2100 channel border areas. This can be done provided that the traffic between the U2100 carriers is equally distributed. G900 neighbors are also only defined in the case of U900 border.

11.3.3 Summary

The layering between different WCDMA carriers within the same band or different bands should try to balance the load between all the carriers. Load here can mean number of HSPA users, voice erlangs, and signaling. It is especially important to monitor the signaling traffic (number of RRC connection requests and cell updates) which can increase rapidly and block one of the carriers. Therefore, equal sharing of UEs among all carriers is mandatory. This means that some parameter tuning needs to be performed from time to time as the total traffic increases.

11.4 RAN Capacity Planning

Capacity planning and system dimensioning is an important aspect when operating a wireless communication service. A balance must be struck between over-dimensioning the system, leading to premature capital expenditures, and under-dimensioning the services provided, which can lead to poor customer experience and even customer churn. Historically, statistical models such as the Erlang B model have been used by wireless operators to estimate the voice service GoS (Grade of Service) they are providing to their customers. Once it is forecasted that the BH (Busy Hour) erlangs of a cell may result in a blocking rate of 1–2%, additional voice trunks or circuits are provisioned to improve the GoS below the targeted threshold. Dimensioning the data services is much more challenging due to their more bursty nature compared to voice and the fact that many different services can run over data channels that have vastly different service requirements. For example, data services such as email or MMS have a much more relaxed delivery delay requirement compared to real time services such as video or music streaming. Unfortunately, RAN (Radio Access Network) KPIs really don't tell the whole story because the counters blend many data services with different service requirements together. As discussed in Chapter 13, the real customer experience should be estimated with new KPIs that measure the performance of each service independently. Since those application specific KPIs are not available yet, this section reviews some of the RAN (Radio Access Network) KPIs that might give us some insight into the average customer experience.

11.4.1 Discussion on Capacity Triggers

For HSPA+ data services, the customer demand can be measured in terms of MB volume transferred over the air interface. The average sector throughput of a cell is an important metric because it shows the potential amount of data that can be transferred during an hour and the throughput speed gives an indication of what services on average can be supported. The average throughput a sector can support is a function of the user geographical distribution inside the cell footprint. In general, if most of the users are near the NodeB, the cells' throughput is higher than in a cell where most of the users are in poorer RF condition on the cell edge. One metric to monitor closely is the cell TTI (Time Transmission Interval) utilization. In HSPA, the length of one TTI is 2 ms, hence there are theoretically 1.8 million TTIs that can have customer payload per hour. Once all the TTIs are utilized the cell is sending the maximum MB volume that it can support. Figure 11.18 (left) plots the DL MB volume vs. the TTI utilization. For this particular cell, the maximum MB customer demand it can support during the busy hour is approximately 700 MB. The slope of the line in the graph is the weighted average HSDPA throughput of 1.7 Mbps.

$$HSDPA\ Throughput\ (Mbps) = \frac{(MBVolume)}{TTI\% * (3600\,s)} \frac{8\,bits}{1\,Byte} \frac{2^{20}}{10^6}$$

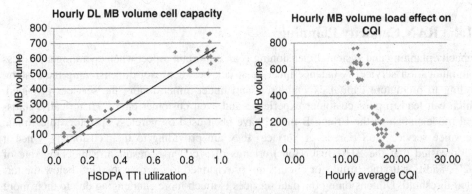

Figure 11.18   HSPA+ cell capacity and loading

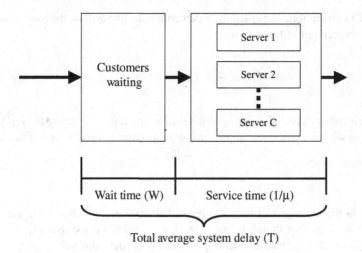

Figure 11.19   Generic queuing model

Figure 11.18 (right) shows the effect on the CQI (Channel Quality Index) beginning reported by the mobiles' as more data is sent over the HSPA+ air interface and the system loads up.

The TTI utilization is an indication that a cell is at its maximum capacity; however, what we really need to know in order to dimension the system is what the average user experience is when the cell is loaded. Since the HSDPA air interface is a shared resource in the time domain it is important to try to measure the throughput a typical user will see as other users are added to the system. Measuring the delay of the entire system is the goal for data networks, so queuing models such as Erlang C may be utilized. There is one scheduler assigned per cell on an HSPA+ NodeB, so essentially there is a single server and we can look at the TTI utilization in terms of erlangs of scheduler utilization and model the total system delay as an M/M/1 queue [5] which is a specific case of Erlang C where the number of servers or channels is equal to one. Figure 11.19 illustrates the concept. For data traffic, we are interested in the queuing delay because it will give us an indication of the entire systems' performance as perceived by the customer. For an M/M/1 queue the average total system delay (T_A) can be expressed as:

$$T_A = W_A + {}^1\!/_\mu$$

$$W_A = \frac{p}{(1-p)\mu}$$

Where W_A is the wait time (queue time), $1/\mu$ is the average service time, and p is the server utilization (TTI%). For an HSPA+ system, W_A is the average time customer data is waiting to be scheduled at the scheduler, the service time is the average time it takes to send the customer data over the air interface. The service time is a function of the total time it takes to deliver data and the TTI utilization. The customer's perception of time will be the total time T_A, while the networks' perception of time is $1/\mu$.

$$\frac{1}{\mu} = T_A(1-p)$$

The same customer data that is sitting in the scheduling queue is also sent over the HSDPA air interface, so we can express the above time domain expression in terms of throughput.

$$R_U = R_S(1-p)$$

Where R_U is the average end user's perceived throughput and R_S is the average sector throughput. A good rule of thumb is that at 90% TTI utilization the user experience is only 10% of the sector throughput, at which point the poor data throughput is probably very noticeable to the customer. Another important insight to the M/M/1 model is that as the TTI utilization approaches 100%, caused by a large amount of users in the system, the user perceived average throughput approaches zero, which implies that there is an infinitely long queue of data waiting to be scheduled.

Figure 11.20 shows the hourly statistics collected over a two-day period for a single highly loaded cell. When the TTI utilization approaches 100% the sector throughput still looks fine, but the user experience from 9:00 to 17:00 is below an acceptable level. Also, the user perceived throughput calculated via the M/M/1 queue model closely matches the HSDPA average user throughput as reported by the RAN counters, which are measuring the amount of data in the scheduler buffer. The degraded CQI that is reported by the mobiles as the system becomes loaded and the high TTI% utilization have a compounding effect on the customer

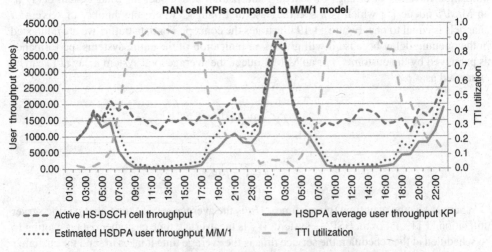

Figure 11.20 Cell and user throughput KPIs in a loaded sector

experience. The former reduces the potential throughputs that a mobile can get when it is scheduled and the latter is an indication of how often it has to wait before it can be scheduled.

11.4.2   Effect of Voice/Data Load

UMTS supports both CS R99 voice and HSPA+ PS data services on the same carrier. The NodeB transmit power is a key air interface resource that is shared between voice and data services. To support the pilot and common control channels, 15–25% of the cell PA (Power Amplifier) is utilized. Downlink power is a shared resource, therefore the voice traffic has big impact on the available power that can be used for HSDPA data services. Often the voice channels consume more than half of the PA power, even though the data services represent the larger portion of the cell capacity.

The following example is extracted from a network with a significant voice traffic load. On the loaded cell shown in Figure 11.21 over 60% of the power is used to support R99 voice during the busy hour. The remaining 40% of the average power of the PA can be used during peak times to support HSDPA. However, from a spectral efficiency perspective the HSDPA payload represents nearly 80% of the cell capacity, while the voice represents only 20%.

The RAN KPIs only report the average HSDPA power over an hour. However, if very few TTI are used during the hour the average HSDPA traffic will be very low, even though when it was in use the power may have consumed all the remaining PA power that was not being used for R99 voice. For the same cell that was analyzed above, Figure 11.22 shows the average hourly HSDPA power vs. the TTI utilization over the same two-day period. Note that no more average power can realistically be allocated to HSPA+ because nearly all the TTIs are in use during the busy hour. For the same cell, not unsurprisingly the DL HSDPA power is also highly correlated to the MB volume of HSDPA data.

In smartphone dominated markets the majority of the capacity on a UMTS cell is represented by the HSPA+ data, even though it may appear that most of the power is being utilized for R99 voice.

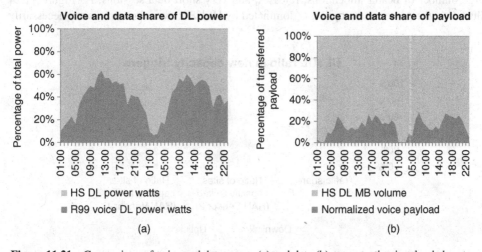

Figure 11.21   Comparison of voice and data power (a) and data (b) consumption in a loaded sector

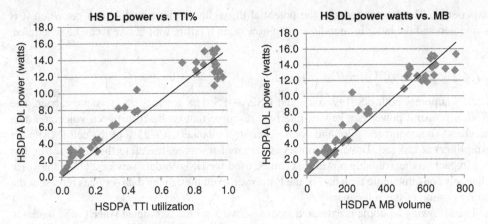

Figure 11.22   Impact of TTI utilization and data payload on HSDPA power consumption

11.4.3   Uplink Noise Discussion

The uplink capacity is also a very important consideration when evaluating capacity solutions. In terms of spectral efficiency, HSDPA Release 8 is 3–4 times more spectrally efficient than Release 6 HSUPA, so theoretically the uplink should not be a bottleneck until the ratio of downlink traffic payload to uplink traffic payload is on the order of 4 : 1. In a smartphone dominated network, the ratio of downlink to uplink traffic volume is on the order of 7 : 1 or even greater because much of the data transferred in the network comes from downlink dominated applications such as video streaming. However, in live networks 30–60% of congested sites experience uplink capacity issues, which isn't consistent with the spectral efficiency expectations. There are also significant differences across infrastructure vendors, with some vendors being more prone to uplink congestion, as illustrated in Figure 11.23.

Due to the bursty nature of the smartphone traffic, the uplink direction will experience a large number of bearer allocations, releases, and very short data sessions. As Figure 11.24 illustrates, the power in the uplink is dominated by HSUPA, even though it is not necessarily

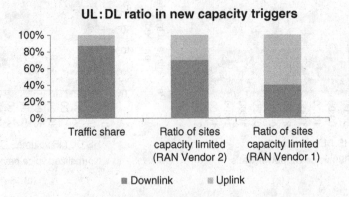

Figure 11.23   Comparison of DL/UL traffic share vs. capacity triggers

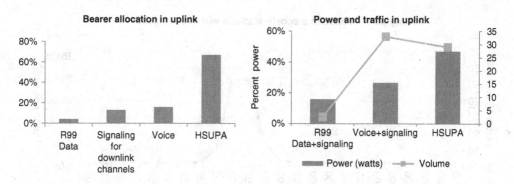

Figure 11.24   Power and payload breakdown in uplink based on service and channel type

the main source of payload. Chapter 12 will provide a deep analysis of the reasons for the excessive uplink noise that is observed in networks with heavy smartphone traffic, and possible optimization actions to mitigate this problem.

As the noise rises in the NodeB, the devices must increase their transmit power to hit the required BLER targets, which in turn generates more uplink noise. There comes a point where the mobile device is at full power but the signal received by the NodeB is too low to be detected, which can lead to access failures and poor uplink performance in terms of throughput.

The goal of determining the uplink capacity for a sector is to estimate the point at which the uplink noise rise becomes so high that uplink throughput is impacted and other real time services, such as voice, begin to experience congestion. When HSUPA 2-ms TTI is deployed it is possible that the uplink noise rise of the system may approach 8–10 dB in heavily loaded cells. When this much noise rise is experienced, system KPIs – such as access failures – begin to degrade. As the noise rise increases, typically data traffic is affected first, followed by voice traffic and eventually SMS services.

In cells with high uplink congestion the number of active users is nicely correlated to the number of data sessions. Three cells with high noise rise in three different areas of the same network were analyzed. There is a ratio of 300–400 data sessions per active user. In Figure 11.25, hourly RAN KPI data is plotted to show the relationship between active users

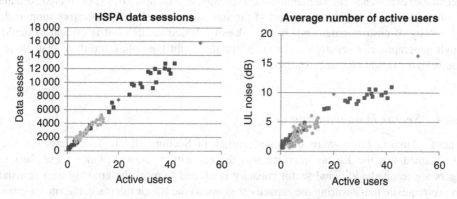

Figure 11.25   Impact of number of simultaneous users on UL noise rise

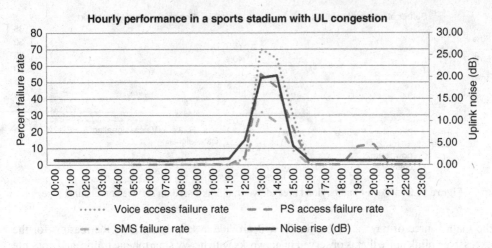

Figure 11.26   Extreme UL noise during mass events

and the uplink noise rise throughout a couple of days. As the average number of active users is trended into the 20–30 range, the uplink noise approaches the 8–10 dB uplink noise threshold.

During events where there is a high concentration of smartphone devices in a small geographical area, such as a sports stadium or convention center, the noise rise can quickly escalate to over 10 dB, after which point the network KPIs and customer experience are dramatically affected. Figure 11.26 shows a cell that serves a sports stadium with thousands of spectators. The chatty traffic generated by the smartphone helps increase the uplink noise, leading to very high levels of system access failures.

Stadiums that serve major sporting events can have a user demand that is nearly 20 times that of an average cell site. They are very challenging from a system capacity planning perspective because the high concentration of smartphones lasts only a few hours, during only a few days a month. A very high capacity system must be deployed to augment system capacity so that the customer experience can be maintained. Typically, neutral host DAS (Distributed Antenna Systems) are deployed with the support of the stadium owners to help operators support the high density of customer demand. In high density device scenarios it is critical to deploy enough geographically separated cells such that they split the uplink traffic so that the full capacity of the downlink can be utilized.

11.4.4   Sector Dimensioning

Capacity planning limits were discussed earlier in Section 11.4.1. These capacity triggers assumed that the RF air interface was the capacity bottleneck. Once these capacity triggers are reached additional sector capacity is needed to meet the growing data demand. As a pre-requisite to measuring the capacity triggers on the RF air interface, the site baseband capacity and IP Iub capacity limits need to also be checked to make sure they are not limiting the RF capacity. These capacity constraints are discussed in the next section and Section 11.5 respectively.

Once it has been verified that the RF capacity is the bottleneck and capacity triggers such as high TTI utilization are measured, as discussed in Section 11.4.1, the next step is to determine if the congested sector throughput is also good. If the sector throughput is good, the quickest and most cost effective solution is to add an additional carrier frequency. If there are spare baseband and power amplifier resources available the activation of a carrier can even be performed without a site visit. In order to mitigate the performance impacts of handling multiple carriers, as discussed in Section 11.3, it is often good practice to add carriers not only to the site that needs it but also to some of the neighboring sites that have high interaction rates with the congested sector.

Congested sectors with poor throughput will also benefit from additional carriers, however carrier adds are not a long-term solution to providing additional network capacity. For these sectors, more investigation is needed to help determine why the sector throughput is substandard. Sector optimization techniques, as discussed in Section 11.2.2, may be needed to improve the SINR in the coverage areas that are generating the traffic demand. If it is determined that the traffic is being generated in poor coverage areas that can't be improved by optimization, or that a sector is already utilizing all the carrier frequencies that are available to the operator, the next option is to build additional macro sites or micro sites to improve the coverage in the poor RF areas and expand the overall network capacity.

11.4.4.1   Optimization of RAN Baseband Resources

When data traffic increases in the network, the throughput offered to the existing users will be reduced. When the throughput levels reach certain quality limits determined by the operator, the capacity of the network will have to be increased.

While this is expected from any technology, and typically fixed adding more data channels (UMTS carriers), in the case of HSPA+ there is an intermediate step that needs to be considered: incrementing the baseband capacity of the site.

In heavily loaded sites, the lack of baseband resources can play a major role in performance. The effect is the same as normal radio congestion – throughput reduction – however in the case of baseband shortage the scheduling rates are typically low. This can be seen during drive tests, which will show a high DTX rate in the transmissions, or in the network KPIs, with a low TTI utilization or lack of HS-PDSCH codes.

Consider the example in Figure 11.27, in which one cluster was presenting low overall throughputs (2.2 Mbps per carrier) even though the radio conditions were decent. It was found that three sites in the cluster required an upgrade on their baseband resources, which resulted in 100% improvement of the throughput in the cluster, to an average of 4.4 Mbps.

One important thing to consider is that baseband shortage can occur even in cases with relatively low amounts of transferred data, especially in networks with a high penetration of smartphone devices. In the example above, even during the night time there were at least 19 HSPA customers connected to the site. The HSPA users in connected mode will consume baseband resources even if they are not actively transmitting data; as will be discussed in Chapter 12, it is possible to adjust the state transition settings to limit this impact, at the expense of longer connection latency.

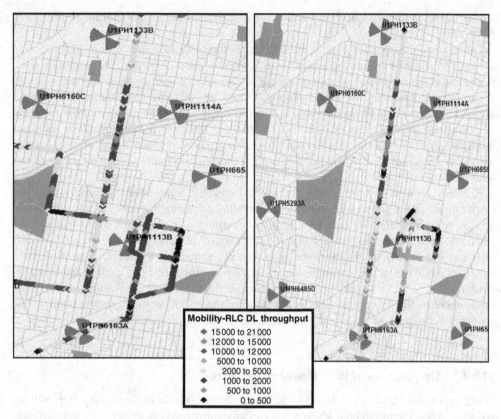

Figure 11.27 Throughput improvement after baseband upgrade. Low baseband case (left) presents a high number of low throughput dots (dark gray)

11.4.5 RNC Dimensioning

The RNC (Radio Node Controller) serves as an anchor for mobility management and radio resource management in an HSPA+ network. The dimensioning of the RNC is important because it will degrade the performance of multiple NodeBs for reasons not related to the air interface capacity, so it can often be difficult to troubleshoot. Below are some common parameters used to dimension the RNC:

- Aggregate Throughput (Mbps);
- Number of NodeBs and Cells;
- Number of Active Users;
- Processing Load.

Originally, RNCs were designed to focus on throughput capacity as their main bottleneck; however, in smartphone dominated networks it became clear that the signaling capacity could rapidly become a bottleneck as well. Figure 11.28 shows that there are situations in which the signaling capacity (Y axis) can be exhausted well before the RNC has been able to achieve the designed throughput capacity (X axis).

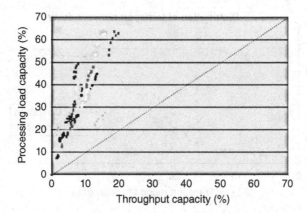

Figure 11.28 Illustration of RNC capacity bottlenecks: Y axis shows control plane capacity, X axis shows user plane

As RNCs are added to the network to keep up with increasing smartphone penetration, sites need to be "re-homed" from their original RNC; this leads to more RNC handovers, and an additional operational complexity derived when a large number of NodeBs need to be logically "re-built" and configured on new RNCs.

There are important differences in vendor RNC design, with some vendors providing a more flexible architecture that enables the "pooling" of resources for either the control or user plane, thus facilitating the balance of the processing capacity and delaying the need to increase the number of RNCs as traffic increases.

11.5 Packet Core and Transport Planning

While not as complex to plan and tune as the radio network, it is important to properly plan the core and transport networks to ensure they provide the required bandwidth and latency for the data services being offered. In this section we discuss some considerations regarding the planning of the Serving GPRS Support Node (SGSN) and the Gateway GPRS Support Node (GGSN).

Modern HSPA+ networks typically implement the GTP one tunnel solution, which tunnels the user data transfer directly to the GGSN, effectively leaving the SGSN to primarily handle data signaling traffic. GGSNs, on the other hand, will be the main node dealing with user data transfer and are specialized to cope with large amounts of data volumes.

When dimensioning the packet core, the number of SGSNs will be related to the amount of sessions being generated, while the number of GGSNs will depend mostly on payload and number of PDP contexts that can be sustained.

SGSN configuration and dimensioning will also be impacted by the PS paging load in the network. If the paging load is too high, the routing areas will need to be made smaller.

SGSNs can be configured to distribute the data sessions in one area across multiple GGSNs. This is known as "pooling" and can be used as a way to avoid over-dimensioning; however, it may have negative effects on user experience as it will be discussed later on this section.

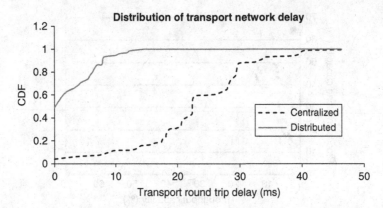

Figure 11.29 Example transport network delay in a centralized vs. distributed deployment

In addition to SGSN and GGSN, many networks deploy a wireless proxy, whose goal is to improve data performance and reduce network payload. There are many flavors of proxies, however they typically implement the following features:

- TCP Optimization, which provides advanced methods to compensate for typical TCP problems in wireless environments.
- Image and text optimization, which reduces image and text payload by compressing or resizing images.
- Video optimization, targeted at reducing video payload while providing an optimum user experience.

While network proxies can help reduce overall payload, their effectiveness will decline as more traffic is encrypted, which limits the possible actions the proxy can take.

As discussed earlier, another important consideration regarding core network dimensioning is latency. Packet delay is a major factor affecting data user experience: HSPA+ networks will not be able to perform at the highest level if there are long delays. This is due to side effects from the TCP protocol, which mistakenly interprets long delays as if data was being lost in the link.

One of the most important considerations to optimize latency in the network is the placement of the GGSNs. In networks covering wide territories it is recommended to distribute the GGSNs to try and have them as close as possible to where the traffic originates. Figure 11.29 shows an example network in which the overall round-trip delay was reduced by an average 22 ms when the GGSNs were distributed, as compared to a centralized deployment.

11.5.1   Backhaul Dimensioning

As the RAN (Radio Access Network) has evolved from sub 1 Mbps to multiple Mbps speeds it has become clear that legacy T1 (1.544 Mbps) or E1 (2.048 Mbps) speeds that were used to support circuit switched voice would no longer be able to scale and support the data speeds that would be achievable on HSPA(+) and LTE networks. Ethernet backhaul became an attractive replacement based on its common availability, ubiquity in the Internet industry and support of speeds that could scale proportionally at a reasonable cost for the expected increase in data demand. Common Ethernet rates scale from 10 Mbps Ethernet (10BASE-T), to 100 Mbps Fast Ethernet (100BASE-TX) and even 1 Gbps Gigabit Ethernet speeds and beyond.

High speed wireless data networks often advertise peak data rates. These peak rates are the maximum a technology can provide and can realistically only be measured in a lab environment under near perfect RF (Radio Frequency) conditions. It is important when dimensioning the Ethernet backhaul that we look at how these RAN (Radio Access Network) technologies are deployed and what other constraints limit the available speeds that a customer may realistically be able to achieve in a real wireless network if there were no backhaul limitations. Peak data rates provide a ceiling for what is achievable from a wireless technology, but they are almost never realized in a real operating environment. The support of high peak data rates is important from a user experience point of view, since data is sent in bursts; however the application requirements for these bursts are not so high. As referenced in Chapter 13, peak bursts of 11 Mbps were observed when streaming a Netflix movie.

Average spectral efficiency is often used to estimate the capacity a wireless network can provide. However, using average throughput to dimension backhaul may underestimate the amount of time that the customers' throughput is limited by the provisioned Ethernet backhaul bandwidth. The following approach attempts to adjust the peak throughput ceiling to more realistic throughput targets. HSPA+ utilizes higher order modulation, such as 64QAM modulation, to achieve the peak data speeds which requires very good SNR (Signal to Noise). However, network statistics show that the RF environment is good enough to utilize 64QAM modulation typically less than 20% of the time. Also, under more realistic network conditions, more FEC (Forward Error Correction) is needed when the SNR decreases, so the achievable speeds are further reduced.

Table 11.5 shows some target throughputs based on the following assumptions:

- Reduce the peak throughput by a factor of 1.5 (difference between 6 bits/symbol 64QAM and 4 bits/symbol 16QAM).
- More realistic throughput using an effective coding rate of 6/10, which is based on the maximum throughput of 16QAM in the 3GPP CQI tables.
- Using these two correction factors we see the target throughputs scale down to approximately 40% of the peak speeds the RAN can theoretically support.

Table 11.5 –Realistic peak HSPA+ speeds for Ethernet backhaul dimensioning

Technology/downlink spectrum	Downlink theoretical peak (Mbps)	Theoretical without 64QAM modulation (Mbps)	Realistic peak[a] without 64QAM modulation (Mbps)
HSPA+ 21 (5 MHz)	21	14	8
HSPA+ 42 (10 MHz)	42	28	17

[a]*Note:* Realistic peak speeds assume an effective coding rate of 6/10, which is based on 3GPP CQI mapping tables for 16QAM modulation.

Table 11.6 Recommended backhaul dimensioning by technology

Technology/downlink spectrum	Ethernet backhaul typical traffic (Mbps)	Ethernet backhaul heavy traffic (Mbps)
HSPA+ 21 (5 MHz)	20	20
HSPA+ 42 (10 MHz)	20	50
3–4 HSPA+ carriers	50	100

Backhaul is a shared resource between sectors of the same site, so there are trunking efficiency gains that can be realized when HSPA+ is initially deployed and it is assumed that only one sector will be in use during a given TTI (Time Transmission Interval). Table 11.6 shows some example backhaul dimensioning calculations. If we had used the theoretical peak speeds to dimension the system we would need approximately 125 Mbps of backhaul to support a site that had HSPA+ 42 (dual carrier) on all three sectors. Based on the example calculations, for the initial deployment of these high speed technologies they can realistically be supported by as little as 1/7th of the maximum air interface speed summed across a three sectored site. This assumes only one sector per site will be busy initially, but as the traffic grows multiple sectors may be utilized at the same time, so the backhaul will need to be augmented accordingly to maintain an adequate Grade of Service (GoS) to the customers. However, care must be taken to also check that applications that customers are using are actually requesting the peak speeds that the air interface is capable of supporting. Incremental capacity should also be added to support the control plane usage, but it is expected that this only adds as much as an additional 4–6% to the overall backhaul requirement.

11.6 Spectrum Refarming

11.6.1 Introduction

The concept of refarming is borrowed from agriculture, where many scattered plots of land are consolidated and redistributed among the owners. In wireless, refarming refers to the consolidation of spectrum assets in preparation to their use for the new radio technology. The typical case is to deploy HSPA or LTE technology on the existing GSM frequencies at 850/900 MHz and 1800/1900 MHz. The motivation for refarming is to get more spectrum for growing HSPA traffic and to improve the HSPA coverage by using lower spectrum with better signal propagation. The North American and Latin American networks use 850 and 1900 MHz refarming for HSPA, most European operators and some African operators have deployed HSPA on 900 MHz band, and both 850 and 900 MHz frequencies are used for HSPA in Asia. Some of the 900 MHz networks are also new spectrum allocations, like the world's largest UMTS900 network run by Softbank in Japan. The GSM 1800 MHz spectrum is also being refarmed to enable LTE deployments.

Refarming UMTS to the low bands 850 and 900 MHz provides a major benefit in coverage. The typical coverage area of a three-sector base station in an urban area is shown in Figure 11.30. The coverage calculation assumes the Okumura-Hata propagation model, a correction factor of 0 dB, a base station antenna height of 25 m, indoor loss of 15 dB, slow fading 8.0 dB, and required location probability of 95%. The antenna gain at low bands is

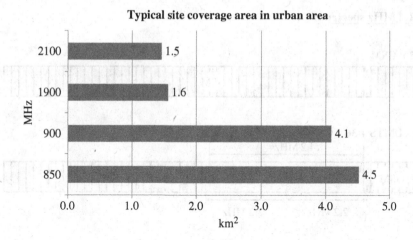

Typical site coverage area in urban area

Figure 11.30 Coverage area of 3-sector cell site in urban area with 25 m base station antenna height and 15 dB indoor penetration loss

assumed to be 3 dB lower than at high bands. The coverage area of 900 MHz is nearly 3 times larger than at 2100 MHz. The difference between 850 and 1900 MHz coverage is also similar. The better coverage is a major motivation for the operators to deploy UMTS on the low spectrum. The coverage improvement helps not only for the rollout in rural areas but also improves the indoor coverage in the urban areas.

Another motivation for the GSM band refarming is the wide support of multiple frequency bands in the devices: in practice all new 3G devices support UMTS900, or UMTS850, or both bands, and also most devices already out in the field support the low bands. Therefore the deployment of UMTS900 or UMTS850 can provide coverage benefits for the existing devices too.

11.6.2 UMTS Spectrum Requirements

The main challenge with refarming has been to clear the spectrum from GSM traffic for the UMTS deployment. The spectrum requirement can be minimized when GSM and UMTS base stations are co-sited. The chip rate of UMTS carrier is 3.84 Mcps, which leads to 3.84 MHz actual transmission bandwidth. The difference between the nominal carrier spacing of 5.0 MHz and the chip rate of 3.84 MHz can be considered as in-built guard band, see Figure 11.31.

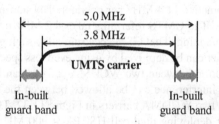

Figure 11.31 UMTS chip rate of 3.84 Mcps compared to nominal carrier spacing of 5.0 MHz

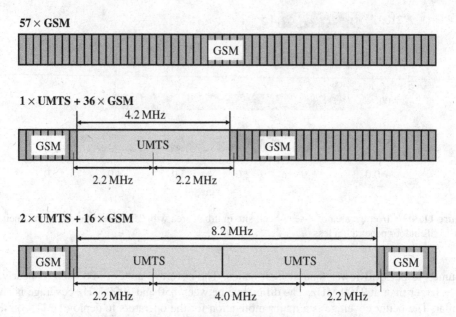

Figure 11.32   Example spectrum migration from GSM to UMTS

The guard band allows there to be two uncoordinated operators next to each other without any additional guard bands. When GSM and UMTS system are co-sited, the signal levels are similar between GSM and UMTS, and there is no need for this in-built guard band and less spectrum can be used for UMTS. Many commercial networks use just 4.2 MHz for UMTS in the case of GSM spectrum refarming, saving in total 800 kHz or 4 GSM carriers. The UMTS spectrum requirement can be selected with a resolution of 400 kHz because the GSM channel spacing has 200 kHz resolution on both sides of the UMTS carrier. The next smaller step below 4.2 MHz is 3.8 MHz, which is also doable but there is already more interference happening between GSM and UMTS. The very narrowband 3.8 MHz deployment can be practical in such cases when the amount of spectrum is very limited. The optimized use of 4.2 MHz and 3.8 MHz requires typically changes in the base station receive and transmit filtering, while no changes are required to the terminals.

Figure 11.32 shows a spectrum migration example where the operator has in total 11.4 MHz. That would be the case with 900 MHz band with three operators: the total 900 MHz band is 35 MHz which leaves in practice 11.4 MHz per operator. The starting point is the GSM-only case with 57 GSM carriers. The UMTS refarming takes 4.2 MHz or 21 GSM carriers, which leaves 36 GSM carriers remaining. That GSM spectrum allows a high GSM capacity.

The second UMTS carrier can be deployed by using even less spectrum. The measurements show that 4.0 MHz separation between two WCDMA carriers can be used in practice, even 3.8 MHz spacing if some interference can be allowed between the two carriers. We assume 4.0 MHz spacing between two WCDMA carriers in Figure 11.32. The operator can still have 16 GSM carriers left after deploying dual-cell HSDPA at 900 MHz band. The combination of good coverage with low band and high data rate with dual-cell HSDPA brings a powerful radio solution.

11.6.3   GSM Features for Refarming

The UMTS refarming becomes smoother with the latest GSM features that improve the GSM spectrum efficiency: the same GSM traffic can be carried but with less spectrum. The most relevant GSM features are

– Adaptive Multirate (AMR) voice coding, where the voice coding rate is changed between 4.75 kbps and 12.2 kbps according to the channel quality. Lower radio channel quality can be tolerated by using a lower data rate and more robust channel coding. Therefore, tighter frequency reuse can be applied in GSM by using AMR codec.
– AMR half rate voice coding uses every second frame for the transmission, which can double the voice capacity and reduce the interference levels.
– Orthogonal Subchannelization (OSC) allows packing even four users per time slot. The solution is to transmit to two users simultaneously and to separate the transmissions by interference cancellation receivers.
– Dynamic Frequency and Channel Allocation (DFCA) uses time synchronized base stations where the Base Station Controller (BSC) allocates slots in time and frequency dimensions so that the inter-cell interference is minimized.

Using all these features can improve the GSM spectral efficiency by more than a factor of four. It is possible to squeeze down the GSM spectrum requirement considerably while still carrying good levels of GSM traffic. Most of the existing terminals in the field support these GSM features, which makes the practical deployment of these features easy.

11.6.4   Antenna Sharing Solutions

The preferable base station site solution utilizes the same antenna both for GSM and for UMTS in order to minimize the number of required physical antennas. Figure 11.33 shows the antenna sharing options. Option (a) uses so-called "over the air combining" where different antenna branches are used for GSM and UMTS. The dual cross-polarized antenna includes in total four ports: two for GSM and two for UMTS. The antenna width is slightly larger than the normal cross-polarized antenna. The benefit of option (a) is that UMTS can be deployed without any changes to GSM and there are no losses in the combining. Option (b) uses cross-polarized antenna and RF combining of GSM and UMTS signals. The in-band RF combining would normally cause some losses, but one option is to use one antenna branch for GSM and another for UMTS without any losses. Such a configuration is possible if MIMO is not used in UMTS and if only one carrier is used in GSM. Option (c) is the most advanced sharing solution with a single RF unit in the base station supporting both GSM and UMTS at the same time. This solution is called RF sharing and it is becoming the typical solution in network modernization cases.

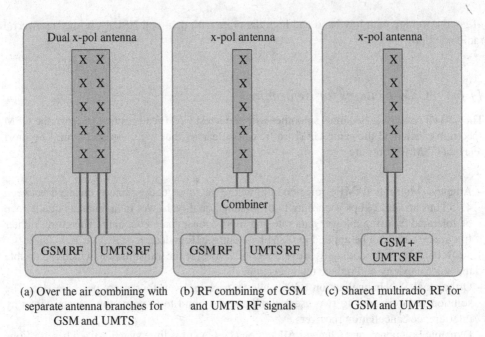

(a) Over the air combining with   (b) RF combining of GSM   (c) Shared multiradio RF for
    separate antenna branches for   and UMTS RF signals   GSM and UMTS
    GSM and UMTS

Figure 11.33   Antenna sharing options for GSM and UMTS

11.7 Summary

In this chapter we reviewed various topics covering key network deployment and dimensioning aspects.

Below are a summary of some of the most relevant considerations discussed:

- HSPA+ networks are typically limited in interference, and special care needs to be taken to control the amount of power transmitted in the network. Lack of coverage should be compensated with different solutions depending on the type of network – urban (typically limited by interference) or suburban/rural (typically limited by power).
- Antenna tilts are a very important tool to configure and maintain RF performance; however downtilting is often abused as the only way to achieve interference containment. Other methods include reducing powers, or even eliminating unnecessary sectors that are not required.
- Multilayer configuration should be done carefully depending on the type of deployment scenario: hotspots or continuous coverage zones, single or multiband, and so on.
- In networks with heavy smartphone penetration there are unexpected bottlenecks at the cell level (uplink congestion) as well as the RNC level (signaling processing capacity).
- As GSM traffic is reduced it should be possible to repurpose that spectrum for HSPA+. New carriers can be allocated in blocks as low as 3.8 MHz.

References

[1] Chevallier, C., Brunner, C., Caravaglia, A. *et al.* (2006) *WCDMA (UMTS) Deployment Handbook: Planning and Optimization Aspects*, John Wiley & Sons, Ltd, Chichester.

[2] Qualcomm Engineering Services Group (ESG), "Aspects of HSUPA Network Planning", Revision A (April 6, 2007) (updated Sept 2008).

[3] Viterbi, A.J., Viterbi, A.M., and Zehavi, E. (1994) Other-cell interference in cellular power-controlled CDMA. *IEEE Transactions on Communications*, **42**(234), Part 3, 1501–1504.

[4] Holma, H. and Toskala, A. (2006) *HSDPA/HSUPA for UMTS. High Speed Radio Access for Mobile Communications*, 1st edn, John Wiley & Sons, Ltd, Chichester.

[5] Bi, Q. (2004) *A Forward Link Performance Study of the 1xEV-DO Rev. 0 System Using Field Measurements and Simulations*, Lucent Technologies.

12

Radio Network Optimization

Pablo Tapia and Carl Williams

12.1 Introduction

To ensure an optimum operation of the HSPA+ technology the operator has to make a multitude of decisions when planning and optimizing the network, considering multiple components: from the device to the network and up to the application servers. In a HSPA+ system there are four main areas that the operator needs to address in order to achieve the desired service performance:

- Radio network planning, such as the location of cell sites, orientation of antennas, transmit power adjustments, and radio network capacity adjustments.
- Transport and core network planning, which will determine the latency and capacity of the backhaul and interconnection links, the location of the core network nodes, and the corresponding core network capacity adjustments.
- Terminal devices. The devices are a major component of the experience of the customer, and care should be taken on the selection of the right hardware components such as CPU and memory, the optimization of transceiver and radio modem, selection and adaptation of the OS, and development and adjustment of apps.
- Soft parameters, which control the adjustment of radio and core functionality such as resource allocation, scheduling, and mobility management procedures.

When planning and optimizing the network, there will often be a tradeoff between quality, capacity, and cost. In some cases, if certain key guidelines are followed, tradeoffs are not necessary and the operator will achieve the maximum potential of the technology with a minimum cost. This chapter will provide the operator with important HSPA+ guidelines, as well as explain some of the tradeoffs that are possible when optimizing the HSPA+ technology from a *radio network* standpoint; planning and optimization aspects for other parts of the system will be covered in other chapters: Chapter 11 (Network Planning) has already discussed

HSPA+ Evolution to Release 12: Performance and Optimization, First Edition.
Edited by Harri Holma, Antti Toskala, and Pablo Tapia.
© 2014 John Wiley & Sons, Ltd. Published 2014 by John Wiley & Sons, Ltd.

strategies to deploy the radio, transport, and core, while specific device optimization will be covered in detail in Chapters 13 and 14.

The latter part of the chapter is dedicated to engineering tools. In the last few years, new tools have been developed that can greatly help engineers with their network optimization and maintenance tasks, especially in today's situation, in which engineers need to deal with increasingly growing traffic levels and deal with multiple technologies simultaneously. The chapter will describe some of the new tools that are available for HSPA+, illustrating with real network examples how an operator can make use of them to optimize their network capacity and performance.

12.2   Optimization of the Radio Access Network Parameters

The principal bottleneck of HSPA+ typically lies on the radio interface. The wireless link is not reliable by nature, with multiple factors affecting its performance: pathloss, shadow and fast fading, load and interference, user mobility, and so on. While the HSPA+ system has special mechanisms to overcome many of these challenges, such as link adaptation and power control, they require special adaptation to the network conditions to operate at an optimum point. For example, the handover triggers in a highway cell need to be set to enable a fast execution of the cell change procedure, otherwise the call may drag too long and result in a drop; on the other hand, these triggers may need to be set more conservatively in low mobility areas with non-dominant pilots, to avoid continuous ping-pong effects that also degrade the service performance.

In today's fast changing networks, radio optimization must be considered a continuous activity: once the network is deployed, it will require constant tuning since traffic patterns change with time, as load increases and new mobile phone applications put the network to the test. Some useful guidelines to tackle network optimization are:

- Recognize when the problem is due to improper planning: trying to compensate for a planning problem through network optimization is sometimes possible, but it will result in a suboptimum performance and capacity situation; whenever possible, planning problems should be identified and tackled before optimization.
- Optimize key traffic areas first: it is impossible to achieve a perfect optimization of the whole network, so where the majority of traffic is should be taken into consideration.
- Don't rely solely on network KPIs: sometimes network KPIs are hard to interpret, especially those related to data services. Driving the network gives the engineer a totally different perspective of what's happening in certain areas.

To support the optimization activity it is recommended to drive test the areas with a certain frequency, since this is the best way to figure out problems in the radio layer. Alternatively, as the network becomes more mature, the operator could use other tools to minimize drive testing, as will be discussed later in Sections 12.3.1 and 12.3.2.

There are multiple functional areas that can be optimized in HSPA+, including antenna parameters, channel powers, handover parameters, and neighbor lists, among others. This section will provide an overview of possible optimization actions on each one of these areas; it also provides a detailed analysis on causes and solutions to optimize uplink noise rise, one of the major capacity problems on HSPA+ networks with high smartphone traffic penetration.

12.2.1 Optimization of Antenna Parameters

Once deployed, HSPA+ networks will be in a state of continuous change: cell breathing due to increased traffic, creation of new hotspot areas, introduction of new sites, and so on; when necessary, antenna parameters should be adjusted to improve network capacity or performance. Normally, the adjustment of the antennas post-launch will focus exclusively on tilts: antenna height changes are very rare, and azimuth modifications have very little impact unless a significant rotation angle is required. Since the adjustments of the antenna tilts can be quite frequent, it is recommended to use Remote Electrical Tilts (RET) capable antennas as much as possible to avoid visiting the site location, which is normally an expensive and time consuming thing to do.

As with other wireless networks, in HSPA+ the throughput and capacity are heavily tied to interference. The main source of interference in HSPA+ is the power coming from other sectors, which is especially harmful in areas where there is no clear dominant pilot (pilot pollution). In addition to interference, pilot polluted areas are prone to "ping-pong" effects during HS cell changes, which also result in lower throughputs and an overall poor service experience.

Figure 12.1 illustrates the interference (left) and corresponding throughput (right) of a single carrier HSPA+ network. As can be observed, the areas with higher interference – darkspots corresponding to low Energy over Noise and Interference ratio (Ec/No) – is where the HSPA+ throughput becomes lower (mid-gray colors on the right figure).

To mitigate these spots, the antenna tilts in the cluster were adjusted to try and ensure proper pilot dominance. The key metrics to monitor in this case are:

- The best serving cell at each location, to ensure that each area is being served by the nearest sector, and that proper cell boundaries are attained.
- Geolocated traffic and Received Signal Code Power (RSCP) samples, to avoid degrading the customers with lower signal strength.

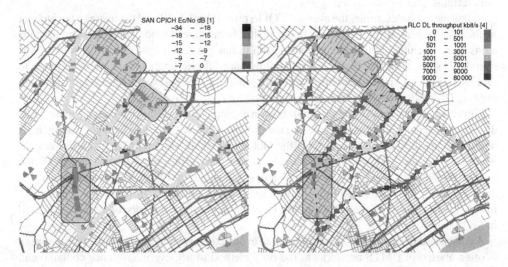

Figure 12.1 Correlation between interference (indicated by Ec/No) and throughput in a live network

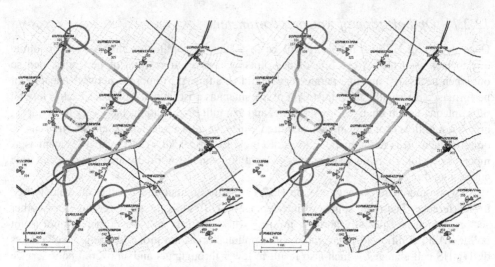

Figure 12.2   Best server plot, before and after antenna optimization

- Signal quality metrics, such as Ec/No and Channel Quality Indicator (CQI), which will provide an indication of the interference situation in the cluster.
- Throughput, captured by a drive test tool.
- IRAT handovers and 2G traffic, to ensure that the optimization is not achieved by pushing traffic to the underlying GSM system.

A good method to analyze this exercise is to display the best server plots: Figure 12.2 illustrates the best server plot before and after the antenna adjustment. As the figure shows, while not perfect, the optimization exercise improved the pilot dominance in the border areas. Perfect cell boundaries are hard to achieve in practice, especially considering urban canyons, hilly terrains, and so on.

In this particular example, the average CQI improved by 1.5 dB and the average throughput in the area increased by 10%, as a result of the better interference situation. The antenna adjustments included uptilts and downtilts, and in one particular case one of the sectors was shut down since all it was doing was generating harmful interference. It is not uncommon to find useless sectors in the network, especially when the HSPA+ sectors are co-located with existing GSM sites that may have been deployed for capacity reasons.

Figure 12.3 provides an antenna optimization example using geolocated data. Figure 12.3a displays the sector traffic samples before the optimization, with darker spots indicating areas with high interference (low Ec/No); as the plot shows, there are a number of traffic samples being served quite far away from the sector, in fact getting into the footprint of sectors that are two tiers away – this is known as an overshooting situation. Figure 12.3b shows the traffic after the sector has been optimized (in this case through downtilt), in which the samples are mostly contained within the cell footprint. Geolocation tools will be discussed in more detail in Section 12.3.

As discussed in Chapter 11, it is important to avoid excessive antenna downtilts. Downtilting is often abused in UMTS networks as the only method to achieve interference containment, however doing it in excess can create significant coverage issues, especially in sectors with

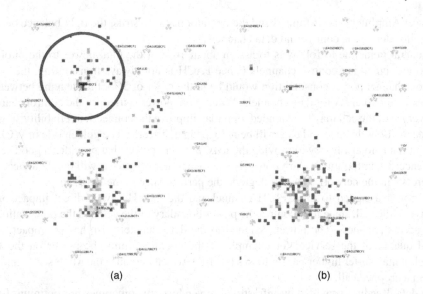

Figure 12.3 Analysis of geolocated samples from one sector before (a) and after (b) tilt optimization

low antenna height. Cell power adjustments can be used as a complement to antenna tilts to achieve a good delimitation of cell boundaries.

When downtilting or reducing the CPICH it is possible that the main KPIs show a performance improvement, however the sector may be losing coverage. It is important to monitor the traffic and IRAT levels to prevent these situations, and if possible validate the new coverage footprint with drive test data or with the help of geolocation tools.

12.2.2 Optimization of Power Parameters

One of the key parameter sets that need to be tuned in a HSPA+ network is the power assigned to the different physical channels. Typically in UMTS networks the operator can control the power level assigned to the following channels:

- Common Pilot Channel (CPICH)
- Downlink Common Control Channel (CCCH): Broadcast Channel (BCH), Acquisition Indicator Channel (AICH), Paging Channel (PCH), Primary Synchronization Channel (P-SCH), Secondary Synchronization Channel (S-SCH), Paging Indicator Channel (PICH), and Forward Access Channel (FACH).
- Release 99 (R99) Dedicated Data Channels.
- R99 Voice Channels.
- HSDPA Channels: High Speed Shared Control Channel (HS-SCCH) and High Speed Downlink Shared Channel (HS-DSCH).

12.2.2.1 Adjustment of Common Control Channels

The power allocation of the control channels needs to be done carefully since it will have implications for both coverage and capacity planning. The total amount of power available in the Power Amplifier (PA) is limited and the operator needs to strike the right balance between power allocated to the control and data channels.

A general guideline to follow is to assign about 10% of the total power to the pilot, and 15% to the remaining control channels (since FACH is not always transmitting, the typical common channel power consumption would be in the order of 5%), thus leaving between 75 and 85% of the total PA for data channels. When more power is used for the control channels, the coverage of the cell may be extended, or it may improve the connection reliability, however the capacity of both voice and data will be reduced in that sector. The golden rule in WCDMA optimization is to always try to provide the *least possible* power that provides a good service experience: when too much extra power is used when not needed, it will create unnecessary interference in the network that will degrade the performance in the long run.

The power assigned to the CPICH channel and the CCCH have a direct impact on the coverage of the cell, since they control the power boundary of the cell and the power allocated to the access channels. The power assigned to the data channels also has an impact on the offered quality of the service: for example, if the operator limits the power on the voice channel, it may suffer higher packet losses (BLER), which degrades the MOS score, and may result in a dropped call.

The default values provided by infrastructure vendors are sometimes not optimum for the needs of the operator. In some cases, the values are set too conservatively to meet certain contractual KPIs, at the expense of network capacity. It is therefore recommended to review and tune the power parameters based on the specific situation of the operator. Table 12.1 provides some parameter ranges for the most relevant CCCH. With the introduction of Self Organizing Networks (SON) tools, it is possible to adjust these settings for each particular cell,

Table 12.1 Typical parameter ranges for CCCH power settings

Channel	Description	Network impact	Typical values (in reference to CPICH)
CPICH (common pilot)	Transmits the pilot of the cell	Accessibility Retainability	0
AICH (acquisition indicator)	Indicates a data channel to be used by the UE	Accessibility	−7 to −8 dB
BCH (broadcast)	Provides information about the network and cell	Accessibility	−3 to −5 dB
PCH (paging)	Alerts the UE of incoming communication	Accessibility	−0.5 to 0 dB
P-SCH (primary synchronization)	Enables UE synchronization with the network	Accessibility Retainability	−3 to −1.5 dB
S-SCH (secondary synchronization)	Enables UE synchronization with the network	Accessibility Retainability	−4 to −3 dB
FACH (forward access)	Transport for small amounts of user plane data	Accessibility Retainability	0 to +1.5 dB
PICH (paging indicator)	Provides information about the transmission of the PCH	Accessibility	−8 dB

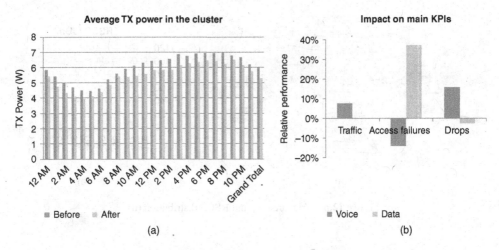

Figure 12.4   Average power reduction in the cluster (a) and main network KPIs (b)

unlike the typical situation in which these are configured equally across the whole network. SON tools will be discussed later in Section 12.3.3.

The following example shows a case in which the network CCCH has been optimized, adjusting the default power values to a lower level while keeping the same network performance. The following changes were implemented:

- BCH: from −3.1 dB to −5 dB
- AICH: from −7 dB to −8 dB
- P-SCH: from −1.8 dB to −3 dB
- PICH: from −7 dB to −8 dB.

Figure 12.4 summarizes the impact on the cluster, both on capacity (Tx power, on the left), as well as on the main network quality metrics (right chart).

This resulted in an average power reduction of 12% in the cluster, with a higher reduction on the top loaded sectors in the area, thus providing an important capacity saving in the network. The main KPIs in the network were improved, or didn't suffer, with the exception of the voice access failures, which were slightly degraded.

12.2.2.2 Adjustment of Pilot Power

The selection of pilot power is normally one of the first tasks to be done when preparing the network planning. The power of the pilot is normally chosen to be equal in all sectors of the same cluster, and a small percentage of the maximum desired carrier power – typically 10%. Keeping homogeneous pilot power simplifies the network deployment; however, in mature networks it may be a good idea to adjust the pilot powers depending on traffic load and coverage needs of each individual sector. The best way to achieve this in practice is with the help of SON tools, which will be discussed later on Section 12.3.3.

While in the initial UMTS days the power amplifiers typically had a maximum power of 20 W, today Power Amplifier (PA) technology has evolved and existing units can transmit

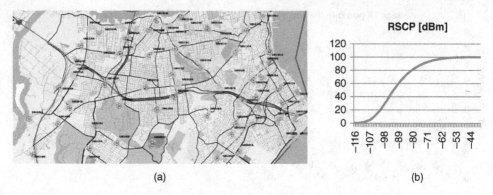

(a)   (b)

Figure 12.5   Trial area (a) and RSCP distribution (b)

80 W or more. This power can be split into several carriers, thus providing the possibility of adding more sector capacity with a single piece of hardware – provided the operator has the spectrum resources to allow for an expansion.

While it is often tempting to make use of all the available power resources in the PA, the operator needs to consider future possible capacity upgrades before deciding how much power to allocate to each carrier. Furthermore, in the case of UMTS, having more power does not provide any additional advantages beyond a certain point, once the basic coverage is ensured in the area.

The following example illustrates the case of a real network that was originally configured to operate at 40 W per carrier, and was converted to transmit with a maximum power of 20 W in all sectors. The trial area included a mixed urban and suburban environment with an average intersite distance of 1.2 km; the average RSCP reported by the network was −92 dBm, which may be considered as coverage challenged area. The trial area and RSCP distributions can be observed in Figure 12.5.

In this exercise, both the maximum transmit power of the sector and the pilots were modified. Each of the pilots was adjusted depending on its specific traffic and coverage needs. Some antenna tilts were also adjusted to compensate for power losses, if necessary. At the end of the exercise, the pilots in the area were reduced by an average of 1.8 dB, and the average transmit powers were reduced by 2.25 dB. As a consequence of the lower power used, the drive test results indicated an improvement of Ec/No of 1.5 dB in the area. The lower powers did not result in traffic loss, neither for voice or data, as can be observed in the data traffic volume trend in Figure 12.6b.

As Figure 12.6 illustrates, the power reduction did not have a negative impact on network KPIs. A small throughput reduction was observed in the cluster (5%) due to the lower power available. The rest of the KPIs were maintained, with some improvement observed on IRAT performance, probably due to the mentioned Ec/No improvement. A similar exercise performed in an urban environment with better coverage revealed further capacity gains, with 3.5 dB reduction in overall transmit power, and no impact on network KPIs.

In summary, the adjustment of the pilot power can be an effective tool to optimize the interference situation in the network. The operator needs to decide what the right tradeoff is between offered capacity, throughput, and cost, and configure the pilot power accordingly.

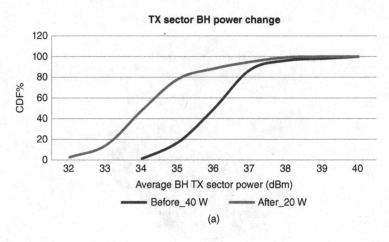

(a)

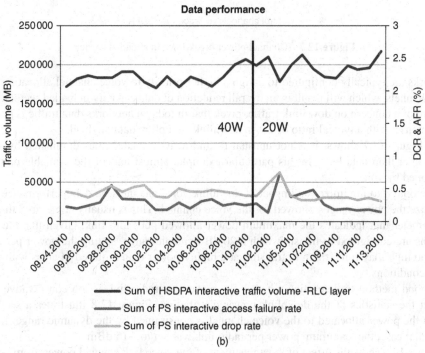

(b)

Figure 12.6   Average transmit power (a) and data KPIs (b), before and after the change

12.2.2.3   Adjustment of Voice Channel Power

In HSPA+ networks the spectrum resources are shared between voice and data channels, in addition to the control channels previously discussed. While voice payload is typically very small when compared to data, the dedicated nature of the voice channel allows it to utilize a fair amount of power resources in the system, therefore increasing the overall noise in the network and contributing to data congestion in hotspot areas. Furthermore, HSPA+

Power allocation in a loaded carrier

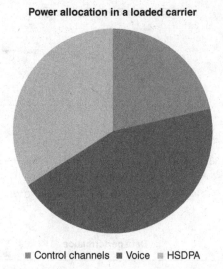

■ Control channels ■ Voice ■ HSDPA

Figure 12.7   Channel power breakdown in a loaded carrier

networks are typically configured to assign a higher priority to voice channel allocation over data channels, which will result in an overall reduction of data capacity in loaded sectors. This is more of a concern on downlink traffic, given that in today's networks data traffic is heavily asymmetric, with a typical ratio of 7:1 in downlink vs. uplink data payload.

As Figure 12.7 shows, it is not unusual to find heavily loaded sites dominated by voice load rather than data load. In this particular example, almost half of the available power is consumed by voice.

One option to minimize the impact of downlink voice channels on data performance is to optimize the power ranges allowed for the voice channel. This is usually configured through a parameter that indicates the maximum power allowed per radio link. Given that the base stations are configured to transmit with a 20 dB dynamic range, setting the power per radio link too high may force them to transmit at a higher power than required in areas with good radio conditions.

A good method to determine whether the power per radio link is too conservative is to extract the statistics of the downlink power distribution. Figure 12.8 illustrates a sector in which the power allocated to the voice channel is constrained by the dynamic range. In this particular case, the maximum power per radio link was set to −15 dBm.

Figure 12.9 show the effect of the adjustment of the voice link channel power, from −15 to −18 dB. As can be observed on the voice channel power distribution (left chart) the network is operating in a more optimum manner after the adjustment, which also results in overall capacity savings. The trend on Figure 12.9 (right) shows that this optimization resulted in about 1 dB average power reduction per link.

12.2.3   Neighbor List Optimization

The optimization of UMTS neighbor list is one of the most frequent network maintenance tasks that field engineers need to tackle. The intra-frequency, inter-frequency, and IRAT neighbor

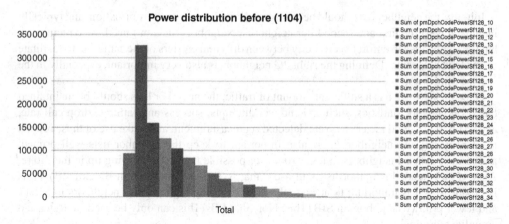

Figure 12.8 Suboptimal voice power distribution in a sector (leftmost bins correspond to lower transmit power)

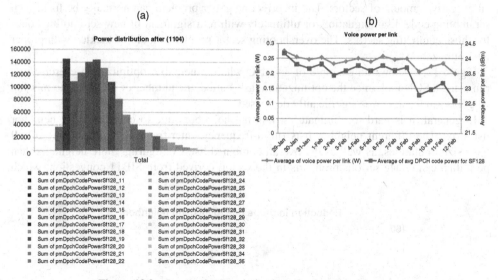

Figure 12.9 Results from optimization of voice channel power

lists should be continuously modified to cope with changes in traffic and to accommodate new sites that may be coming on air. If this task is not diligently performed on a regular basis, the network performance will be impacted via an increase of drops, typically revealed as drops due to missing neighbors, or UL synchronization. In some extreme cases, the drops due to missing neighbors can account for up to 20% of the total number of drops in cluster.

Typically the engineers will have some computer tool available to facilitate the configuration of neighbors. These tools range from basic planning tools, in which the neighbors are configured based on RF distance, to Self-Organizing Networks (SON) tools that will automatically configure the neighbors on a daily basis. SON tools will be discussed in more detail in Section 12.3.3.

Initially, the neighbor lists should be configured based on planning information, and typically include at least the first and second tier neighbors. Neighbor lists don't need to be reciprocal, but it's a good idea to enforce reciprocity between different carriers of the same site to facilitate traffic load balancing. Defining the right 2G neighbors is also very important, especially in the case of rural sites.

Once the network has a sufficient amount of traffic, the neighbor lists should be maintained based on network statistics, such as handover attempts, success and failures, drop call rate, and so on. It is a good idea to activate detected set measurement, and make use of these reports to identify missing neighbors. It is also important to keep the neighbor lists well pruned, and remove unused neighbors to leave room for possible new cells showing up in the future. Also, if an extended neighbor list is in use – that is, System Information Block (SIB) 11bis is broadcast – it is important to arrange these lists to keep high priority neighbors in SIB11 and lower priority neighbors in SIB11bis. Note that SIB11bis can only be used by Release 6 compliant terminals.

The engineer should also be on the lookout for possible handover problems due to wrong neighbor selection, as well as overshooting sectors, which will tend to generate handovers to an excessive amount of sectors. The incorrect neighbor problem can normally be fixed with scrambling code disambiguation, or ultimately with re-assignment of new scrambling codes for those conflicting sectors. The overshooting sector problem should be tackled with proper RF optimization.

Figure 12.10 shows an example of an RNC where a neighbor optimization exercise was run using a SON tool: once the optimization task started, the number of drops due to missing neighbors in the area was significantly decreased.

In general, it is hard to eliminate all missing neighbors due to constraints on neighbor list sizes. The different neighbor lists (intra-frequency, inter-frequency, and inter-RAT) all have size limitations, as well as the overall combined list. Configuring more neighbors than permitted can create serious problems in the sector, derived from SIB11 congestion, which

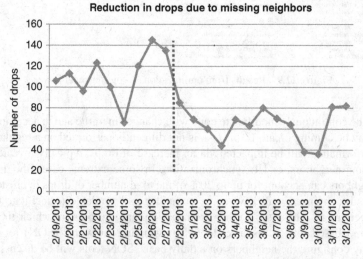

Figure 12.10   Neighbor optimization exercise in one RNC

in extreme cases can put the site off service and will require a site reset to be fixed. For this reason it is a good idea to maintain a pruned list size, eliminating the neighbors that are not strictly required as previously discussed.

Some infrastructure vendors provide special reporting of the detected set neighbors, which are neighbors not contained in the neighbor list that are properly decoded by the UE using soft handover event triggers. These special reports provide aggregated information containing the RSCP and Ec/No value of the detected scrambling code, and can be useful to identify missing neighbors in live networks.

12.2.4   HS Cell Change Optimization

The mobility performance in HSPA+ presents significant challenges due to the nature of HS cell change procedure, which is significantly different from the handover method of the original UMTS dedicated channels. The R99 UMTS systems included soft handover mechanisms to avoid the harmful effects from adjacent cell interference that are inherent in systems with a single frequency reuse: the user plane transmission was never interrupted, and links from multiple sectors could be combined to improve the performance at the cell edge.

In the case of HSDPA, and to a certain extent of HSUPA, the transition between sectors can result in a reduction of user throughput due to the interference levels endured at the cell boundary: since the device is only transmitting and receiving data from one sector at a time, the new target sector will become a source of interference until the cell change procedure has been completed. In addition to this interference problem, it is not uncommon to observe "ping-pong" effects, by which the UE bounces back and forth between two cells that have similar power levels, which sometimes further degrade the offered HSPA throughput.

More recently, the HSDPA performance at the cell edge has been significantly improved thanks to the wide adoption of Type 2i and 3i receivers which are able to cancel the strongest sources of interference in the HS-DSCH channel. Advanced receivers were presented in Chapter 8, and practical performance aspects were discussed in Chapter 10.

While HSUPA was designed to use Soft Handover (SHO), the cell change procedure suffers from practical limitations that impact its performance. For example, some infrastructure vendors limit the SHO utilization to the control plane alone to limit excessive baseband consumption for users in border areas. Furthermore, the HSUPA cell changes will follow the change of serving cell decided on downlink, and thus will also be impacted by ping-pong effects in the HSDPA cell change.

The HS cell change procedure can be triggered by an explicit event (1D), by which the UE notifies the network of a more suitable serving cell, or implicitly following a soft handover event (1A–1C). The event triggers can be based on RSCP or Ec/No, or both, depending on vendor implementation. Table 12.2 summarizes the relevant UE events that can trigger a cell change procedure.

These events are configured on the network side, and are typically defined with a level threshold and a time to trigger. Very short trigger times can lead to an excessive number of cell changes and frequent ping-pong, while very long times can result in decreased quality (CQI) due to a longer exposure to interference. The metric used to trigger the event is also relevant: for example, events triggered based on Ec/No can result in higher fluctuations in the case of HSDPA data.

Table 12.2   UE Events that can trigger a HS cell change

Event	Impact
1A	SHO radio link addition
1B	SHO radio link removal
1C	SHO radio link replacement
1D	Change of best cell

It is important to realize that macro level network KPIs can sometimes mask some of these problems that occur only in specific areas, such as cell boundaries. Customers that are located in cell boundary areas may be experiencing a consistent poor data experience, but these problems will go unnoticed because these stats will be combined with the overall sector KPIs. Therefore, to tune the performance at these locations the best course of action is to rely on drive test exercises, or alternatively on geo-located KPIs. Section 12.3.2 discusses a set of tools that can be very useful to troubleshoot and optimize this type of problem.

In the following example we analyze the performance of a cluster that presented a high number of ping-pong effects, which resulted in poor data performance in those particular locations. The performance degradation could be noticed even during stationary conditions when located in the cell boundary areas.

The objective of the exercise was to reduce the amount of unnecessary cell changes, with the ultimate goal to improve the throughput in those challenging areas. Table 12.3 summarizes the original parameter settings controlling the HS cell change, as well as the optimized settings adopted during the trial.

Figure 12.11a illustrates the trial area, in which the sectors with higher data loads are indicated with a darker color. It shows the stationary test locations, as well as the selected test drive route, which was chosen to go through equal power boundary areas where there was no clear cell dominance. As discussed earlier, these are the most challenging conditions for HSDPA traffic channels. The Figure 12.11b shows the throughput distribution along the drive route, with baseline (light gray) and new parameters (dark gray).

As the throughput distribution in Figure 12.11b shows, the performance for the customers in the worse performing areas was significantly improved with the new parameters. For an equal percentile (20% on the chart), user throughputs increased from 100 kbps to 1.5 Mbps. As Figure 12.12 illustrates, the tests performed in stationary locations also showed improved

Table 12.3   Results from HS cell optimization exercise

Parameters	Baseline	Optimized	Unit
Addition window	1.5	0	dB
Drop window	3.5	6	dB
Replacement window	1.5	3	dB
Addition time	320	640	ms
Drop time	640	1280	ms
Replacement time	320	1280	ms
HSDPA CPICH Ec/No threshold	−5	−9	dB

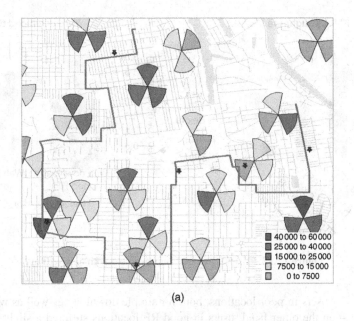

| 40 000 to 60 000 |
| 25 000 to 40 000 |
| 15 000 to 25 000 |
| 7500 to 15 000 |
| 0 to 7500 |

(a)

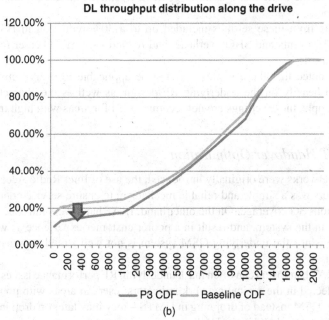

DL throughput distribution along the drive

— P3 CDF — Baseline CDF

(b)

Figure 12.11 HS cell change optimization area (a) and throughput distribution (b)

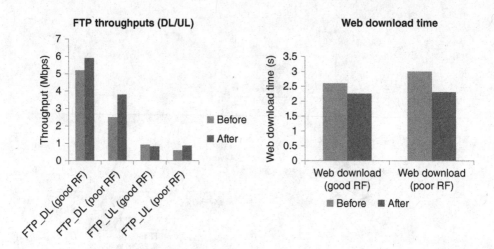

Figure 12.12   User experience improvement with new settings

performance for users in poor locations, both for simple download as well as web browsing performance – on the other hand, users in good RF locations suffered a slight reduction in uplink throughput.

Other benefits from these settings included up to a 30% reduction in HS cell changes, reduction in SHO events and SHO overhead, and reduction of radio bearer reconfigurations as well as R99 traffic.

It should be noted that these results may not be applicable to every network, since the settings depend heavily on the underlying RF design, as well as the characteristics of the cluster. For example, these settings are not recommended for areas with high traffic mobility.

12.2.5   IRAT Handover Optimization

When UMTS networks were originally introduced, the use of Inter Radio Access Technology (IRAT) handovers was a simple and reliable mechanism to ensure service continuity in areas with discontinuous 3G coverage. On the other hand, IRAT handovers introduce an additional point of failure in the system, and result in a poorer customer experience as well as reduced system capacity since the underlying GSM system is not well suited to transfer high speed data connections.

The use of IRAT handovers can often mask underlying RF performance issues in the network that are not reflected in the typical network indicators: users in areas with poor performance will handover to GSM instead of dropping in UMTS – they may later on drop in GSM as well, but this drop is not considered in the UMTS optimization and this results in a poorly optimized network exercise.

Frequent IRAT also increases the probability of missing paging. Call alerts may be missed when the devices are in the middle of a transition between technologies and the location area update has not been completed. Although this problem can be mitigated with MSS paging coordination, this is a complicated problem to troubleshoot, since there are no counters that keep track of these events.

In general, as networks mature and a sufficient penetration of 3G sites is achieved, operators should try and minimize the utilization of IRAT handovers whenever possible. It is recommended to try and disable IRAT in areas with 1 : 1 overlay; if there is not 100% overlay, the default parameters can be adjusted with more aggressive settings at the expense of a higher drop rate in UMTS – however this increase is often compensated with a drop reduction in the underlying GSM layer.

Figure 12.13 illustrates an example in which the voice IRAT thresholds were reduced to a very aggressive setting (RSCP level of −120 dBm, Ec/No of −15 dB). Note that these values were used for testing purposes and may not necessarily be appropriate for a widespread use in commercial networks.

The test cluster was the same one referenced in Section 12.2.2.2, where there was approximately a 70% overlay between UMTS and GSM sites. In the example, the IRAT ratio was reduced from an original 10%, down to a level close to 1%. The 3G dropped call rate in the area subsequently increased from 0.5 to 1%. On the other hand, the 2G drop rate improved, because there were not so many rescue calls going down to the GSM layer. At the end of the day, each operator will need to decide their own specific tradeoffs between 3G traffic share and 2G/3G traffic quality.

In the case of HSDPA data, the service is more forgiving to poor radio conditions and thus permits even more aggressive settings to retain the traffic on the HS channel. There are two sets of parameters controlling this, the PS IRAT parameters as well as the parameters controlling the transition from HS to DCH. Figure 12.14 shows the case of a network in which these

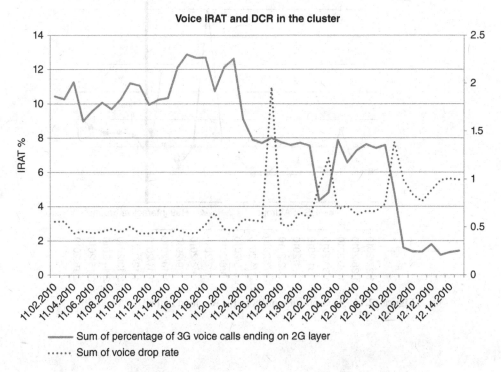

Figure 12.13 Optimization of voice IRAT settings

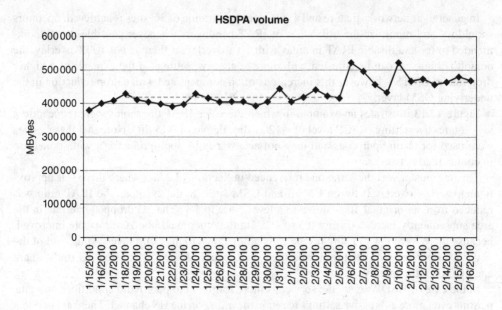

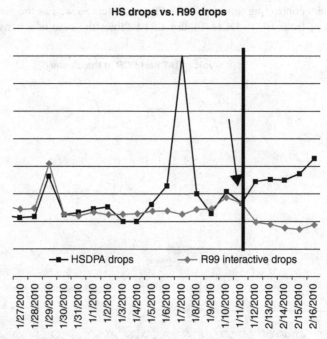

Figure 12.14   Optimization of PS traffic retention in the HSPA layer

settings have been aggressively adjusted:

- HS to DCH transitions due to poor quality OFF.
- Ec/No service offset for Event 2D (used by compressed mode) from −1 dB to −3 dB.
- RSCP service offset for Event 2D (used by compressed mode) from −6 dB to 0 dB.

By adjusting these settings the users will stay longer in HSDPA, thus avoiding unnecessary transitions to R99 – which is a less efficient transport – or ultimately 2G. Figure 12.14 shows the impact of the change on the overall HSDPA traffic volume (left), which increased by 12%; the chart on the right shows the impact of the change on PS drops. The overall packet drop rate remains stable, although there was an increase in HSDPA drops that was offseted by a reduction in R99 drops.

Nowadays, many operators have completely removed IRATs for data and some operators also for voice, which has typically resulted in better KPIs, assuming that the UMTS coverage is sufficiently good. If IRAT is used for data, then it is normally triggered only by RSCP and not by Ec/No.

12.2.6 Optimization of Radio State Transitions

Section 4.10 in Chapter 4, and Section 10.3 in Chapter 10 discussed how the delay in the radio state transitions has an important impact on the HSPA+ latency, and therefore on the final data service experienced by the customer. While packets on the DCH state enjoy a fast transfer speed and a protocol architecture that minimizes latency, those packets transmitted over the FACH channel will be transferred at a very slow speed. Even worse is the case where the device is in a dormancy state, either IDLE or PCH, in which case there is an additional delay to set up the corresponding channel.

The following diagram (Figure 12.15a) illustrates the typical RRC state transitions in today's networks. The diagram does not include the IDLE state, since nowadays it is widely accepted that the default dormancy state should be Cell_PCH (either Cell_PCH or URA_PCH). This is due to the fact that the transition from Cell_PCH is much faster than from IDLE, and involves a significantly lower amount of signaling messages, as illustrated in Figure 12.15b.

When the UE is in the PCH state, a new data packet that needs to be transmitted can either be sent via DCH – if the direct transition feature is activated and certain conditions are met – or via FACH. To trigger a direct DCH transition the data volume should be larger than a certain threshold (B1, on the diagram); in certain implementations there is also a quality criterion, by which the direct transition is only triggered if the Ec/No is better than a certain threshold. Figure 12.16 illustrates the result from a trial network in which different quality thresholds were tried. As the threshold is relaxed, the amount of direct transitions increase, but there's also an increase in upswitch failures.

When the UE is in FACH state, it will transmit at a very low bitrate, typically 32 kbps. If the application data volume increases beyond a threshold B1, a transition to DCH will be triggered; this promotion time is quite slow and can take around 750 ms, so in general the operator should avoid entertaining continuous FACH<->DCH transitions. When there is no data to be transmitted, the UE will fall back to PCH (or IDLE) state after an inactivity time (T2).

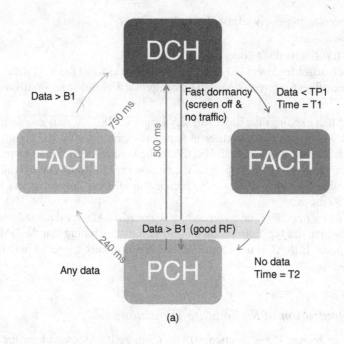

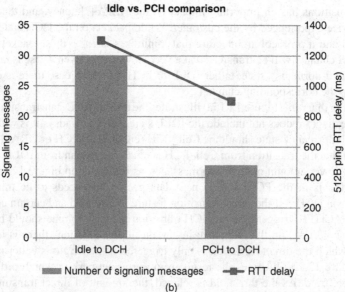

Figure 12.15 Illustration of RAN state transitions for HSPA (a) and comparison between idle and PCH performance (b)

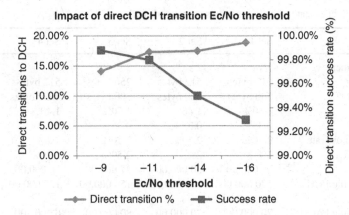

Figure 12.16 Impact of different Ec/No thresholds on the PCH to DCH direct transition (buffer size = 512 B)

The DCH state is the fastest state, where the user can enjoy minimum latency (as low as 25 ms) and high-speed data transfer. The UE will fall back from this state into FACH state in case the transmission speed falls below the TP1 threshold for a certain amount of time (T1). There is also a possible DCH to PCH direct transition for mobiles that implement the Release 8 fast dormancy feature that is described in Chapter 4. This is typically triggered when there is no additional data to be transmitted, and the phone screen is off.

The configuration of the HSPA+ radio states parameters is a complex task as it affects many different factors. For example, a large T1 timer will have a good effect on user experience, but may create capacity issues in the cell. Similarly, a low B1 buffer to transition to DCH would reduce latency, but can degrade some network KPIs. The following summarizes the impact of these parameters on different aspects of the network:

- User experience: small buffers will minimize the connection latency; long inactivity timers can reduce the number of transitions to DCH, thus reducing connected-mode latency; facilitating direct transitions from PCH to DCH will also reduce connection latency. Section 13.5.5 in Chapter 13 provides additional details on the impact of state transitions on end user experience.
- NodeB baseband capacity: longer timers will increase the consumption of channel elements, since there will be an increase in number of simultaneous RRC connections.
- RNC processing capacity: large buffer thresholds will minimize the number of state transitions and reduce the signaling load in the RNC. Very short timers will increase the amount of signaling due to continuous switching between channel states.
- Network KPIs: aggressive settings for direct transition can increase the channel switching failures; small channel switching thresholds can increase the access failure rates.
- Battery drain: long inactivity timers will result in a faster battery drain.
- Radio capacity: long timers will keep more users in DCH state, thus increasing the uplink noise (see Section 12.2.7.2 for more details); very short timers generate too much signaling, which can also create harmful effects on network capacity.

Table 12.4 summarizes an optimization exercise in a live network in a dense urban environment, where various buffer and timer settings were tried.

Table 12.4 Effect of state transition settings on network KPIs

Parameter/KPI	Set 1	Set 2	Set 3	Set 4	Set 5
T2 inactivity timer	5 s	3 s	3 s	5 s	5 s
UL RLC buffer	512 bytes	512 bytes	256 bytes	512 bytes	1024 bytes
DL RLC buffer	500 bytes	500 bytes	300 bytes	1000 bytes	1000 bytes
PS AFR	1.40%	1.40%	1.70%	1.40%	1%
PS DCR	0.35%	0.35%	0.35%	0.30%	0.40%
Downswitch failure rate	2.50%	2.50%	2.50%	2%	3%
Upswitch failure rate	2%	2%	2.50%	2.20%	1.50%
FACH traffic	38 000.00	30 000.00	27 000.00	38 000.00	40 000.00
Channel switching PCH to FACH	130 000.00	150 000.00	150 000.00	125 000.00	130 000.00
Channel switching FACH to HS	70 000.00	70 000.00	80 000.00	65 000.00	50 000.00
RNC processor load	15.50%	15%	16.50%	14.50%	14.50%

As previously discussed, the results show that there is a tradeoff between user experience and network resources. Shorter buffers and timers result in higher signaling and processor load, and also revealed an increase in access failures due to upswitch failures. On the other hand, as Figure 12.17 shows, the user experience – measured in this case as web download time – is significantly improved with short buffers (33% web download time improvement with 512 bytes as compared to 1024 bytes).

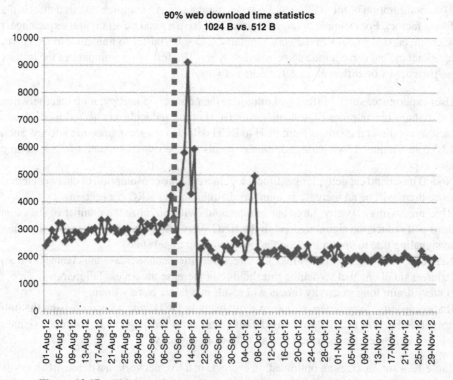

Figure 12.17 Web browsing user experience improvement with shorter buffers

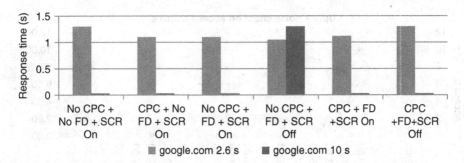

Figure 12.18 DNS response time for two different T2 settings: 2.6 s (light gray) and 10 s (dark gray). CPC = continuous packet connectivity; FD = fast dormancy; SCR = phone screen

Longer timers result in better user experience due to the extended use of the DCH channel; however, in practice the timers can't be set to values beyond 5 s because they have a significant impact both on battery drain and in network performance – in particular, the uplink noise. A follow up exercise to the one presented, where the inactivity timers were set to 10 s, revealed a significant increase in PS access failures.

When the penetration of Continuous Packet Connectivity (CPC) devices achieves significant levels, it should be possible to further tune these settings to maximize the data user experience with a minimum impact to network resources or battery life.

The exercise in Figure 12.18 shows that the DNS response time would be reduced by over 1 s in case of longer timers; furthermore, thanks to CPC the battery drain would only be 7% higher comparing a baseline 2.6 s timer with an extended 10 s timer. Furthermore, the Uplink Discontinuous Transmission (UL DTX) function of CPC would minimize the amount of uplink interference.

12.2.7   Uplink Noise Optimization

Even though data uplink traffic is only a fraction of the volume transmitted in the downlink direction, the current UMTS systems have been unexpectedly suffering from uplink capacity limitations, typically detected in the form of excessive uplink interference or uplink noise.

In most of the cases, operators will try to maintain the network capacity when uplink capacity limits are approached, in the same way that they deal with traditional 2G cellular network congestion by:

- implementing cell splits;
- carrier additions;
- downtilting of antennas;
- traffic redirection between layers and cells;
- and incorporating DAS systems into their networks where applicable.

Although these methods are able to resolve uplink network congestion, they are all costly solutions and do not tackle the root cause of uplink noise rise.

With the introduction of HSUPA, particularly HSUPA with the 2 ms TTI feature, uplink noise rise in many networks was seen to increase. Many operators expected this to occur since

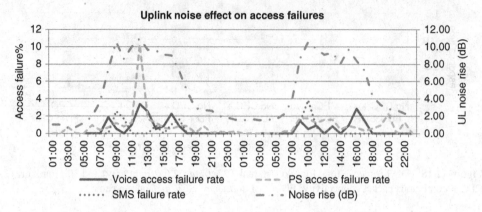

Figure 12.19 Impact of UL noise on access failure rate

the inclusion of HSUPA traffic in a cell allowed for an 8 to 10 dB noise rise (dependent on the network design) above the thermal noise floor to fully obtain the benefits of HSUPA.

Furthermore, the Release 6 HSUPA standards were designed to transfer high bitrate data streams for extended periods of time; however, with the explosion of smartphone traffic in the last few years, it became clear that the existing methods were not optimized for the chatty traffic pattern that is typical of smartphones. This chattiness generates a continuum of small packets flowing to the NodeB from multiple simultaneous locations, which makes it hard to control the interference level in the uplink.

The main impact of high uplink noise is the reduction of the cell coverage, which will be reflected in an increase of packet access failures, as Figure 12.19 illustrates. In this example, when the uplink noise reaches average hourly values beyond 8 dB, the number of access failures increases significantly, both for voice and data connections.

The increase in uplink noise will not only affect the performance of that particular cell, but also of adjacent cells, since the mobiles that are attached to it will transmit at a higher power, thus increasing the interference floor of nearby sectors.

Given the importance of uplink noise in smartphone HSPA+ networks, this topic is discussed in various chapters in the book: Chapter 13 provides further details on smartphone chatty profiles, Chapter 11 provided some insight around the relation between the number of connected users and uplink noise, and Chapter 4 discussed some of the HSPA+ features that will help mitigate uplink noise problems in the future, such as the use of CPC and enhanced FACH/RACH. This section provides further analysis on the causes of uplink noise, and some practical methods to mitigate this problem in current networks.

12.2.7.1   Uplink Noise Rise Causes

There are two main reasons for excessive uplink noise in HSPA+ networks:

- a power overhead due to the inefficient design of the uplink channel (DCH), which will cause high levels of interference even in situations where there is no data transmissions; and
- power race scenarios that can be triggered by various causes, including near-far effects and system access with too high power, among others.

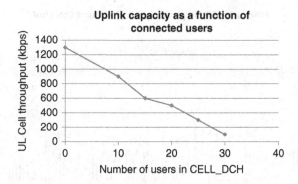

Figure 12.20   Simulation of the reduction of cell capacity with number of connected users (no data). Data from [1]

Figure 12.20 summarizes an analysis conducted at 3GPP in the context of Continuous Data Connectivity (CPC), which shows the overhead caused by the DPCCH of inactive HSUPA users. In this example, a sector with 30 users in DCH state would deplete the available capacity, even when they are not transmitting any data at all. This analysis is related to the analysis shared in Chapter 11, Section 11.4.3, which shows the relation between UL noise and the number of active users in a sector.

The most effective method to combat this problem is to adjust the RRC state transition parameters; in particular, reduce the T1 timer as will be discussed later in Section 12.2.7.2.

The introduction of HSUPA users in a sector will also generate a power race between these and the existing voice traffic in the sector: the increase in noise causes the R99 users to no longer meet their uplink EbNo, SIR targets, which results in the failure to meet the BLER targets for the bearers – the system response is to increase the uplink power for each R99 bearer, which causes an increase in the overall uplink system noise. Figure 12.21 shows an example of a congested sector, in which it can be observed how the power of the R99 channels stays typically around 2 dB above the designed uplink load threshold.

Following a similar trend, uplink RACH power for UEs will also be higher in a cell with E-DCH bearers, and given the fact that the RACH power does not have an effective power control mechanism, this may cause instantaneous uplink power spikes.

In general, the HSUPA system is highly sensitive to uplink power spikes, which may generate power race conditions such as the one described before. The following describes some of the scenarios that can generate these power spikes.

A wireless system based on spread spectrum will suffer from inter-cell interference in both the uplink and the downlink. When a UE approaches a cell boundary, the power received by the non-serving cell will be within a few dBs of the serving cell. This problem is more pronounced with the introduction of HSUPA, where soft handover is not widely utilized. One approach to dealing with this issue is to downgrade the uplink scheduling TTI based on an Ec/No threshold from 2 to 10 ms, allowing for less power to be used in the UL. Another option is to control the soft handover parameters for HSUPA service, expanding the triggering of cell addition as well as the time to trigger additional windows, in order to control the interferer. This will, however, have an impact on the NodeB baseband resources.

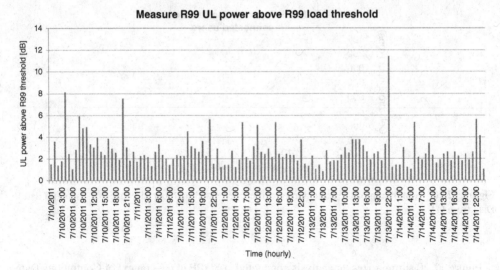

Figure 12.21   Additional uplink power used by R99 channels when HSUPA traffic is present

A similar problem is caused by building shadowing, where a UE is connected to one cell but due to mobility and shadowing (such as moving around a building) finds itself within the footprint of a new cell. This results in a significant amount of power being received in the new cell before the UE can enter into soft handover and the interference brought under control.

Figure 12.22 shows the results of this effect – a cell carrying no UE traffic and experiencing high uplink noise in the beginning part of the trace before the UE changes cells. A plot of the UE TX power during the cell change is shown in Figure 12.23.

In this example, prior to cell change the UE TX power ramps up to 14 dBm before reducing to −15 dBm post cell change.

In many cases, it is not possible to avoid those uplink spikes; however it is possible to control how the system deals with these spikes. Section 12.2.7.3 presents an example in which the

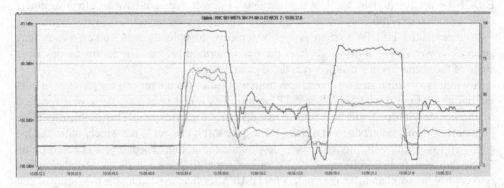

Figure 12.22   Power measurement from live sector showing near–far problem

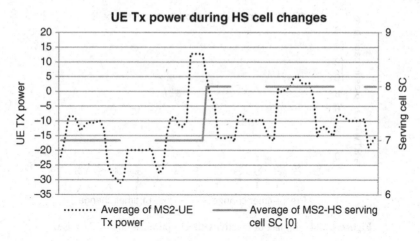

Figure 12.23   Illustration of UE Tx power peak during HS cell change

system parameters are adjusted to limit the reaction of the devices in the cell, thus effectively mitigating the consequential increase in uplink noise.

12.2.7.2   Adjustment of State Transition Timers

As previously indicated, an effective way to reduce uplink noise in congested cells is the optimization of the RRC state transitions parameters, in particular the reduction of the T1 timer, which controls the transition from Cell_DCH to Cell_FACH state. A reduction in T1 timers will lead to fewer erlangs which, as discussed in Chapter 11, Section 11.4.3, is helpful in reducing the noise rise, but will also increase the signaling load in the network. Furthermore, as indicated in Section 12.2.6, a reduction in these timers will result in a longer connection delay, which will impact the user experience. Unfortunately, in most vendor implementations these settings are configured at the RNC level and it is therefore not possible to reduce the timers only in sectors with extreme congestion.

Figure 12.24 illustrates the impact of modifying the T1 timer from 8 s down to 3 s in a live RNC. The chart represents the uplink Received Total Wideband Power (RTWP) distribution for the top 5 loaded cells in the RNC, before and after the change.

As the figure indicates, there is a significant reduction of uplink noise after the change, as can be seen in the reduction of samples between −98 and −83 dBm. As a consequence of the optimization, the following aspects were improved in the network:

- HSUPA packet session access failure rate was reduced from 10.8 to 3.8%.
- The usage of the UL R99 channel was reduced (80% reduction in triggers due to reaching the limit of HSUPA users).
- Reduction in NRT SHO overhead from 25.7 to 23.8%.
- 13 to 45 mA reduction in battery utilization, depending on device type and user activity.

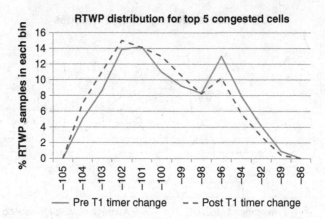

Figure 12.24 Impact of inactivity timer optimization on UL noise

The drawbacks of this change include:

- Higher (2–3%) RNC processing load due to increased signaling.
- An increase in utilization of FACH/RACH channels.
- Increase in connection latency.

12.2.7.3 Adjustment of Filter Coefficients

In some infrastructure vendors, one effective method to reduce the uplink noise rise caused by these effects is to modify the filter coefficient length used by the NodeB to aggregate the uplink noise rise measurement. By extending the filter coefficient, significant improvements in the uplink noise rise can be obtained. As shown in Figure 12.25, the uplink noise spiking caused by the shorter filter coefficient is eliminated and a smoother, less "spiky" noise measurement is obtained.

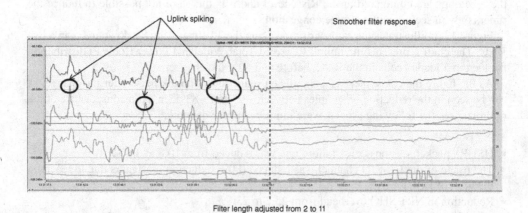

Figure 12.25 Filter coefficient adjustment and impact on the measured noise floor

The modification of the filter coefficient has a direct impact on the PRACH outer loop power control, and can introduce more stability into the system. If we refer to the initial PRACH power calculation:

Preamble_initial_power = Pathloss + UL interference + Prach Required Received CI

The filter coefficient controls the way the *UL interference* component of this formula is calculated by the NodeB and transmitted to the mobiles via the SIB7 information block. A conservative setting of the filter coefficient makes the system less susceptible to instantaneous modifications in uplink noise.

Statistical analysis of a real network has shown the following improvements in the RTWP with the introduction of changing this value, as shown in Figure 12.26.

12.2.7.4 Adjustment of PRACH Parameters

Care should also be taken to tune the initial PRACH target settings. Lab testing has shown that, when these values are not optimally adjusted, excessive spiking occurs when there are small coupling losses between the UE and the network as shown below in Figure 12.27

Many network operators approach PRACH power settings with the goal of ensuring reduced call setup times. Their approach is therefore to maximize the power used in the initial PRACH settings which ensures an access grant on the first attempt. In a spread spectrum network this approach has a negative impact on the uplink shared resource.

In a study conducted on a live network cell, adjusting the initial PRACH parameters showed immediate relief from excessive uplink noise rise (Figure 12.28).

Table 12.5 defines the parameter sets that were used in the above study. PRACH optimization needs to be done cautiously as it has an impact on cell coverage and very aggressive values could result in missed calls. When analyzing different impacts, special care should be taken to ensure that traffic is not being lost due to the smaller sector footprint. This can be checked through drive tests or with the help of geolocation tools.

In this study, the biggest improvement occurred after step 1, the contributor to this change is the required PRACH carrier to interference target. The other steps deal with further decreasing the required carrier to interference ratio; however, not much further improvement is obtained for this change.

Combining the filter coefficient together with the PRACH settings, a significant amount of the noise that exists in the uplink can be controlled.

12.3   Optimization Tools

With the increase of technology complexity the planning and maintenance of the network has become increasingly difficult; however, new and powerful engineering tools have been designed to help the engineers in their daily tasks. In this section we discuss three of the most interesting tools that are available today, which can help operators improve their network operational efficiency and performance: geolocation, user tracing tools, and self-organizing network (SON) tools.

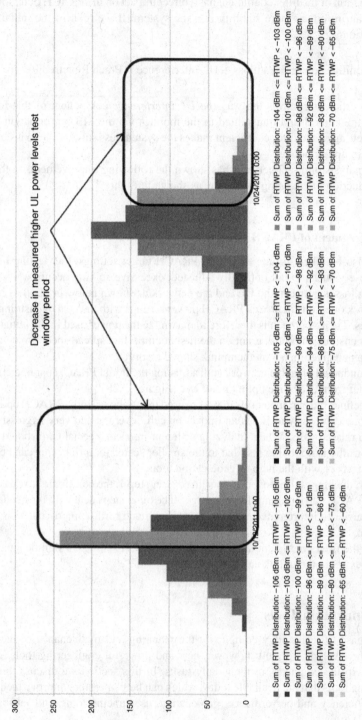

Figure 12.26 RTWP distribution in the test cluster. Left: with default (shorter) coefficients; right: after parameter change

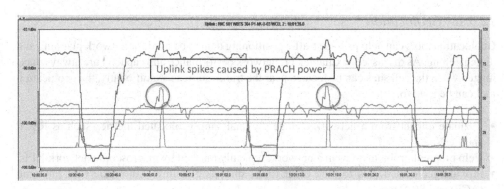

Figure 12.27 Spiking in uplink noise due to initial PRACH access attempts; circles show noise spiking effects

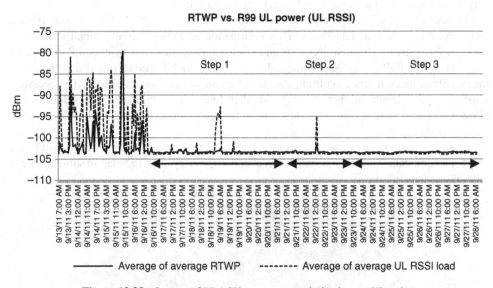

Figure 12.28 Impact of PRACH parameter optimization on UL noise

Table 12.5 Summary of PRACH parameter changes

Parameter	Baseline	Step 1	Step 2	Step 3
Required PRACH C/I	−25 dB	−28 dB	−30 dB	−33 dB
PRACH preamble step size	2 dB	1 dB	1 dB	1 dB
RACH preamble retransmissions	32	16	16	16
RACH preamble cycles	8	1	1	1
Power offset after last PRACH preamble	0 dB	2 dB	2 dB	2 dB

12.3.1   Geolocation

Geolocation tools can help provide traffic positioning data to be used for network planning and dimensioning. As discussed in Chapter 11, network performance and capacity are always maximized when the cell sites can be located near the traffic sources. Additionally, the geolocation tool can be used to:

- Evaluate impact from a network change, by analyzing geolocated metrics such as RSCP and Ec/No.
- Help pinpoint problem areas in a network, typically masked by macroscopic network KPIs.
- Provide geolocated data to feed other tools, such as planning, Automatic Cell Planning (ACP) and SON tools.

There are two types of tool that can provide geolocation information: the first type requires a special client embedded in the handset device that reports statistics collected at the application layer with GPS accuracy; one example of this tool is carrier IQ. This first type of tool provides very good geolocation accuracy, but limited radio information.

The second type of tool can provide detailed RF metrics, such as pilot power, Ec/No, and RRC events such as handovers. This information is gathered from special RNC trace interfaces that stream UE measurements and events, such as Ericsson's General Performance Event Handling (GPEH) or NSNs MegaMon. Since the mobiles don't report their location, these tools estimate it based on triangulation information and timing advance. The accuracy improves with a larger number of neighbors; therefore there is typically higher accuracy in dense urban environments (around 100 m), as compared to suburban (around 300 m) and rural environments (up to 1000 m).

Figure 12.29 shows an example network in which a signal strength map has been built based on geolocation data. Traffic information has also been overlayed on the map to show the areas with higher network traffic.

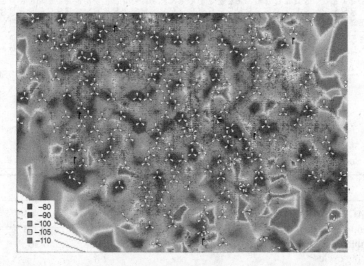

Figure 12.29   Example of a geolocation exercise. The shades of gray represent the pilot signal strength, and the black dots the location of the traffic samples

12.3.2   User Tracing (Minimization of Drive Tests)

The user tracing tools permit one to visualize RF metrics and Layer 3 messages from any mobile in the network, providing a subset of the functionality than that of a drive test tool, with the convenience of remote operation.

A similar use case has been standardized for LTE systems, and is referred to as "Minimization of Drive Tests" or MDT. In the case of UMTS, the functionality is non-standard, but most infrastructure vendors provide a wealth of trace information through proprietary interfaces – the same one that was mentioned in the case of the geolocation tool. Consequently, often both tools – geolocation and user tracing functionality – are offered within the same framework.

The main uses of the MDT tools are:

- Substitution for drive testing, which can result in significant savings in engineering resources.
- Generation of real-time alerts based on network or user KPIs (dropped calls, access failures, etc.).
- Troubleshooting support to customer care. These tools can typically save historical trace data, which can be used in case a specific customer issue needs to be analyzed.
- Support for marketing uses, based on customer behavior.

Figure 12.30, taken from one of the leading tools in this space, shows the typical capabilities of this type of tool. As can be observed, the engineer can monitor specific geographical areas, identify issues, and drill down into the trace of the specific subscriber that has suffered the problem. Historical data from the subscriber is also available to understand what happened to the connection before the specific problem occurred.

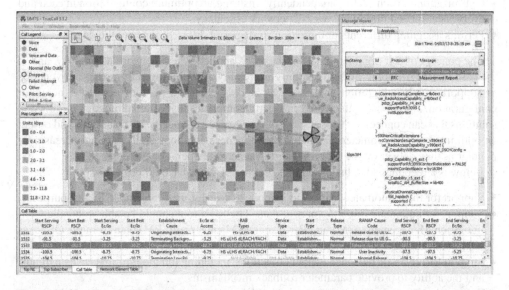

Figure 12.30 Snapshot of remote tracing tool, showing user events and traces. Reproduced with permission of Newfield Wireless

As indicated earlier, these tools nowadays need to operate on closed, proprietary interfaces that have been defined by the infrastructure vendors. This complicates the integration and maintenance of such tools, and makes them dependent on the willingness of the infrastructure vendors to share their format information.

12.3.3 Self Organizing Network (SON) Tools

While network technologies have been evolving very rapidly, the operational tools that the field engineers utilize in their daily activities are still very labor intensive. A significant amount of the engineer's work day is spent in manual tasks, reports, and troubleshooting activities, which prevents them from concentrating on proper optimization activities, which are often very time-consuming.

As discussed in many instances in this chapter, in order to achieve optimum efficiency and performance, the network should be configured on a cell by cell basis. In some cases, it should be configured differently at specific times of the day, since traffic distributions can change dramatically in certain conditions. However, it is practically impossible to manually design and manage so many changes, and it risks human error due to the complexity of the task.

This challenge was widely discussed during the introduction of the LTE technology, and considerable effort has been put into a new concept called "Self Organizing Networks," whose ultimate goal is to improve the operational efficiency of the carriers through increased network automation. Some of the benefits of utilizing SON techniques are:

- Permits operators to optimize many aspects of the network at once, with minimal or no supervision.
- Provides fast and mathematically optimum configurations of the network.
- Continuous network optimization, adapting to changes in the environment.
- Optimization of each sector independently, ensuring an optimum configuration in each and every location and improving overall network efficiency.
- Permits assigning different configuration strategies at different times of the day.

The concept of self organizing networks can be applied to any network technology, and in the case of HSPA+ some SON algorithms are readily available from infrastructure vendors, to support mainly integration tasks such as NodeB plug and play, or automatic neighbor configuration. Additionally, there are third-party companies that offer automatic optimization solutions that can connect to the operator's data sources to extract the relevant KPIs and output an optimized set of parameters for that network. These solutions are becoming increasingly popular as they deliver on the promised performance and operational efficiencies.

Below is a recommended list of features that the SON tool should provide:

- Support for fully automated, closed-loop mode as well as open-loop mode for test purposes.
- Support for automatic import/export of data. The SON tool should be able to handle Configuration Management (CM) and Performance Management (PM) data from the OSSs, as well as streaming trace data such as GPEH and Megamon feeds.
- Support for advanced data management functions, including functions to analyze data trends, and the ability to provide parameter rollback mechanisms.
- Fast processing and reaction time to near-real time data.
- Coordination mechanisms for different algorithms and parameters.
- Configurable optimization strategies.

Table 12.6 Table with automated 3G strategies

Automated planning, configuration, and integration

NodeB integration support	3G automatic neighbors
Automatic NodeB rehomes	Parameter consistency enforcement

Automated optimization

Cell outage detection and compensation	Energy savings
3G mobility optimization	Automatic load balancing
Automatic CCCH optimization	Coverage and interference optimization

Automated performance monitoring and troubleshooting

Automatic performance reports	Real-time alerts
Detection of crossed antenna feeders	Cell outage detection and compensation
Automatic sleeping cell detection and resolution	

12.3.3.1 SON Algorithms for HSPA+

The optimization algorithms will typically try to find a balance between coverage, capacity, and quality, depending on a certain policy that can usually be configured by the operator. To achieve this, the SON tool typically modifies network parameters that are configurable through the OSS, while for example, the Automatic Cell Planning (ACP) tool deals primarily with RF parameters such as antenna tilts, azimuth, height, and so on. However, as SON tools have become more sophisticated, the boundary between these tools has become blurred, and now it is possible to find SON algorithms that optimize certain RF parameters, such as pilot powers and antenna tilts. Table 12.6 summarizes some of the HSPA+ SON strategies that can be found in today's tools.

The following section provides a high-level description of the most interesting algorithms for automatic 3G optimization.

UMTS Automatic Neighbors (ANR)

This algorithm configures neighbors of a 3G cell to try and minimize dropped calls due to missing neighbors and UL sync issues. Most existing modules will configure intra-frequency, inter-frequency, and inter-RAT neighbors to GSM, and will keep the neighbor list within the limits imposed by the technology or vendor implementation. In certain implementations, this module can operate in near real time taking input from RAN traces, which can be very helpful during cell outages.

Automatic Load Balancing

These algorithms will try and adjust the sector footprint, handover weights, and inter-layer configuration parameters to distribute the load in case of high congestion in one specific hotspot area. The goal is to maximize service availability (limit access failures) and increase throughput fairness for all customers, irrespective of their geographical location. There are two flavors of load balancing mechanisms, one that reacts quickly to sudden changes in traffic, and one that tries to adjust the network configuration based on long-term traffic characteristics.

Automatic Sleeping Cell Resolution

Sleeping cells are sectors which stop processing traffic for no apparent reason, and for which no alarm is generated. The sleeping cell detection algorithm identifies a sleeping cell based on certain performance criteria and resets the cell accordingly.

Coverage and Interference Optimization

This module adjusts the footprint of the cells by modifying antenna tilts (in case of RET capable) and powers to (i) eliminate coverage holes in the area, (ii) minimize inter-sector interference, and (iii) minimize the amount of power used in the cluster. In some cases, the algorithm will use geolocation information to improve the accuracy of the decision. If antenna ports are being shared by more than one technology, then decisions for tilts need to be assessed for all network layers that share that path.

12.3.3.2   Results from SON Trials

HSPA+ SON tools are commercially deployed in a number of networks today. The examples in this section are extracted from test results of these tools in a live network.

The following example illustrates the use of a SON tool during a network recovery situation. The network presented was impacted by a heavy storm which caused a very large number of cell outages in the area; when this happened, the number of drops due to bad neighbor configuration skyrocketed and the engineers were spending a large amount of their time configuring neighbor relations, which took away time from other network recovery efforts. Three days after the event, a SON tool was activated in the area, which took care of automatic neighbor configuration in the network. Figure 12.31 shows the relevant network statistics and time savings.

As can be seen, the number of voice drops was dramatically reduced immediately after the activation of the ANR solution. Due to the dramatic cell outage situation, the amount of drops due to missing neighbors accounted for almost 20% of the total number of drops; after the tool was activated, this percentage was reduced to less than 4%. And most importantly, as indicated on the chart on the right, all this improvement was obtained with no intervention from the engineering crew, except for the initial configuration required.

In the next example, a SON tool was activated in two live RNCs in a market for a month. Several SON algorithms were run continuously, including:

- automatic parameter enforcement;
- plug and play (configuration of new sectors);
- ANR;
- coverage and interference optimization.

During the trial period, other RNCs in the market were used to track any possible changes in the network, such as natural traffic growth, seasonality effects, and so on.

Figure 12.32 shows the reduction in drops in one of the RNCs (RNC4) once the SON tool was activated. It should be noted that the tool runs autonomously, gathering KPIs every 15 min interval, and is capable of implementing parameter modifications at that rate if necessary. The

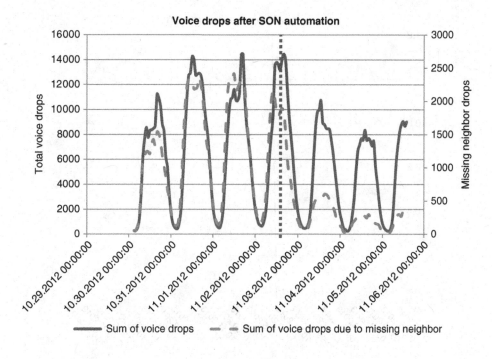

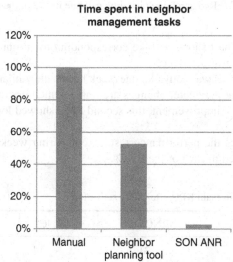

Figure 12.31   Results from SON ANR execution during massive cell outages situation

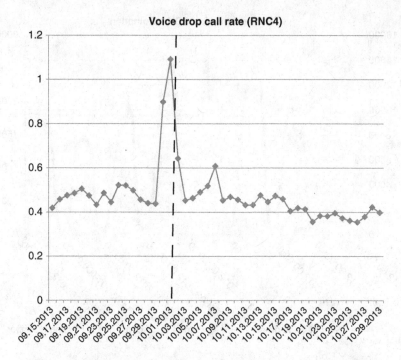

Figure 12.32   Reduction in voice drop call rate in one of the trial RNCs

Drop Call Rate (DCR) chart shows a spike corresponding to an outage in the network that happened before the tool was activated.

The second RNC (RNC2) was activated one week later with similar performance improvements. Given the fact that parameter changes are implemented gradually, which results in a slow slope in performance improvement, this second RNC showed lower gains at the time of the analysis.

Table 12.7 summarizes the performance results, comparing weekly statistics before and after the tool was activated in each of the RNCs.

Table 12.7   Improvement in main KPIs after SON activation

KPIs	RNC4 (%)	RNC2 (%)	Other RNCs (%)
Voice drop call rate gain	21	14	0
Voice access failure rate gain	30	10	−14
Voice traffic gain	8.2	9.2	7
Voice IRAT gain	11	15	2.2
Data drop call rate gain	14.5	19	−1.3
Data access failure rate gain	56.3	2.4	11.7
Data traffic gain	5.6	7.3	5.1
Data IRAT gain	2.1	−2.9	2.4
Data throughput gain	6.3	9.9	3.5

As the results show, the SON tool was able to improve in all main KPIs, while still capturing more traffic than the rest of RNCs in the market. Even more interesting is the fact that calls to customer care in the trial area were reduced by about 15%.

12.3.3.3 Evolution of the SON Architecture

In the long run, SON tools will become a central piece in the network operations process, and will likely evolve into a more complex architecture in which it interacts with other engineering tools, in addition to the OSS. The leverage of big data technologies will help facilitate this paradigm shift. The following aspects are facilitated by an evolved SON architecture:

- Increased scope of data sources and network elements, expanding from OSS to operator specific units, such as work order and inventory databases.
- Simplified data extraction and consolidation that is used by the multiple tools to pull data and configuration information, and push new configurations as required by the automated tasks. It should provide advanced filtering capabilities as well as data storage, and the ability to work with historical data as well as near real-time.
- The SON tool will interact with existing engineering tools that add value to the system, but don't operate in an automatic fashion. For instance, the tool may receive special alerts that are generated in real time by the MDT tool, and react accordingly.
- New, custom automation tasks can be defined and created in the form of SON plugin modules, that can be scheduled to run in background, either periodically or reacting to upcoming events from the SON framework or other tools.

Figure 12.33 illustrates this concept.

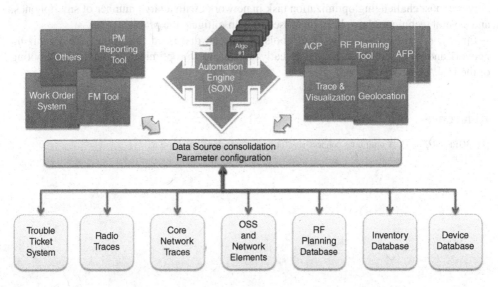

Figure 12.33 Advanced SON management architecture

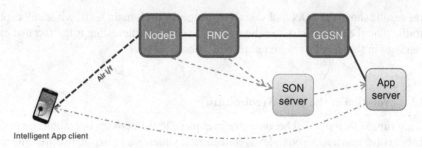

Figure 12.34   Example of advanced SON interactions

The above mentioned architecture would facilitate an increased level of network automation. It will enable a transition from network-centric SON actions, which are typically based on aggregated node KPIs, to customer-centric specific actions, including app-level optimization.

Figure 12.34 illustrates an example interaction that would be possible with this architecture. In this example, the SON system detects congestion in certain areas, and notifies an application server that is generating excessive traffic in the area – the app server will modify the service quality for the users in the affected areas, which will in turn alleviate the congestion in the network.

12.4   Summary

This chapter has presented the main focus areas for optimization in commercial HSPA+ networks, including neighbor management, antenna and power configurations, handovers, and radio state transitions.

A special section has been devoted to analyzing the uplink noise rise problem, which is probably the most challenging optimization task in networks with a large number of smartphones, and several techniques have been discussed to help mitigate the problem.

Finally, several useful engineering tools have been discussed, including self organizing network and MDT tools, which would greatly help during the optimization and troubleshooting of the HSPA+ networks.

Reference

[1] 3GPP TR 25.903, "Continuous connectivity for packet data users", Release 8.

13

Smartphone Performance

Pablo Tapia, Michael Thelander, Timo Halonen, Jeff Smith, and Mika Aalto

13.1 Introduction

In recent years the wireless industry has witnessed an impressive surge in smartphone and data services adoption: in 2013, smartphones accounted for nearly 50% of worldwide handset sales; however, in the USA alone this figure was even more dramatic, surpassing the 90% mark [1].

While the wide adoption of data services is quite positive to the operator from a revenue standpoint, there are big challenges associated with the amount of traffic being generated by modern smartphones. Smartphones are powerful computing devices with fast Internet access speeds that enable a similar connection experience to that of typical home computers. Such performance levels permit wireless customers to enjoy a wide range of wireless services, from typical web browsing to live streaming of high definition video, which entails a higher consumption of data content overall. Furthermore, as will be analyzed later on this chapter, the better the device, the more data it is bound to consume, which puts operators in the difficult situation of juggling the desire to satisfy their customers and the need to limit their impact on network resources.

The resource consumption associated with increased smartphone penetration is not only tied to the amount of data transferred, but also involves an even higher increase of network signaling: unlike typical computers, smartphones are always connected to the network, contributing to the overall traffic with a myriad of small, intermittent packet flows that have put in check many of the design fundamentals of the original HSPA equipment, creating bottlenecks in the control plane rather than in the user plane, as everyone expected. This chapter will analyze in detail how this signaling is generated, establishing comparison across different operating systems and applications, which are ultimately responsible for these harmful effects.

The second part of the chapter is focused around smartphone customer experience. Most consumers' experience with mobile broadband stems from their use of a smartphone. As such,

HSPA+ Evolution to Release 12: Performance and Optimization, First Edition.
Edited by Harri Holma, Antti Toskala, and Pablo Tapia.
© 2014 John Wiley & Sons, Ltd. Published 2014 by John Wiley & Sons, Ltd.

from a mobile operator's perspective they should place concerted effort on ensuring that the experience is a good one. One common belief in the industry and among consumers is that a "faster network" provides a better user experience, meaning that a consumer's experience with a smartphone will be inherently superior in an LTE network than an HSPA+/DC-HSDPA network. The reality can be somewhat different.

The operator with the fastest network, as determined by measurements that focus solely on throughput, will have an inherent advantage when it comes to marketing the service offering. However, the higher data speeds may not translate into a better user experience. In this case, the user experience is defined by the parameters that the mobile data subscriber actually observes while using the smartphone. Examples include:

- Video playback start time and the amount of buffering that occurs during the video playback.
- The battery drain on the device along the day.
- The time required to begin interacting with a social networking service.
- The time to load a certain web page, to name a few.

To varying degrees, the actual throughput capability of the underlying cellular network impacts these parameters. However, there is a diminishing return and beyond a certain threshold higher throughput is inconsequential. There are also other factors that influence the user experience, and in many cases these factors have a greater influence on the outcome, meaning that they undermine the potential benefits of having a faster network.

In this chapter we analyze in detail some of the key aspects that influence the end user experience, and provide some guidelines about how smartphone performance can be tuned to take full advantage of the HSPA+ technology.

13.2   Smartphone Traffic Analysis

HSPA networks were not originally designed to carry traffic types such as those observed from smartphones today. Both downlink and uplink channels were conceived to transfer large volumes of data very efficiently at high speed, however the traffic patterns observed in today's networks are somewhat different. Understanding the nature of smartphone traffic is important to try and improve the efficiency of data transmission across the network.

Figure 13.1, from a network with a large penetration of smartphones, illustrates the share of data consumption compared to the number of session requests, for uplink and downlink sessions. This chart shows the large asymmetry in data consumption between the uplink and downlink directions, with most of the sessions initiated at the mobile device side.

In terms of actual content, it is worth differentiating between data volume and frequency of access, which is tied to the popularity of one particular application. Figure 13.2 illustrates this.

As Figure 13.2a illustrates, a large part of the data volume is consumed by video content; however, although this service is quite popular, it is not as dominant in terms of frequency of access. This is due to the nature of the video service, in which one single session will generate a significant amount of traffic, typically in the order of 200 MB per hour. On the other extreme, Internet searches are quite frequent but generate only a very small amount of traffic. One interesting aspect to note is the popularity of social networking services, such as Facebook and Instagram, and how they are increasingly generating larger amounts of volumes due to embedded multimedia sharing capabilities.

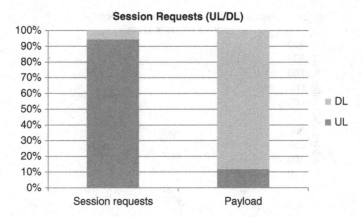

Figure 13.1   Downlink vs. uplink radio traffic characteristics

Smartphones tend to transmit a significant number of small packets in quite short sessions, which is referred to as "chattiness." This device chattiness is principally caused by smartphone "apps" that operate in the background and periodically retrieve small amounts of data with status updates; examples of these apps are weather widgets, social network updates, tweets, and so on. This is not reflected in Figure 13.2 because those URLs are hard to classify within a single content group.

Figure 13.3 compares the amount of data that is sent within every HS-DSCH session for three types of HSPA+ networks: a smartphone-centric (left), a USB-dongle centric (right), and one with a mix of devices. As can be observed, smartphone-centric networks tend to transmit smaller amounts of data each session, which is related to the device "chattiness" previously discussed.

Considering that smartphone networks transmit an average of 100 kB per HS-DSCH session, with default inactivity timers set to 8 s this results in about 2% efficiency during data transfer, assuming a downlink data transfer of 5 Mbps. Figure 13.4 illustrates this idea.

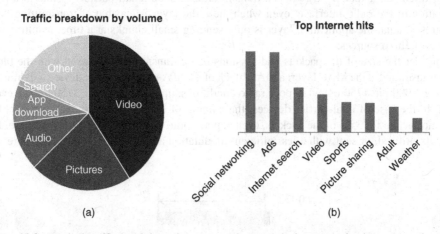

Figure 13.2   Typical traffic breakdown in an example network, in terms of volume (a) and service usage (b)

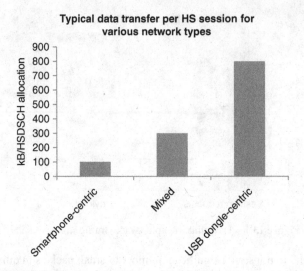

Figure 13.3   Data transfer per HS-DSCH allocation for various network types

This explains in part why networks with high smartphone penetration suffer from uplink interference effects: smartphone devices keep transmitting unnecessarily in the uplink direction while the channel is up. Uplink noise problems are discussed in further detail in Chapters 4, 10, and 12.

Figure 13.5 shows the distribution of packet sizes transmitted in uplink and downlink, in the same network discussed earlier. As can be seen, the use of small packets is quite dramatic in the uplink direction, mostly due to TCP acknowledgements sent for downlink data; due to the asymmetrical nature of the data, the number of acknowledgements sent in the uplink direction is much larger than in the downlink direction.

Note that even in the downlink direction, about 25% of the packets transmitted are quite small, often fewer than 90 bytes. The transmission of such small packets results in a poor utilization of the radio interface: even when the radio layer is capable of transmitting large amounts of data, the application layer is only sending small chunks at a time, resulting in a waste of radio resources.

Consider the size of the packets and sessions being transmitted compared with the block sizes transmitted at the MAC layer: a MAC block of a HSPA+ single carrier device will transfer between 6000 bits (750 bytes) in poor radio conditions, up to 30,000 bits (3750 bytes), every 2 ms. In the case of dual-carrier devices, the amount of downlink data transferred doubles. This means that for 35% of the packets, even at poor radio conditions the MAC block is either not fully utilized, or sent with a less efficient modulation and coding scheme. Therefore, the

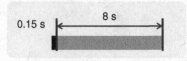

Figure 13.4   Illustration of inefficient use of the DCH channel for smartphone traffic

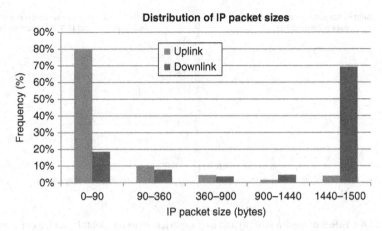

Figure 13.5   Distribution of IP packet size in both directions (uplink and downlink)

spectral efficiencies attained in real networks with a large amount of smartphone traffic will be far from the theoretical figures, which can only be reached when operating in full buffer mode. Code multiplexing can help improve efficiency in networks with small packets by allowing multiple users in parallel per TTI.

Due to the characteristics of smartphone IP traffic, including asymmetry and chattiness, there's a great potential to tackle this problem at the upper layers, including optimizing the individual applications and services; and creating smart methods at the operating system level to control transmission at the packet level. Some of these potential optimizations will be discussed in this chapter.

13.3   Smartphone Data Consumption

The exponential increase in network data consumption can be distilled into three main causes that will be analyzed along this section:

- Increased penetration of smartphones in the device base.
- Increasing consumption with newly released smartphones.
- Excessive resource consumption from a small percentage of customers.

Smartphones typically consume over ten times the amount of data of regular feature phones. The reason for the increased data consumption is the improved usability and connection speed that smartphones present, as compared to typical feature phones that are cumbersome to use and have limited data capabilities. Consider the example of a web browsing experience in a smartphone, featuring a browser capable of rendering full web pages in a high resolution display, as compared to a clumsy browser in a feature phone which can possibly only show WAP content and which is hard to navigate and scroll: naturally, the customer with the smartphone will tend to watch web content more often than the one with the feature phone, even if the feature phone was able to offer equal data access speeds.

Another example of higher content consumption comes from the increased popularity of online video services. One hour of online video can consume between 200 and 350 MB,

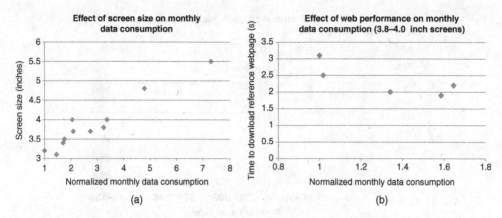

Figure 13.6   Effect of screen size (a) and user experience (b) on monthly device consumption

depending on the service. Consequently, as smartphone penetration increases in the base, massive uptake in overall network data volumes is expected, thanks both to the increased amount of active data plans – which are typically required by operators from smartphone users – as well as the difference in monthly consumption between smartphones and feature phones.

The increased data consumption is also noticeable across different smartphone devices: newer smartphones present better characteristics than previous versions, which in turn result in higher data consumption. Two key factors play a role in this: the size of the screen and the Internet user experience of the device. Figure 13.6 illustrates the impact of these two factors in monthly data consumption, from statistics collected in a Tier 1 US operator.

Figure 13.6a illustrates the relation between screen size (in inches) and monthly data consumption. The results show that a device with a 5.5-inch screen consumes more than seven times the amount of data of a device with a 3.2-inch screen, and about twice as much data as a device with a 4-inch screen. There are two main reasons for the increased consumption with larger displays: first, customers with large displays are more prone to using their devices to access multimedia content; and second, larger screens demand larger images and videos, which account for the majority of the data payload in the device.

Improved Internet data experience also plays a role in driving data consumption: phones that offer a better data user experience tend to consume more data, as illustrated in Figure 13.6b. The chart compares phones with a similar screen size and different web browsing speeds, and shows how the devices with a faster web browsing speed (lower time to download a reference webpage) can consume over 50% more data than similar smartphones. It is important to note that the correlation is related to the web performance rather than the device category: the download and upload speeds are irrelevant if these are not translated to an improved user experience.

Another interesting aspect to note with regards to smartphone data consumption is the role of the operating system. There are notable differences in the methods used to control the way the devices access the network: some OS permit the transmission of background data, while others don't; some OS permit the customer to configure the quality of the video services, while others don't, and so on. Furthermore, there have been observed significant differences between different versions of the same operating system, as illustrated in Figure 13.7 which compares the average consumption of different Windows mobile devices with versions 6.5, 7,

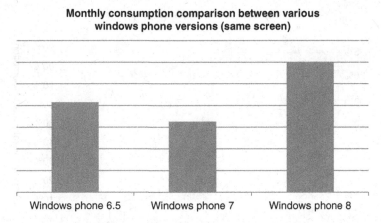

Monthly consumption comparison between various windows phone versions (same screen)

Windows phone 6.5   Windows phone 7   Windows phone 8

Figure 13.7 Data consumption differences based on OS version

and 8 – all of them with the same screen size (4.3 inch). In this case it can be observed that Windows Phone 8 devices tend to consume much more data than their predecessors – in this particular case, mostly due to the default configuration of the video service quality.

Increased data consumption isn't always good news for the customer nor the operator. In the case of customers, if they are not careful they can quickly consume their monthly data quota if they access multimedia content frequently. For example, a typical Netflix movie can consume about 500 MB in a regular smartphone, which means that customers with data plans of 2GB per month will run over their monthly quota after having watched just four movies.

For operators, increased data consumption will force additional network investments to cope with the capacity demands, or else face customer complaints due to performance degradation in the network. One interesting thing to note when analyzing customer consumption patterns is the fact that not all customers access the network in the same manner; therefore it is good idea to classify customers based on typical monthly usage. Figure 13.8 illustrates the consumption patterns in an example network in which there are no data consumption limits.

As the figure shows, there is a minority of customers that consume an excessive amount of data: in this example, 2% of the users account for over 20% of the overall data transferred in the network. In order to minimize capacity investments, and at the same time preserve the service satisfaction for the majority of the customers, the operator can implement different data control mechanisms to prevent harm from these excessive users, including data bitrate capping, blocking of certain operations such as peer-to-peer transfers, tethering, and so on.

In addition to volume consumption, the operator needs to analyze other forms of resource consumption by the smartphone devices, such as signaling. Sometimes it's less harmful for the network to transmit a large volume of data than a myriad small packets, as will be discussed in detail in the following sections.

13.4 Smartphone Signaling Analysis

The increased penetration of smartphone devices has caused a rapid increase in the signaling load at the network level. As Figure 13.9 illustrates, the growth in signaling traffic in many

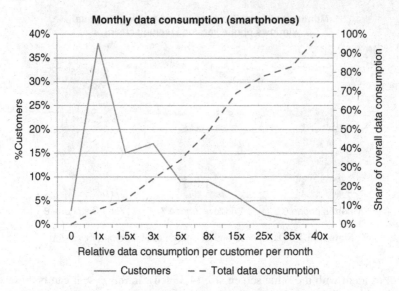

Figure 13.8   Distribution of monthly data consumption (smartphone users)

cases has been faster than the data volume growth, leading to congestion, dropped calls, and increased access failures.

On the device side, their excessive connectivity has also led to poor battery life, unresponsive user interface, slow network access, and non-functional applications for many smart device users.

End users do not notice the impact of increased signaling directly; however, overall this results in a decreased quality of experience (QoE). To overcome these problems, network operators have increased their investments beyond their original capacity expectations.

The operators can, however, try and minimize the impact of poor performing devices by analyzing the behavior of their smartphones. This section describes a methodology to analyze network impact from smartphones, mobile applications, and mobile operating systems.

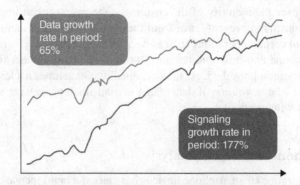

Figure 13.9   Volume vs. signaling growth example from a live network

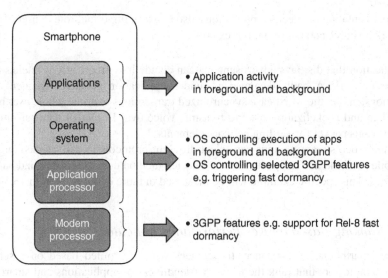

Figure 13.10   Components impacting smartphone performance

13.4.1   Smartphone Profiling

Smartphone profiling consists of the analysis of multiple protocol layers and their inter-related behavior in complex end-to-end systems. Promoting network friendly smart devices and applications can help operators avoid unnecessary capacity expansions, as well as improve user experience – which will in turn result in better customer loyalty. As will be discussed in more detail in the next section, the user experience depends on various factors, including network performance and the specific application being used. Even the battery life time of a mobile device can be impacted by network features, as well as by well designed and implemented applications. Figure 13.10 illustrates the various components that need to be considered when analyzing smartphone performance.

Some of the features that need to be taken into account when analyzing smartphone per-formance are standardized by 3GPP. Those features are implemented at the modem chip set level and are either fully controlled by its firmware or partially by the mobile OS. On the net-work side, the 3GPP features are split between access stratum (i.e., radio access network) and non-access stratum (i.e., core network). A good example of one such feature is the so called "network controlled fast dormancy" or 3GPP Release 8 fast dormancy: when properly utilized, this feature can help optimize the amount of signaling caused by RRC state changes, at the same time that the device battery life is extended by making a better use of the CELL_PCH state.

There are also important features that are implemented outside of the 3GPP domain, such as the operating system of the device. One example of this is the implementation of OS keep-alive messages required for push notification services. Always-on applications use push notification services, which send messages from the network servers to clients in mobile devices. These notifications require persistent TCP connections, which again require periodical keep-alive messages to keep the TCP connection established. The lower the keep-alive frequency is, the fewer transactions that generate additional RRC state changes and signaling.

The implementation of specific applications also plays an important role in user experience and network resource consumption. For example:

- An application that displays advertisements can decide how frequently it fetches new ads from the network. A badly designed ad-funded application generates frequent network transactions, and in the worst case synchronized transactions over the whole user base.
- The coding and packetization used to transmit Voice over IP (VoIP) data can impact the load in various packet forwarding network elements.
- Audio and video streaming apps can select different protocols to deliver the content over the mobile network, and depending on the selected method user experience and battery life time varies. This specific example will be discussed in more detail in Section 13.5.6.5.

13.4.2   Ranking Based on Key Performance Indicators

To analyze smartphone performance, the devices will be profiled based on a set of key performance indicators that rank the network-friendliness of applications and smart devices with regards to key network domains like mobile device, radio, packet core, transport, and IP edge (firewalls, NATs, etc.).

The advised ranking-based methodology presented in this section enables mobile operators to find the ideal balance between the best user experience and the lowest impact on network performance.

Smart devices are tested with real applications or services such as Voice over IP (VoIP), mobile video, and social media in order to emulate as close as possible the behavior of a typical smartphone user. The smartphone profiling provides essential insight into the interaction between factors such as smartphone battery life and applications or operating systems. For example, Figure 13.11 illustrates a comparison between two different smartphones running a social networking application.

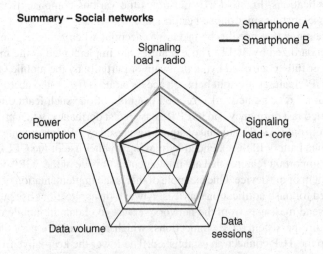

Figure 13.11   Example profiling of a social networking app in two different smartphones

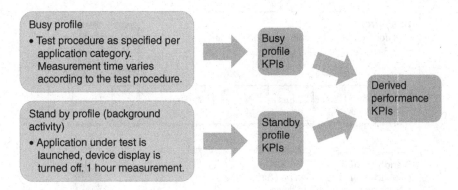

Figure 13.12   Both busy and standby profiles are used to determine overall device performance

The profiling of smartphones provides a relative ranking for applications or devices, based on their behavior in a real environment. This enables mobile operators to make performance comparisons systematically and quantitatively.

13.4.3   Test Methodology

The profiling of a smartphone device is performed both for single applications and with multi-application profiles – in which the device is loaded with multiple commercial smartphone applications such as Facebook, Skype, email, and so on. Both test methodologies will utilize busy – typical use of the applications – and standby test profiles. The standby mode is used to characterize background activity, that is, when the device is powered on, but not actively used. Both profiles are equally important to understand the performance of the devices and applications, as illustrated in Figure 13.12.

- Standby testing reveals specific issues related to OS or applications, like too frequent keep-alive messages or lack of synchronization with keep-alive messages. It also helps to better understand why certain smartphones drain their battery faster than others.
- Tests with busy profiles provide information about busy hour signaling and power consumption of the device as well as protocol behavior of the applications.

The busy profile test cases are application specific and emulate the typical usage of the application. The example in Figure 13.13 is taken from a VoIP application testing with alternating Mobile Originated (MO) and Mobile Terminated (MT) calls, with periods of inactivity between calls to capture all necessary signaling. Call length can be operator specific, but the breaks need to be long enough to capture inter-call signaling.

The standby profile test is simply a one hour test in which the application is activated, although not actively used. This same test case can be used for any application category, and provides understanding of the keep-alive process of the application, as well as useful insight about specific app behavior; for example, in the case of a VoIP call, it explains how presence signaling works in the background.

Figure 13.14 shows an example of a multi-application test.

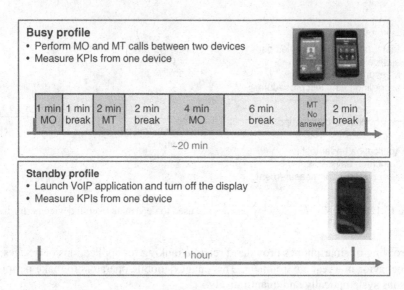

Figure 13.13 Example VoIP application activity for busy and standby profiles

The multi-application test case in this example includes six different applications, and its main goal is to emulate the typical smartphone user's busy hour behavior. Applications can be operator specific (e.g., top 10 applications), but this profile should not change too often to be able to compare several smartphones over time.

13.4.4 KPIs Analyzed during Profiling

Figure 13.15 illustrates the most typical measurement points and KPIs used for smartphone profiling. It is important to define a set of standard KPIs together with standard test procedures, to facilitate the comparison of applications within the same application category.

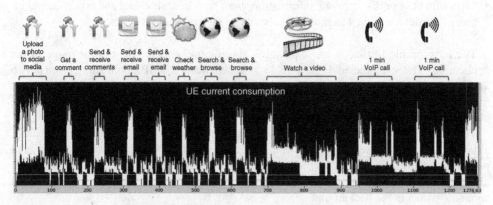

Figure 13.14 Device testing with a multi-app profile

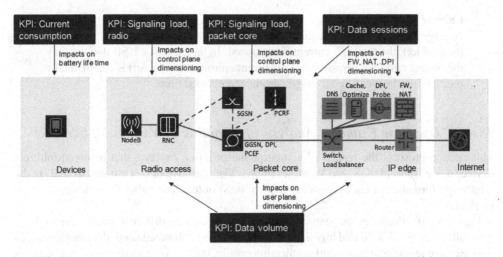

Figure 13.15 Measurement points and KPIs collected during smartphone profiling

The following KPIs are typically used to measure network impact and end user experience. All network KPIs are collected for both busy and standby profiles to understand the impact to different network domains.

Total radio state change signaling frequency in RNC

This KPI is used to evaluate the impact on the RNC control plane load. The total number of external (Uu/Iub, Iur, Iu) signaling messages due to radio state changes is measured over time and reported as frequency (msg/min).

RAB establishment frequency

This KPI is used for evaluating the impact on RNC and SGSN control plane load, as well as the impact of direct tunnel. The total number of successful RAB establishments per minute (RABs/min) is measured.

Note that this KPI is not that significant with modern Release 8 fast dormancy smartphones that tend to stay always RRC connected. However, if there are still old devices in the network, then legacy fast dormancy impacts on this KPI.

Total IP volume

Total IP volume (bytes) is measured separately for uplink and downlink. When the measurement period is taken into account, the average IP throughput (bits/s) can be calculated. The data volume includes the UE endpoint-specific IP headers.

TCP/UDP data sessions

This KPI is relevant for network elements and functions that handle IP flows such as firewalls (FW), network address translation (NAT) and deep packet inspection (DPI) functions. Total number of TCP and UDP sessions is measured.

Current consumption

Average current (mA) is measured over the test period. In the case of a smartphone (or a tablet), the total UE current is measured. In the case of a USB dongle, only the dongle current is measured. The current consumption (mAh) is calculated by multiplying the measured average current and elapsed time.

13.4.5   Use Case Example: Analysis of Signaling by Various Mobile OSs

This section presents the results of a smartphone profiling exercise that involved multiple devices using Android, iOS, and Windows Phone. The objective of this test was to analyze the relative performance of each Operating System (OS) in terms of radio network signaling, in particular.

Figure 13.16 illustrates the standby traffic pattern of three different smart devices from three different mobile OSs and highlights the impact of synchronized keep-alive messages. All devices were tested against a multi-application profile, that is, there were several applications installed and activated in each of the tested smartphones.

During the standby tests, most times a new packet needs to be sent, it will trigger a transition in the channel state involving a number of RRC message exchanges. Frequent signaling will

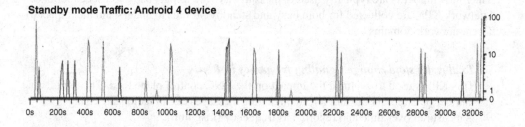

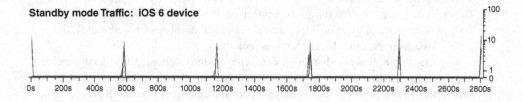

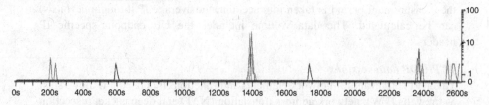

Figure 13.16   IP packet transfer during standby tests for different operating systems

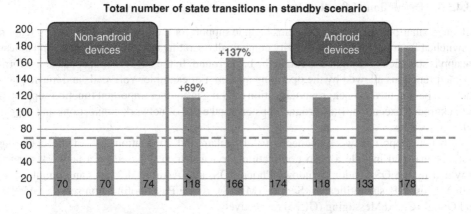

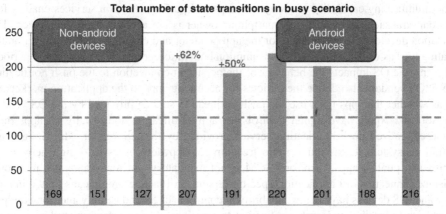

Figure 13.17   Radio signaling comparison of Android vs. non-Android devices

impact device current consumption negatively, because every transaction will require several seconds of RRC connection in the Cell_DCH or Cell_FACH channel.

As Figure 13.16 illustrates, there is a significant difference in the standby transfer profile of the operating systems analyzed: iOS and Windows Phone devices tend to transmit less data, and the data transfer is performed in a synchronized manner. By aligning the messages, more data can be transmitted within the same state, for example, Cell_DCH, and fewer RRC state transitions are required.

Figure 13.17 illustrates the number of state transitions for the same multi-application test case. Test results are grouped by non-Android (cases 2 and 3 from the previous example) and Android devices to highlight the difference between operating systems. Both "stand by" and "busy" profiles are analyzed.

As explained earlier, the main difference can be explained by the lack of coordinated keep-alive messages in the Android operating system. The difference is obviously bigger in the "standby" case where keep-alive messages are contributing most of the traffic, but it can be also seen in the "busy" case. The performance variation within Android devices can also be explained by the lack of use of Release 8 fast dormancy in some of the devices.

13.4.5.1 Discussion of the Results

All major smartphone operating systems include support for multitasking: when an application is switched from the foreground activity control, all operating systems suspend the app unless the app is specifically designed to run in the background. In the background, apps can perform selected functions allowed by the operating system. For example, VoIP calls and music playback can be executed in background; location tracking is another important function supported in background execution. Background apps can make use of network connections and trigger data transactions.

To better support network connectivity of background applications, all of the operating systems analyzed include a push notification service, which is based on a network service provided by the OS vendor. Windows Phone, iOS, and Android devices can communicate with Microsoft Push Notification Service (MPNS), Apple Push Notification service (APNs), and Google Cloud Messaging (GCM) respectively.

The multitasking capability of always-on apps and push notification services enables frequent data transactions even if the smartphone owner is not actively using the device. The application developer is responsible for the app behavior and can significantly impact on the signaling load as well as UE current consumption both in foreground and background periods.

The specific OS impacts the behavior of the persistent connection to the push notification service. OS vendors also define the policies for accepting apps to the application markets, as well as whether the apps must use the push notification service provided by the OS vendor. Microsoft and Apple have tighter controls for app certification and usually all apps must use the centralized push notification service provided by the OS vendor. The Google and Android ecosystem leaves much more freedom for app developers to design networking features and publish apps, therefore GCM is not mandatory and many apps use their own persistent connections to application-specific servers. In practice, this means that Windows Phone and iOS devices have better capability to control background activity and, for example, synchronize keep-alive and push notification transactions. Android devices tend to generate more signaling traffic, because different always-on apps may generate independent, non-synchronized keep-alive and push notification transactions in background. At the time of writing this chapter, Google was actively working on improving these aspects with various GCM and Android initiatives.

13.5 Smartphone Performance

All network operators strive to offer the best possible service experience to their customers; however, this goal is becoming increasingly challenging as data services become more prominent. Unlike in the case of voice, which was a relatively simple service to analyze, data services are more complex and often typical network KPIs do not provide the required observability to understand their performance. As will be discussed later in this section, there is not a single "data service" experience, rather a service experience linked to each individual data service: web browsing, video streaming, social networking, and so on.

Understanding the end user experience requires the analysis of multiple factors, from devices to networks and services. Table 13.1 provides a list of some of the factors that play a role in the customer experience with smartphone devices, broken down into two major categories: battery life and Service Quality of Experience (QoE).

Table 13.1 Summary of key factors affecting smartphone performance

E2E domains User experience	Smartphone HW/SW design by OEM	Network capabilities and conditions	Applications and services Client and server design and conditions
Battery life	Battery capacity HW component selection (e.g., display) Chipset energy saving features Standby activity by OS and OEM SW package	Energy saving features in radio RF conditions impacting on Tx power RF condition, load and QoS policies impacting on upload/download times	Application background activity Content delivery solutions (e.g., frequency of ads)
Application/ service QoE	Display quality, touch screen performance CPU performance, memory size Modem performance (category) OS multitasking IP stack	RRC state transition performance RF condition, load and QoS policies impacting on packet loss and upload/download times TCP and content optimization	Server location, load and capacity App UI design App background behavior Application layer protocol design Content optimization IP stack

At a very high level, the main areas to be considered are:

Battery Performance. One of the main sources of grief for smartphone customers is the poor battery life of these devices, compared to the previous generation of feature phones. Smartphones are more heavily utilized than their precursors; however, there is still room for improvement at the cellular, OS, and app level.

RF Coverage. A fast performing network is meaningless if it is being accessed from an area where there is poor or no coverage. The quality of the RF signal is intimately tied to the maximum speed that can be delivered by the network. When accessing the service outdoors and/or under lightly loaded conditions the networks are typically capable of delivering the promised data rates; however, under more challenging conditions, such as indoors, the measured RF signal levels (RSCP, RSRP) can be quite low and this situation results in a poor user experience.

Network Connection Times. Before the first byte of application data is sent, the smartphone must connect to the network and perform a DNS query to determine the location of the host web site. Although in many cases the total connection time is relatively inconsequential, in some cases it can easily be on the order of several seconds, which can be equivalent to the actual transfer time of a modest amount of data.

Effects from TCP Protocol. Most of today's applications are based on the TCP/IP protocol stack. Although TCP is the most widely used protocol in the Internet, it presents significant challenges in wireless networks, as the multiple versions of the protocol that are "optimized" for wireless give testimony[1]. The variable nature of the air interface interferes with the control mechanisms in TCP, which often results in reduced bitrates or data stalling situations.

Operating System Influences. The various operating systems have different ways to control data transfer, which can affect network and device resource consumption. Also, in many cases the same application behaves entirely differently based on the underlying operating system. For example, a Facebook photo upload, operating on iOS, is faster than Facebook in Android due to the fact that the iPhone app resizes or compresses the picture before uploading it to the network.

Application Characteristics. The characteristics of the application can be of paramount importance when it comes to defining the user experience. "Chatty" applications, for example, can require lots of back and forth interactions between the smartphone and the network with only small amounts of data being transferred during each interaction. This situation can negate the benefits of an otherwise fast network. In this case, network latency could have more influence on the user experience. The size and layout of a web site can also influence the time required to download a web page. On the other hand, an application can also mask the true performance of a poor performing network and give the user a far better experience than would otherwise be the case. For example, an application could upload content, such as a picture, in the background so the user isn't aware how long it takes. Alternatively, the application could use compression to reformat the content of the transferred data and ultimately reduce the size of the uploaded picture or the downloaded video.

Applications can also be optimized in other ways to provide a better user experience or to shorten the total user transaction time. For example, a robust predictive text algorithm or the use of voice recognition software could reduce the time required to perform a search function or to enter a destination in a maps application. This capability would be entirely independent of the network's capabilities but from the user's perspective it would still lead to a better experience.

Understanding the behavior of each of these applications and their performance over the wireless network is an important step towards the optimization of the smartphone experience. In this section we analyze two of the most common applications in use today, web browsing – which is the basis for most common "apps" – and the video streaming service. As discussed earlier, these services represent the most popular traffic in terms of number of transactions and volume, respectively.

13.5.1   User Experience KPIs

In a world dominated by data services, defining the right measurements is a key step towards understanding and maintaining proper customer experience. Previously used KPIs, such as access failures and drops, are not so meaningful when referring to data services, as in many cases a data drop will not even be noticed by the consumer. It is therefore important to define a new set of metrics that "make sense" from a data utilization point of view; these metrics will

---

[1] TCP Vegas, New Reno, Hybla, BIC, Westwood, among others.

have to be defined at the application level, such as the "time to access video content," "time to download a web page," or the "number of rebuffering events" during a video playback.

These complex KPIs cannot be captured using typical counters, and require the deployment of new elements, such as network probes, that can capture per-session data to extract the necessary information. To complicate things further, the KPIs should be able to tell, in the case of a performance issue, whether the problem lies in a specific radio network element or in other part of the communication path. There are various alternatives to implementing these new KPIs, including the use of network probes, leveraging special UE reports provided by applications (Real User Measurements – RUM), and the use of custom UE clients, such as carrier IQ.

Network probes require significant infrastructure and expense to instrument. To manage scalability and cost, they are often deployed near the GGSN/PGW, or only at a few locations. Due to this placement well before the radio interface, they are often limited to connected user experience and cannot measure the impact of connection setup latency or failures. This lack of visibility is significant as the setup latency and failures may have a much large impact on user experience than round-trip delay. Core network probes are therefore good at detecting service failures, or massive numbers of errors, but do not provide an accurate picture of user experience. Alternatively, probes can be deployed at the last mile transport connection, which can capture the effects from the radio access network. This information can be combined with control plane traces from the RAN elements to provide an end-to-end picture of the data connection.

UE reports, also known as Real User Measurements (RUM), are often collected by web sites, applications, and video services. These reports are often very specific to the service and can include time spent on a web page, application launch time, web page launch time, video start time, and various client parameters, to list a few. Since these are done directly on the device, they sometimes include both connecting and connected latency; however, it really depends on the service's goals. This information is not normally available to the carrier so they must instrument their own RUM measurements.

Special UE clients, such as carrier IQ, can also be used to capture the customer experience at the device side. This type of client can provide unique information, such as battery performance, that is otherwise not visible to the operator; these clients can also be developed to capture certain data level KPIs; however, they are typically limited in the amount of information they can capture.

13.5.2   Battery Performance

The battery consumption in smartphones is one of the key aspects affecting customer perception of the device. No matter how fast the device performs, if it can't survive at least one day with a full charge, it will be a nuisance to the consumer, who will need to be continuously looking for ways to charge the phone. Such was the case with early HSPA+ smartphones, and has happened as well to current LTE devices. Apart from the fact that customers tend to use their smartphones much more extensively than previous feature phones, there are clear outliers that have a significant impact on battery performance, some of which can be mitigated by proper device and operating system design.

Figure 13.18 illustrates the result of lab test performed by the T-Mobile validation team, considering a 24-h drain on an active user profile. The key elements involved in battery drain in this particular example are ordered from higher to lower.

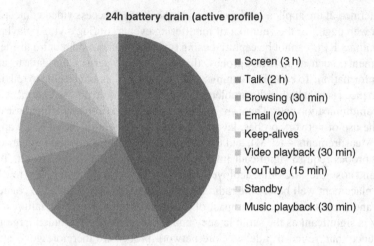

24h battery drain (active profile)

- Screen (3 h)
- Talk (2 h)
- Browsing (30 min)
- Email (200)
- Keep-alives
- Video playback (30 min)
- YouTube (15 min)
- Standby
- Music playback (30 min)

Figure 13.18   24-h smartphone battery drain for an active profile (lab)

As the figure illustrates, the primary source of battery drain is the phone screen. Unfortunately, the screen technology plays a major role here and there are no quick wins to improve this item, apart from playing with dimming and other areas that are really not related to the cellular technology. Next, in this case, is the talk time, which has been extensively optimized as compared to early 3G phones and it is expected to see little improvement in the future.

On the other hand, the chart shows some opportunity to improve the cellular performance when using certain applications, such as web browsing, email, and YouTube. Section 13.5.6 will discuss the impact that app design can have on battery life, through the analysis of various streaming services.

As discussed in Section 13.4.5, it should be noted that keep-alive packets can have much more relevance than is reflected in this chart, due to the fact that lab profiles tend to be less "chatty" than real life situations. There are several possible techniques to mitigate the impact from chatty apps, including special mobile clients that can be used to reshape the signaling behavior of the device. Table 13.2 illustrates the test results of one of these clients, from SEVEN Networks, that shows the potential to optimize the standby mode battery drain. In this particular case, a mixed profile of apps was used and left in background (standby) mode for two consecutive hours in a live network. The device using the optimization client experienced a battery drain of less than half as compared to a regular device.

Another method to improve battery drain is to optimize the use of RRC states: Cell_PCH, Cell_FACH and Cell_DCH, for which the use of Release 8 fast dormancy can play an important

Table 13.2   Impact of handset optimization client (2 h test)

KPI	Normal	Optimization client
Number of RRC messages	2091	414
Max time between state transitions	1 min	15 min
Av current drain	30 mA	12 mA

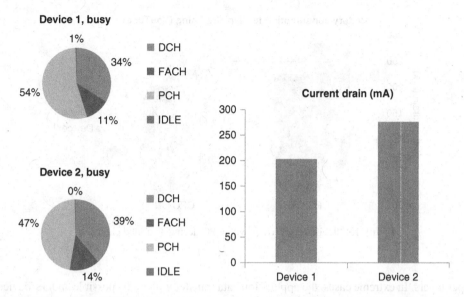

Figure 13.19 Usage of RRC states and battery consumption for two comparable smartphone devices

role. In particular, devices that spend more time in PCH state tend to fare better in terms of battery life. The charts in Figure 13.19 illustrate the difference in RRC state utilization in two comparable high-end smartphone devices; as the figure on the right shows, the device that makes better use of the RRC states consumes around 25% less battery than the non-optimized device.

The Continuous Packet Connectivity (CPC) feature can also be utilized to reduce the battery drain on smartphones, thanks to the Discontinuous Transmission (DTX) and reception (DRX) features. The gain of CPC depends to a large extent on the type of traffic that is being transferred: for example, it can be quite effective for bursty, interactive applications, such as web browsing or video streaming, but it will not help much in the case of FTP downloads. Figure 13.20 shows the test results from a lab experiment using YouTube video streaming for two different devices.

As the results illustrate, CPC in this case is providing overall battery savings in the order of 6–10%. The apparent low gains from the feature are due to the higher drain caused by the screen, as previously discussed in Figure 13.18.

Finally, it should be noted that the most optimized devices aren't always the best in terms of battery life as perceived by the customer. During our analysis it was surprising to see less optimized devices that were equipped with high capacity batteries, which in the end resulted in a better battery life in the eyes of the consumer.

13.5.3 Coverage Limits for Different Services

The availability of proper signal coverage is a key factor to offer a good service experience: if the signal is too weak, or there's too much interference, the radio layer will drop packets frequently and the connection may be intermittently dropped, which results in delays in the

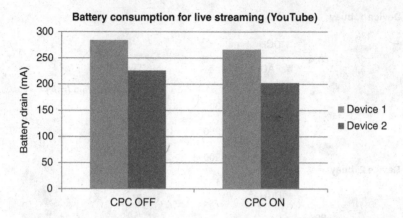

Figure 13.20   Battery savings with CPC during YouTube playback

upper layers. In extreme cases, the application data transfer will not be possible and the device will show an error or the application will simply not respond to user actions.

As discussed earlier, not all data services are equal, and some are more resilient than others with regards to radio coverage. In Figure 13.21 we illustrate the results of a lab test in which we executed various smartphone applications at different combinations of signal strength (RSCP) and interference (Ec/No). The objective of the tests was to find the minimum RSCP and Ec/No at which each service could operate. The app behavior at this point depends on the type of service: for example, for web browsing it would mean that the browser provides an error or doesn't show the page at all; for YouTube, the point where extreme rebuffering occurs, and for Skype, the point at which the call is dropped.

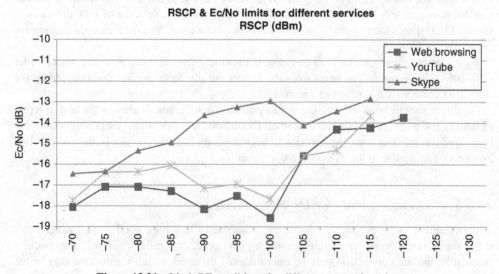

Figure 13.21   Limit RF conditions for different types of services

As the test results show, data services can generally operate at very weak levels of signal strength, as long as the interference levels are low; on RSCP levels under −100 dBm, the services require a more stringent Ec/No. It should be noted that these were lab tests, and the absolute values can only be taken as indicative.

In general, web browsing is the most resilient data service. It can operate at the lowest signal levels, and can also endure higher levels of interference: it can operate at about 1 dB lower than YouTube, and several dBs below the Skype levels in certain conditions. The behavior of Skype is quite interesting, since it requires a cleaner signal than the other services; this difference is clearer in areas with mid signal strength, in which other services can operate with 3–4 dB higher interference levels than Skype can endure.

This information is particularly interesting to operators when planning the network, or during troubleshooting efforts. Typically, operators can get a relatively good estimate of the RF conditions of specific locations; however, it is much harder to understand what those mean in terms of customer experience. By using simple methods like the one described here the operator can get a feeling for what services are available at which locations.

13.5.4   Effect of TCP Performance

The TCP protocol is the most popular transport protocol used in the IP stack. This protocol ensures lossless, in-order delivery of data between two endpoints, and incorporates error recovery mechanisms, as well as congestion avoidance, to limit the packet loss in periods of heavy load conditions in the link. The TCP protocol was not originally designed for wireless environments, characterized by long and variable packet delays, and if not properly optimized can limit the potential offered by the HSPA+ technology.

To ensure correct packet delivery, TCP requires sending packet acknowledgements from the receiver, which also indicates to the sender the available buffer size in the receiver, called the "receive window." Similarly, the sender keeps a "transmission window" also known as "congestion window" on its side until the packets have been properly acknowledged, in case they need to be retransmitted. In networks with high latency, a small transmission window can result in a slower speed than the actual link could provide, since packets can take a long time to be acknowledged by the receiver, thus resulting in the queuing of packets in the transmitter. To help avoid this situation, the TCP parameters (in particular, the TCP window size of the receiver) should be adjusted according to the "bandwidth-delay product," which indicates the effective amount of bytes that the network is able to transmit at any given time. Some OS/TCP implementations, such as Windows Vista and Windows 8, have a built-in TCP windows autotuning function which adjusts the TCP window size based on the end-to-end TCP flow and RTT. This approach has positive aspects, but it can also have side effects: such algorithms do not take the physical medium into account and could conflict with some radio resource scheduling algorithms which try to optimize the data throughput.

Another important effect introduced by TCP is the adaptation of the transmission window to help mitigate congestion situations. To achieve this, the protocol includes two different mechanisms: a probing mechanism called "TCP slow start" by which the transmission window is increased as the initial set of packets get proper acknowledgment, and a congestion avoidance mechanism that adjusts the transmission window when packet losses are detected. The TCP slow start is used at the beginning of the transmission and after a packet has timed out. Figure 13.22 shows how the mechanism works during the FTP download of a file in a HSPA+ network.

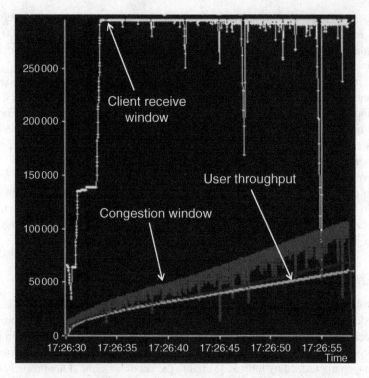

Figure 13.22 Illustration of TCP slow start during FTP downlink transfer

Considering the latency in typical HSPA+ networks, TCP slow start will limit the actual bitrates, especially in the case of transmission of small objects in which there is not sufficient time to increase the window size before the file is transferred.

The TCP congestion avoidance mechanism, which is triggered when packet losses or timeouts are detected, can also introduce unwanted effects in wireless. Packet losses, and especially packet timeouts, are frequent in wireless environments, in which small fluctuations of signal or load can cause significant delay variations on the link. When these occur, the TCP protocol responds by reducing the bitrate, however the transient effect may already be gone and the connection speed will suffer unnecessarily.

In the last few years there has been a significant effort to optimize the performance of TCP over wireless, and multiple TCP versions have been developed: TCP Vegas, New Reno, Hybla, BIC, Cubic, and Westwood, among others. In addition to these protocol implementations, many operators have installed wireless proxies that implement proprietary TCP versions specially designed for wireless. Given that operators do not have control over the TCP stack in the Internet servers, the use of wireless proxy is a practical approach to mitigate the challenges from TCP. Furthermore, these proxies have shown better performance in practice than the aforementioned protocols, as the next analysis illustrates.

Figure 13.23 summarizes a study performed in a live HSPA+ network with dual carrier. The test consisted in a single download of a HTTP file under good and poor radio conditions, using various versions of the TCP protocol: BIC, Cubic, Hybla, and FIT. A wireless proxy, implementing a proprietary version of TCP, was also tested as one additional test scenario.

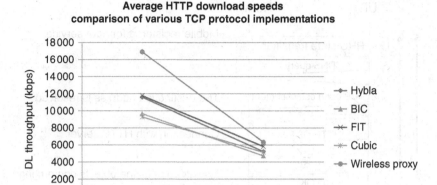

Figure 13.23 Performance comparison of different TCP protocols in a live HSPA+ network

As the results show, the wireless proxy provides the best results in both test scenarios, with a wide difference in good RF conditions and only marginal improvement in poor RF. Additional tests were conducted in a congested area, in which the proxy also offered the best performance, only with a marginal gain. Another interesting TCP protocol is TCP Westwood, which showed good potential in other tests conducted, beating the proxy performance in poor radio conditions.

In addition to the TCP version, an important parameter to adjust is the maximum TCP window size in the terminal device. As discussed before, this window can effectively limit the amount of data that is transmitted to the device, wasting the potential of networks with higher speeds. Figure 13.24 shows the result of a test campaign in a city equipped with dual-carrier HSPA+, in which three different window sizes were tested. The tests were performed in 35 different locations to ensure that they covered many possible combinations of load and signal strength.

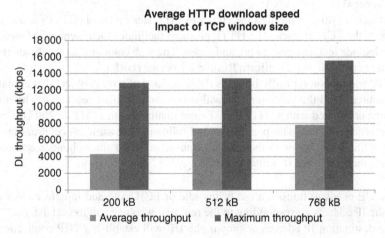

Figure 13.24 Impact of TCP window size on HTTP downloads

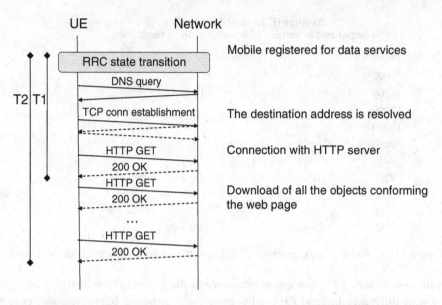

Figure 13.25   Diagram of a typical web content download in a smartphone

In this particular example, the results suggest that the higher window sizes work better for dual-carrier devices. Beyond a certain point (512 kB) the performance differences were not significant, especially in terms of average throughput.

13.5.5   Web Browsing Performance

Figure 13.2b in Section 13.2 illustrated how HTTP traffic accounts for the majority of the data transactions in the network, either directly through the browser or through a specialized app. It is therefore very important to understand the behavior of these services and their performance in a wireless network.

Web services are offered through the Hypertext Transfer Protocol (HTTP) that is typically built on top of the TCP/IP stack. The HTTP protocol facilitates the download of web content, which may include text, images, audio, and video. The web content is uniquely identified and placed in the network using a Uniform Resource Locator (URL).

The HTTP transaction typically involves the download of a main object that contains links to other resources, and the subsequent download of each of those resources. The transfer of each resource is initiated with a "HTTP GET" and finalized with a "HTTP 200 OK" message. Each of the objects on the main page download following the same procedure. Figure 13.25 shows a simplified diagram with the relevant transactions in a smartphone, assuming that in these devices the PDP context is always on:

- First, the UE needs to transition out of the idle or PCH state and initiate a DNS query to resolve the IP address corresponding to the required web page address (URL).
- Once the destination IP address is known, the UE will establish a TCP connection, which is performed using a three-way handshake: the UE sends a "connection request" packet

(SYN), the server responds with a "connection response" (SYN-ACK) and the UE confirms that the server's response has been received (ACK).

- After the TCP connection is established, the UE requests a web page object (HTTP GET) and the network sends multiple TCP packets with the relevant information. When the full web page is sent, the server sends a "web page complete" message (HTTP 200 OK).
- The multiple objects indexed in the original web page will also be downloaded following the same HTTP GET/200 OK scheme. The download of these objects can be sequential (in older versions of HTTP), or in parallel if the browser and server supports pipelining (HTTP 1.1).

Figure 13.25 highlights two instances that are relevant to the consumer, a first time (T1), when the first object is downloaded – typically corresponding to the browser indicating some action in the screen – and T2, when the web page is completely loaded, typically corresponding to the time the browser has finished rendering the full page on the screen.

In order to provide a satisfying user experience, the different actions that take place should happen relatively quickly. The study in [2] provides some high-level guidelines that could be applied to web browsing response times:

- Provide some feedback to the user within 2 s. By this time, the browser should have indicated that an action is occurring
- The customer will perceive a good service experience if the task takes less than 5 s. If it takes longer than that, the browser should provide some feedback to let the user know that the system is still at work
- Beyond 10 s, even when there's feedback, users' patience wears thin and the system is perceived as slow.

Considering these guidelines, it is very important to ensure that the network and device are well tuned to provide a responsive web browsing experience. The main challenges to achieve this in HSPA+ are both the *connection latency*, which is the time to transition from IDLE/PCH to DCH, and the *connected latency*, or the packet round-trip delay once in DCH state.

The connected latency affects the speed at which packets are acknowledged in the network: the lower the latency, the faster the TCP ACK process, which results in an overall faster download time. Interestingly, in many high speed wireless networks the packet latency can become the bottleneck for interactive services such as web browsing, rather than the actual link speed.

The connection latency plays a major role in the perceived web experience. When the phone is in idle or PCH state, the system needs to spend some time to transition into the high speed mode, which is only possible in the DCH state unless the E-FACH/RACH feature is active. As web browsing transactions are initiated by UE, the very first packet is sent in the uplink direction. When the UE requests radio resources for uplink transmission, it can set a "traffic volume indication," which is a trigger used for direct transition from PCH to DCH. Since the web browsing session starts with either a DNS query or TCP SYN packet, both of which are small packets, the traffic volume indication is not set. Therefore, transition normally happens in two steps via FACH, which takes a longer time. Chapter 12 provides some measurements of the various transition times, and Figure 13.26 shows the impact on web browsing times on a few example sites.

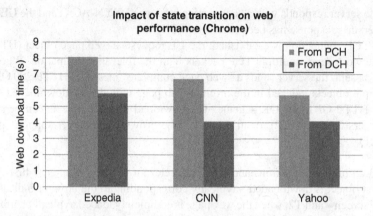

Figure 13.26 Impact of RRC state transition of web performance. Light gray: session starting in PCH; Dark gray: session starting in DCH

As Figure 13.26 shows, the additional delay caused by RRC state transitions can be up to 2 s, often representing a significant amount of the overall web download time. This impact can be minimized using different techniques:

- Extending DCH hold timers, however this has negative impacts on battery life and potentially degradation of network KPIs, such as baseband resource consumption and access failures. The use of CPC should help counter these negative impacts.
- Extending FACH hold timers and activating the E-FACH/E-RACH feature if available.

RNC can furthermore use traffic profiling techniques and predict browsing transactions based on recent user activity, and use direct transition from PCH to DCH even if the UE does not set the traffic volume indicator.

Other important factors affecting the web experience are the performance of the device, the operating system (including the default browser), and, to a lesser extent, the terminal category. Figure 13.27 shows a web download comparison for various popular web sites across multiple mobile operating systems: Android 4, iOS 5, and Windows Phone 8. Note that, at the time of the tests, the iPhone could not be tested with dual carrier in the trial network.

The web download results show how the operating system has a significant influence on performance, often beyond the impact of the device category. In this example, the iOS device tested didn't support dual carrier, however it offered a better experience than the dual-carrier devices while in good radio conditions. The effect of dual carrier can be noticed in the poorer RF conditions, where throughput becomes a limiting factor. The Windows Phone 8 device is the one offering a better performance when considering both scenarios.

There are various reasons why Android presented a slower response in these tests, probably the most relevant is the fact that the amount of data downloaded with Android was larger than with other OS since different contents were served for different browsers. Also, the default browser included in Android phones (WebKit) is not well optimized for JavaScript, as Figure 13.28 (left) shows. On the other hand, Android OS offers an alternative browser, Chrome, which presents a significantly improved performance, as the same figure on the right illustrates.

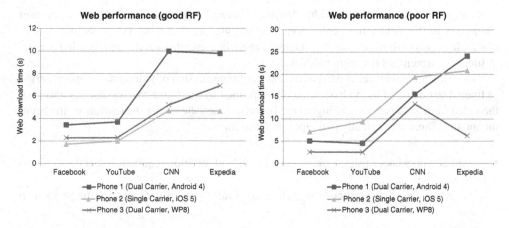

Figure 13.27   Web performance comparison across devices in good and poor RF conditions

In summary, analyzing and optimizing the web experience is foundational to provide an optimum service experience on smartphones, since many of the apps are based on web content. Achieving the best web browsing experience is not just a matter of optimizing the network, or selecting the best modem for the device; there are other factors like the OS and the browser that play a very important role and also need to be carefully analyzed and optimized.

13.5.6   Video Streaming

Streaming services – in particular video streaming – are perhaps the services that were truly enabled with high-speed radio technologies such as HSPA+. These services require a significant bandwidth that older technologies could not offer, at least not consistently across the network coverage area. Furthermore, as discussed in Section 13.2, video streaming services today represent the main source in terms of data volume in networks dominated by smartphone traffic, with increasing relevance in the coming years – as forecasted by Cisco in their Visual Networking Index analysis [3]. As such, this service represents one of the most relevant and, at the same time, challenging traffic type to be served by the HSPA+ network.

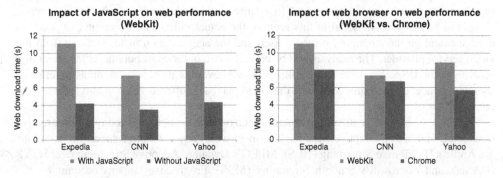

Figure 13.28   Analysis of web performance for Android devices

In addition to the high bandwidth demands, streaming services also present significant challenges to the battery life of the device. While playing streaming video, the device will make use of two of the main sources of battery drain in a smartphone, during extended periods of time: the screen and the radio modem.

Video streaming services consist of the constant transmission of video frames over a period of time. The quality of the video source is determined by the frame rate and the resolution of the video frames: in general, the higher the frame rate, and the higher the resolution, the better the quality. There are typically two types of video frames:

- *p frames*, with a larger frame size, contain the full information required to decode the video image.
- *i frames*, of a smaller size, contain differential information in reference to the previous p frame.

During a video stream, *i frames* are typically sent with a higher frequency than *p frames*; this results in a variable bitrate profile, with instantaneous peaks – corresponding to the transmission of *p frames* – followed by a lower speed transfer of the subsequent *i frames*.

To allow time to decode the frames, and to account for possible fluctuations on the air interface, the receiver client implements a decoding buffer that can introduce several seconds of playback delay. This ensures a smooth video playback compensating for differences in delay arrival between packets or temporary packet loss.

13.5.6.1 Streaming Protocols

The media content can be transferred over TCP or UDP. Transport protocols built on top of UDP, such as RTP, are more suitable for the content of real-time data since they involve less overhead and lower latency. Since these protocols don't provide error protection or sequential delivery, any degradation in the radio layer will be reflected in the video quality, and will result in "blockiness" or even periods of video blackout.

On the other hand, video services that use the TCP protocol will ensure the correct and in-order delivery of all data frames; however, problems at the radio layer will trigger TCP retransmissions or timeouts, which can result in data stalling. With TCP, a temporary data loss will normally be masked to the user thanks to the receiver buffer; however, in areas with poor radio conditions or high load it can result in a pause of video playback or "rebuffering" events.

As will be analyzed in detail in this section, the actual video user experience depends to a large extent on the streaming protocol used and the actual data transfer strategy from the video service provider. The analysis will be based on two of the most popular video streaming services in the USA, YouTube and Netflix, and will cover the most relevant areas in terms of user experience: elapsed time to initiate the video playback, offered video quality, and battery consumption.

Two of the most popular streaming schemes seen on wireless networks today are progressive downloads and HTTP adaptive streams. The most well-known adaptive streaming solutions are Apple's HTTP Live Streaming (HLS), MPEG's Dynamic Adaptive Streaming over HTTP (DASH), and Microsoft's Smooth Streaming (MSS). Progressive streams account for the majority of video volume today, due to the popularity of YouTube; however, this is quickly shifting as even YouTube is changing their content to HTTP adaptive streaming, in particular HLS.

Progressive video streams are generally encoded into different formats (such as 240P, 360P, 480P, 720P, 1080P), but only one quality and bitrate is downloaded at a time. This does not automatically change or adjust based on congestion or the user's available bandwidth.

HTTP Adaptive Streaming (HLS, Dash, MSS) video streams are encoded in many different qualities and are delivered in chunks of 2 to 10 s of video. The video quality is adjusted both up and down as conditions or available bandwidth fluctuates.

A common feature of all streaming services is an initial buffering of multimedia content at the client, which tries to ensure smooth playback in the presence of bandwidth fluctuation and jitter. This buffering is visible to the user as start-up delay and referred to as "fast start." The name comes from the fact that this initially buffered data is typically downloaded using all the available bandwidth, while the rest of the video is downloaded using one of these techniques according to the study in [4]:

- encoding rate streaming;
- throttling;
- buffer adaptive streaming;
- rate adaptive streaming;
- fast caching.

Encoding rate streaming means that after the fast start the client is able to receive data at the encoding rate. For example, YouTube uses the encoding rate technique for HD video in the case of Flash player in Android.

In the case of throttling, the server limits the rate. Throttling can be used, for example, with Flash players on Android or the YouTube app in iOS. Different throttling factors are used depending on the client.

In buffer adaptive streaming, the client stops downloading when the receive buffer is full and continues when the buffer drains to a certain threshold. The YouTube app and HTML5 player in Android use the buffer adaptive technique.

In rate adaptive streaming, the client can switch dynamically between different video qualities depending on available throughput. HTTP Live Streaming (HLS) is one such technology used by, for example, Netflix.

Fast caching downloads as fast as possible even the whole video. This is used, for example, in the YouTube client in the Windows Phone.

13.5.6.2   Video Start Time

Twenty percent of mobile users will abandon a video if it does not start within 10 s [5]. The faster the network, the higher the abandonment rate, so in order to maintain customer satisfaction, it is important that video start time decreases as mobile network speeds increase.

Video Start Time (VST) is an important KPI, but is difficult (costly) to measure from a carrier perspective due to the complexity of the transaction, and it has to be tailored to the specific service that one wants to be monitored. In this section we will discuss two of the most popular video services: YouTube and Netflix.

A YouTube progressive video stream has between four and eight transactions before the video starts. A delay in any of these transactions can have a significant impact on video start time so it is important to include all the transactions in the video start time metric:

1. DNS query for redirector.c.youtube.com.
2. HTTP request for video from redirector and HTTP 302 redirect to nearest YouTube cache.
3. DNS query for redirected address.
4. HTTP request for video from redirected address and either starts downloading the video or receives another HTTP 302 redirect.
5. Steps 3 and 4 can be repeated multiple times depending on the popularity of the video and the location of the YouTube caches.

At the moment of writing this book, the Android YouTube player required at least 5 s of video in buffer before the video will start playing. While this is the minimum required to start the video, the client can store up to 30 s of video buffered by the time the video starts playing. This is due to how YouTube bursts 1 MB (240P) before throttling the connection to 1.25x the video encoded bitrate.

If, however, the initial burst of video is throttled or is slowly downloaded, the client will start the video as soon as it has 5 s of video in buffer. This situation will often trigger video buffering events as the client does not have enough video in buffer to handle the variations in conditions and video bitrate. If buffering occurs, the video start buffer increases to 8 s for the remainder of the video; however, this value is reset when the next video is started.

Figure 13.29 illustrates the YouTube video start time measured with different devices under good and poor radio conditions. As can be observed, the main factors affecting the YouTube performance are the operating system and the device type (single vs. dual carrier).

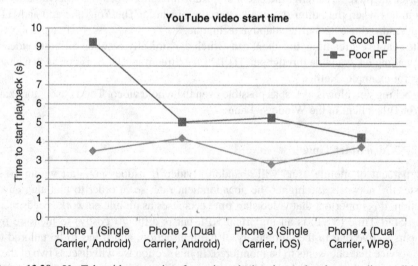

Figure 13.29 YouTube video start time for various devices in good and poor radio conditions

Netflix uses an adaptive stream: it adjusts the quality automatically based on measurements taken by the client while downloading chunks of video data. The delivery mechanism for Netflix varies based on the operating system of the smartphone; this example illustrates the case of an Android client:

1. DNS query for Netflix Content Distribution Network (CDN).
2. HTTPS session where client sends capabilities.
3. HTTPS session where server sends video bitrates and qualities based on #2.
4. DNS query for Netflix CDN.
5. Multiple HTTP sessions for isma and ismv files – isma are audio files and ismv are video files.
6. HTTP session for index file (.bif), which contains image previews used for fast forward and rewind.

The Android Netflix player requires at least 120 s of video in the buffer for the video to start.

13.5.6.3 Impact of Network Proxy on Video Start Time

Network proxies are often introduced to manage and improve data user experience; however, they can have a degrading impact on video performance. Since video services can vary greatly across the same delivery technology, and proxies often apply the same policy per technology, these policies can negatively impact certain users, devices, and sometimes whole services or technologies. Testing for all scenarios to prevent this is not practical with such a broad range of variables so issues often go undetected for extended periods of time.

Two popular ways to manage video from the carrier side are just in time delivery (JIT) and transcoding. These methods can be used together or separately and both aim to reduce overall video volume, but often have unintended tradeoffs.

Just in Time (JIT) delivery throttles the rate at which a video is downloaded to the client so that just enough video is held in the client's buffer to theoretically prevent buffering. Just in Time delivery's main goal is to manage video "wastage," or video downloaded and not consumed, but has tradeoffs such as higher battery consumption, more rebuffering, longer video start time, and less efficient delivery leading to spectral inefficiencies.

An extreme example of this is shown in Figure 13.30. In this example, video start time was delayed up to 30 s for longer videos as the proxy was applying buffer tuning before the video metadata was fully transferred. This caused over 50% of the impacted videos to be abandoned before the videos were able to start. It also caused buffering every few seconds once the video started as the buffer tuned rate did not allow for any variation in video bitrate. This is an extreme example but highlights the challenges faced when trying to apply a blanket policy across technologies.

As revealed by the study in [5], JIT does not play an important role in high-speed networks, therefore its gains may be questionable in modern HSPA+ networks.

Transcoding reduces the video bitrate while attempting to keep a similar perceived quality. There are various techniques to accomplish this, including dropping frames, re-encoding the video into multiple formats, and so on. Transcoding requires significant hardware to process

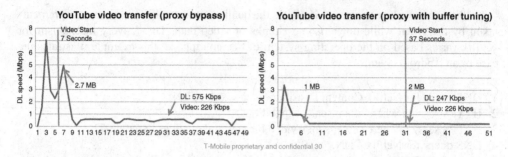

Figure 13.30 User experience degradation with proxy buffer tuning on long videos

the videos and store the content, but can help maintain a decent level of video quality under varying network conditions.

Video services have recognized the need for managing wastage and have implemented client and server buffer management schemes for progressive streams. As mentioned before, many services are also transitioning to an adaptive streaming format, which will capture the benefits of current transcoding techniques. These new schemes of managing wastage, the move to adaptive streams, and the ever increasing speeds of networks have made traditional proxy techniques less relevant for video delivery on HSPA+ networks.

13.5.6.4 Video Quality (MOS)

In addition to video start time, the quality offered during the video playback is a key element to characterize the video service. Video quality is typically measured in terms of video Mean Opinion Score (MOS), and considers items such as frame rate, artifacts due to packet losses, and video interruptions.

One of the challenges associated with video MOS measurements is that, as is the case with voice MOS measurements, the measurement setup requires the transmission of a pre-defined pattern file that will be evaluated by the MOS tool upon reception. This means that video MOS measurements are not possible with on-the-fly content, and the measurements are restricted to spot tests performed by the operator. There are tools that try to estimate video MOS based on real-time measurements from the network, however we won't discuss those in this chapter.

Figure 13.31 illustrates the lab setup configured by T-Mobile's product realization team to perform video quality measurements.

This setup can be used both for video streaming and video calling services. The received video is captured with a camera from the screen of the receiving device, and processed at the video controller device. The test setup provides a controlled environment in which to test various aspects of the video service. Figure 13.32 shows the impact on video MOS from encoding the content with a higher frame rate (left); the chart on the right shows that hardware codecs are able to deliver better performance than software codecs for video calling services.

Another important factor affecting video quality is the transfer rate, which is limited by the radio conditions. Figure 13.33 illustrates the impact on MOS at various test conditions; the chart shows a dramatic video quality degradation beyond a certain point.

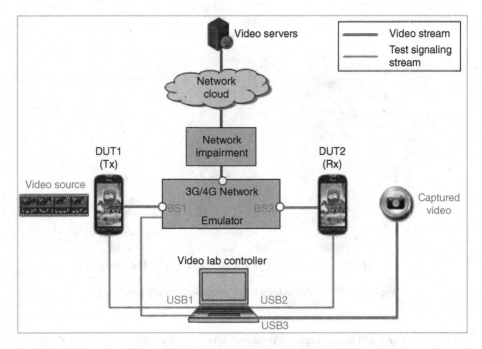

Figure 13.31 Lab setup to measure video MOS

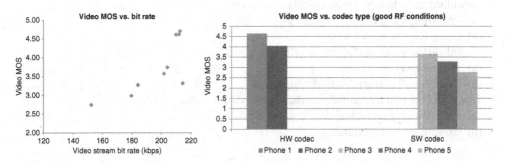

Figure 13.32 Effect of frame rate and codec type on video MOS

13.5.6.5 Streaming Resource Consumption

Until recently, video delivery on mobile networks was treated the same as wired networks – often keeping the radio in an active state for the whole duration of the video consumption. Considering that video is about 50% of network traffic (and growing), developing efficient video applications and delivery is key for both the carrier and user experience: video delivery that is not optimized for mobile networks will increase battery consumption and decrease spectrum efficiencies.

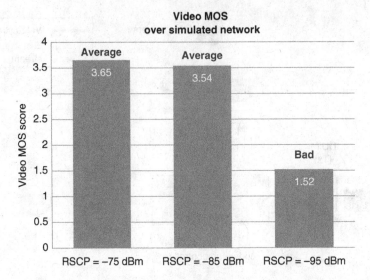

Figure 13.33   Video calling MOS under various test conditions

Each video service and technology has their own unique way of delivering video. Progressive streams are delivered in three basic ways – each with their own benefits and tradeoffs:

1. Download the whole video as quickly as possible
    (a) Pros: Battery, spectrum, signaling
    (b) Cons: High potential wastage.
2. Download the video in chunks every few minutes
    (a) Pros: Battery, spectrum
    (b) Cons: Small potential wastage, signaling.
3. Download the video at slightly above the encoded bit rate.
    (a) Pros: Wastage, signaling
    (b) Cons: Battery, spectrum.

Adaptive streaming (e.g., HLS or MSS) downloads each chunk as fast as possible, that is, servers do not throttle the download rate. This is the case, for example, with Netflix (HLS for iOS and MSS for Windows Phone). This results in efficient battery and spectrum usage, while eliminating wastage and ensuring the best video quality given the conditions. Adaptive streams may or may not save on signaling – depending on the chunk delivery pattern and the carrier's T1 and T2 timers. For example, if the video is delivered every 10 s and the T1 timer is 5 s and T2 timer is 5 s, it will create additional signaling for every chunk of data delivered.

Carriers and video service providers often focus on one item while ignoring the others. This leads to non-efficient video delivery for mobile networks. The key for both is to find the right mix to ensure the proper user experience. A video service provider is less concerned with spectrum efficiencies and signaling, and more concerned with battery usage, wastage, and user

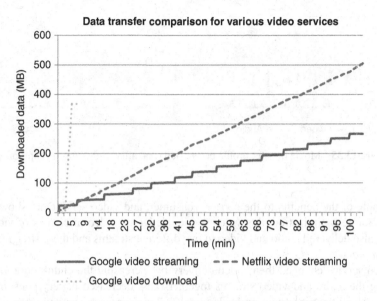

Figure 13.34 Data transfer comparison for different online video services

experience. Luckily, these are related to spectral efficiencies and signaling on the carrier side so there are benefits for both sides from optimizing video delivery.

An example of the above can be seen in Figure 13.34. The same 104-min long movie of standard quality was downloaded in the three ways listed above. Service 1 downloaded chunks of video every 3 to 5 min and consumed 27% of battery on device A. Service 2 downloaded the video throughout the duration of the consumption, which consumed 52% of battery on Device A. By adopting a delivery mechanism like Service 1, the consumer can do more with their device since their battery lasts longer, and the network is able to serve more customers.

Table 13.3 provides the detailed battery consumption for the tested video services.

Adaptive streams may have their video and audio combined into one stream, or they can be delivered in two separate streams. This kind of delivery can have a significant impact on user experience and resource consumption, so there is benefit to both the video service and carrier to working together to optimize delivery.

Table 13.3 Battery drain comparison for two popular video services (Service 1: Download in chunks; Service 2: Continuous download)

Device	Service	Consumption of battery life (%)
Device A	Service 2	52
Device A	Service 1	27
Device B	Service 2	80
Device B	Service 1	35

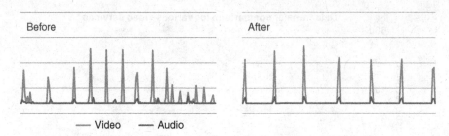

Figure 13.35 Movie transfer profile before (left) and after (right) app was optimized

An example of the benefits to the carrier, consumer, and video service are shown in Figure 13.35. Before optimization, this video service delivered video in chunks of video every 6 to 10 s. It also delivered audio and video in two different streams and these streams were not delivered in sync.

The video service changed their video delivery by increasing the chunk duration to 24 s and syncing the audio and video streams together. This decreased battery consumption by 23%, and reduced the time spent on DCH by 54%. This approach resulted in an increase in radio signaling, but the impact was minimal compared to the battery and spectrum efficiency improvements.

13.6   Use Case Study: Analysis of Smartphone User Experience in the US

In 2012, the US-based research consultancy, Signals Research Group, conducted a large-scale study for a major North American operator to determine the user experience while using typical smartphone applications across a wide range of network technologies and handset/operating system platforms. The tested networks included CDMA2000 EV-DO Rev A, HSPA+, DC-HSDPA, LTE (5 MHz), and LTE (10 MHz) and the operating systems consisted of Android (version 2.3.5 and 4.0), Apple iOS, and Windows 8. For completeness sake, the testing was done in two different major markets and in various locations/times of day in order to gauge the effects of different RF conditions (e.g., loading, high interference, etc.).

Although the study included basic downlink and uplink throughput tests in order to help characterize the performance of the networks as it is traditionally done, most of the testing focused on the measured performance of actual user applications as implemented on the three operating systems and running on the aforementioned network technologies.

The following sample results help demonstrate the point that the typical user experience with a smartphone application depends on many factors that are not necessarily tied to the achievable data speeds of the network. In all figures, the results are a composite across all testing that occurred in the particular market. Further, all times that involved user interaction, such as typing in text or selecting an image, were excluded.

The first figure (Figure 13.36) shows the average downlink throughput that was measured from the various locations and times when the application testing took place. Figure 13.37

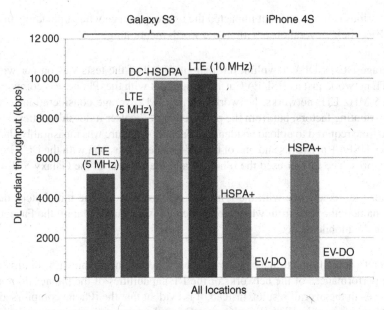

Figure 13.36 Available throughput network technology and operating system. Reproduced with permission of Signals Research Group

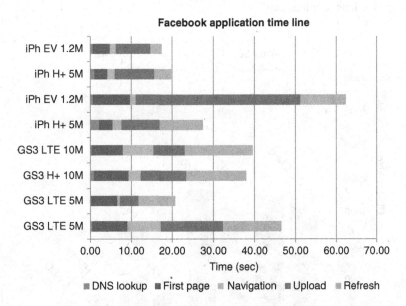

Figure 13.37 Facebook user experience – by network technology and operating system. Reproduced with permission of Signals Research Group

shows the various components that impacted the user experience while uploading an identical picture to Facebook. Several observations are worth mentioning:

- The average DC-HSDPA downlink throughput for all of the tests was on par with the 10 MHz LTE network, just as HSDPA Cat 10 throughput with the iPhone was comparable with the two 5 MHz LTE networks. Network loading and coverage considerations could have been contributing factors, but from the perspective of the user it doesn't matter.
- The total time required to upload an identical Facebook picture was measurably shorter with the legacy HSPA+ networks and one of the EV-DO networks than with the DC-HSDPA and LTE networks. The former used the iPhone 4 and the latter used the Galaxy S III (Android Version 4.0).
- The better user experience with the iPhone 4 also extended to the initial steps during the process, namely the time from when the application was launched until the Facebook page loaded on the mobile device.

The "secret" behind the stronger performance of the iPhone is shown in Figure 13.38. In reality, the performances of the networks or the RF capabilities of the phones do not explain the differences in the overall results. Instead, it is evident that the iPhone compressed the size of the picture before uploading it to the site. From the perspective of the user, the picture quality looks the same when viewed from the screen of the mobile device. However, the user also perceives that the network was "faster," even though it wasn't. The figure also shows that the iPhone on the second EV-DO network didn't compress the picture and this phenomenon also explains the longer upload time.

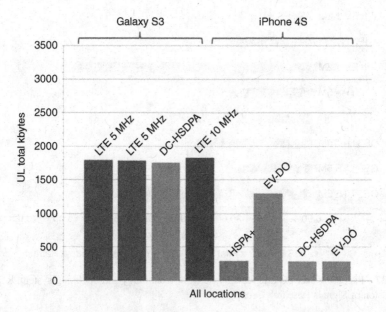

Figure 13.38 Facebook picture upload payload – by network technology and operating system. Reproduced with permission of Signals Research Group

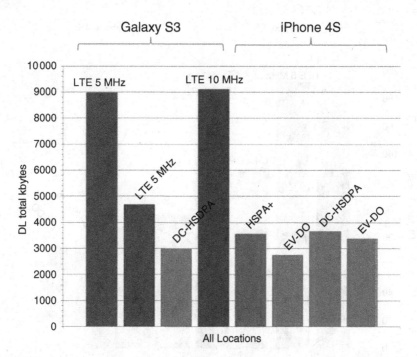

Figure 13.39 YouTube video content vs. network technology – by network technology and operating system. Reproduced with permission of Signals Research Group

It is also worth pointing out that the actual time required to upload the picture didn't necessarily impact the user experience. This situation exists because the picture is uploaded in the background, meaning that the user could switch to another application without having to wait for the video to upload. Facebook does indicate when the picture or video is uploaded and the screen refreshes to reveal the uploaded content, but the user doesn't have to wait for these events to occur.

The study also found differences in the amount of transferred payload associated with downloading and playing an identical YouTube video and loading an identical web site. In the case of YouTube (Figure 13.39), there seemed to be a strong relationship between the capabilities of the network and the quality of the video. With LTE networks, YouTube delivered a higher quality video format, or at least the file size was larger, than it did with the other 3G networks. In the case of one 5 MHz LTE network, the use of a third-party optimization platform could have reduced the video payload. From the perspective of the individuals who conducted the tests, there wasn't any perceived difference in the quality of the videos.

In the case of web page content, Apple's decision to not support Flash in its browser actually had at least one positive impact on the user experience. As shown in Figure 13.40, the size of the downloaded web pages was lower with iOS than it was with Android. Although the web pages still generally loaded faster on the Galaxy S III devices that operated on the faster networks, the differences would have been even greater if the iPhone had downloaded the same amount of web content.

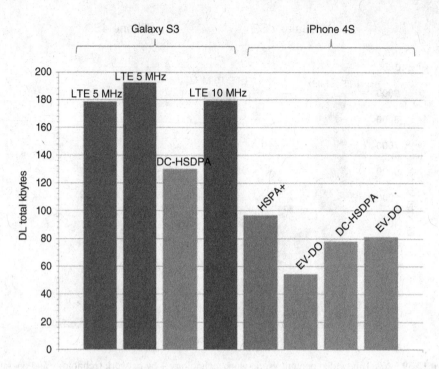

Figure 13.40 Web page data content – by network technology and operating system. Reproduced with permission of Signals Research Group

Generally, the iPhone (iOS) delivered a better user experience than the Android-based smartphones, or the user experience was at least comparable to the Android-based smartphones, despite the latter devices using much faster networks (i.e., legacy 3G versus DC-HSDPA and LTE). This situation, however, didn't pan out with all applications. For example, the study found that the Google voice search application was nearly ten times faster than Apple's Siri when it came to loading the first page of search results – the performance of both applications was still deemed very compelling. Interestingly, the Google search function also used eight times more data traffic, in part due to the incorporation of more graphics and overall content in the results.

13.7 Summary

Below are some recommended steps towards ensuring an optimum user experience in HSPA+ networks:

- *Reeducate consumers on what really matters.* Although easier said than done, operators frequently place too much emphasis on advertising the data speeds that their networks deliver. Instead, they should market their service offering based on the typical user experience that it delivers. This approach will also prove to be beneficial once newly deployed LTE networks experience network loading and previously marketed data speeds are no longer achieved on a consistent basis.

- *Identify non-optimized applications and take corrective action.* Operators need to invest time and energy into studying how applications behave in their network. In many cases, simple fixes to the application are all that is required to transform a poor performing application into a fully optimized application.
- *Mandate new user experience KPIs to ecosystem device and OS partners.* Device and OS partners have a vested interest in delivering a good user experience, but they do not always understand what it takes to deliver a compelling user experience that makes the most efficient use of the available network resources.
- *Minimize poor RF coverage areas.* Leveraging the benefits of SON is perhaps the best way to identify poor coverage areas. In this case "poor coverage" should equate to a minimum user experience threshold not being achieved and not necessarily to a specific RF parameter, although the two could be correlated.
- *Optimize the network parameter settings and the network architecture based on typical user behavior.* Moving popular content closer to the edge of the cell is one example. Adjusting network timer settings (e.g., T1, T2, and T3) in order to minimize RRC state transition times without sacrificing battery life is another example.

References

[1] Sharma, C. (2013), "US Wireless Data Market Update.Q4 2013 and 2013".
[2] Seow, S. (2008) *Designing and Engineering Time*, Pearson Education, Inc., Boston, MA.
[3] Cisco (2012), "Cisco Visual Networking Index: Global Mobile Data Traffic Forecast Update, 2012–2017".
[4] Hoque, M., Siekkinen, M., Nurminen, J. Aalto, M. and Tarkoma, S. (2013) "Mobile Multimedia Streaming Tecniques: QoE and Energy Consumption Perspective", arXiv:1311.4317.
[5] Shunmuga Krishnan, S. and Sitaraman, R. (2012) "Video stream quality impacts viewer behavior: inferring causality using quasi-experimental designs", Proceedings of the 2012 ACM conference on Internet measurement conference (IMC).

14

Multimode Multiband Terminal Design Challenges

Jean-Marc Lemenager, Luigi Di Capua, Victor Wilkerson,
Mikaël Guenais, Thierry Meslet, and Laurent Noël

The mobile phone industry has experienced unprecedented growth since it was first launched over 20 years ago using 2G/ EDGE, GSM, GPRS (EGPRS) technologies. Combined with 3G standards, such as WCDMA/HSPA, and more recently, 4G/LTE (FDD and TDD), mobile phone shipments have continuously beaten records, tripling volumes from 533 million devices in 2003 to more than 1.8 billion devices in 2013. Looking ahead to 2017, it is expected that shipments of mobile phones will then reach 2.1 billion units worldwide [1]. Put into perspective, this means that nearly 70 devices will be shipped every second! Beyond these impressive production volumes, the introduction of smartphones in 2007, unleashed by Apple's original iPhone, is the other major factor that has completely reshaped the definition of mobile terminals. A high-end non touch-screen 2009 device would today be qualified as "feature phone" or "voice-centric phone." The year 2013 was a key turning point in the history of this industry as the "data-centric / do-it-all" smartphone shipments exceeded the volumes for these feature phones. Based on current trends, what are now collectively called "smartphones" are likely to transition to a market segmented into three retail price ranges: low end/cost (<100$), mid-end (<300$), and high-end super-phones (>500$). Given this expectation, it will become difficult to make a distinction between these segments, based on a common classification as "smartphones." This is not unlike the trend seen in the world of PCs. In the high-end segment, the differentiator is likely to translate into a race to be the first to deliver new hardware (HW) or telecom features, such as being the first device to support LTE-Advanced. In the lower-end segment, retail prices will be the primary differentiator.

The other key factor which has influenced the situation is the quick pace at which 3GPP has been delivering new telecom features and air-interfaces, moving from the well established mono-mode EGPRS terminal, then to dual-mode EGPRS-WCDMA, and more recently to

HSPA+ Evolution to Release 12: Performance and Optimization, First Edition.
Edited by Harri Holma, Antti Toskala, and Pablo Tapia.
© 2014 John Wiley & Sons, Ltd. Published 2014 by John Wiley & Sons, Ltd.

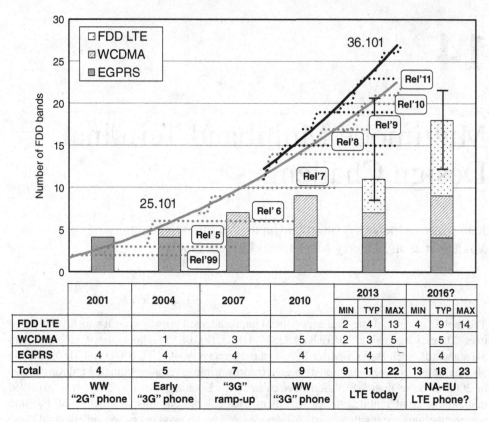

Figure 14.1 Number of frequency bands for WCDMA (TS25.101 -UTRA) and LTE (TS36.101 – E-UTRA) air interfaces vs. typical commercial UE band support per air interface (bar graph)

the triple-mode EGPRS/WCDMA/LTE user equipment (UE). Some of the most recent smartphones also have to support a fourth air interface: TD-LTE. Commensurate with support for this evolution comes the necessity to support more frequency bands. The rate at which bands have been introduced by 3GPP is illustrated in Figure 14.1, which plots the number of bands vs. the initial introductory years for LTE-FDD (36.101) and WCDMA/HSPA (25.101). With the introduction of LTE, the number of bands standardized at 3GPP as of October 2013 is 29 FDD bands and 12 TDD bands (TDD bands are not shown in Figure 14.1). This trend is not expected to abate as the UARFCN formulae are modified to accommodate up to 128 bands in future releases [2]. The table inserted in Figure 14.1 shows that the number of bands supported in mobile phones has followed a similar trend. From 2003 to 2006–2007, dual-mode terminals only needed to support a small number of bands: band I for most networks, plus band II and V in North America. Today, worldwide roaming is achieved in dual-mode terminals by supporting a total of nine bands: four EGPRS (850,900,1800,1900) and five WCDMA bands (I, II, IV, V, VIII).

Figure 14.2 is a depiction of the most common bands deployed worldwide, and helps to better illustrate the complexity of multiple band support for multimode terminals. The map is complemented with a short table listing the band support requirements for three hypothetical geographical regions: worldwide (WW), North America – Europe combined (NA-EU), and EU

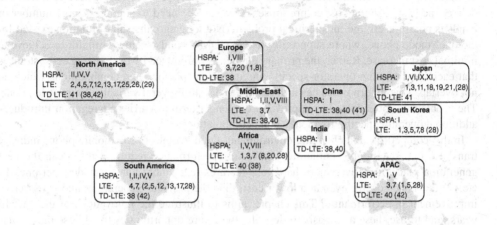

Region, Modes	EGPRS	WCDMA	LTE	Total	Number of co-banded + unique bands*
WW, "quad"-mode	850,900,1800,1900	I,II,IV,V,VIII	FDD: 1,2,3,4,5,7,8,11,13,17,18,19,20,21,25(26,28) TDD: 38,39,40 (41,42)	27	6 + 12 = 18 bands
NA-EU, triple-mode	850,900,1800,1900	I,II,IV,V,VIII	FDD: 2,3,4,5,7,13,17,20,25 (1,8,26)	18	5 + 6 = 11 bands
EU, triple-mode	850,900,1800,1900	I, VIII	FDD: 3,7,20	9	3 + 4 = 7 bands
WW, dual-mode	850,900,1800,1900	I,II,IV,V,VIII	-	9	3 + 3 = 6 bands

Figure 14.2   Main commercial frequency bands and band requirements for multimode terminals. Future LTE commercial bands are shown in brackets. The number bands that can be shared amongst RAT or "co-banded" are underlined

only triple-mode terminals. Global WW dual-mode requirements are provided as a baseline for comparison purposes. These examples highlight the impact of adding LTE support: while 9 bands is all that is required to cover the world in a dual-mode terminal, the triple-mode UE needs to support 24 bands (20 FDD, 4 EGPRS); and this becomes 27 if TD-LTE is to be supported. With 9 bands, the triple-mode terminal can only cover the EU market. Remarkably, the number of bands required to provide just NA-EU coverage in a triple mode terminal[1] is double that of a dual-mode WW phone: 18 bands vs. 9 bands for dual-mode handset. Of course, this situation is somewhat offset by the fact that doubling the number of bands does not necessarily translate into doubling the number of power amplifiers and/or RF filters, thanks to the concept of co-banding[2]. However, even with co-banding, the explosion in the number of bands supported imposes some new constraints. Today, the triple-mode WW "flagship" terminal must be realized using several variants, each covering a particular region of the

---

[1] Reference [15] is one of the first PTCRB certified triple-mode terminals, which nearly covers all of the estimated required bands for EU-NA. It supports a total of 17 bands. The recent tablet [53] is the product that comes closest to the listed WW triple-mode requirements with a total of 22 bands.

[2] Co-banding consists in re-using the RF circuitry common to multiple modes/air-interfaces for a given band of operation, thereby minimizing the duplication of front-end components such as filters and/or power amplifiers. Savings in RF HW circuitry thanks to co-banding is discussed in Section 14.1.3.1.

world/ecosystem. Even dual-mode terminals are frequently produced using several variants to deliver the best roaming/cost compromise. However, the need to support a large number of bands is not, in itself, the most critical problem. Not long ago, the same situation existed for the dual-mode phone, where supporting five WCDMA bands seemed difficult to achieve at minimal cost increase. Rather, the primary challenge introduced by the existing ecosystems is that each have telecom operator-specific and/or region specific band combinations which do not overlap. This in turn translates into having to develop several variants of a mobile phone. The problem is not expected to improve in the longer term as carrier aggregation introduces additional operator specific frequency band combinations.

In this competitive and changing environment, the dynamics of the mobile phone industry translates into seemingly conflicting cost/feature/performance requirements – can the next generation platform deliver more hardware features (e.g., more bands, modes, peripherals, etc.) at the same cost, or even at a lower cost, than the previous generation, and yet offer an improvement in performance? This chapter aims to illustrate the innovations of the past 10 years, and to use these as a basis to describe the future opportunities to address these ever-increasing challenges. A discussion on the tradeoffs between cost and performance resulting from these requirements is presented using two metrics: those of component count and PCB footprint area for cost constraints, and power consumption for performance needs. This chapter invites readers to take a journey to the center of a multimode multiband terminal through the selected subsystems of Figure 14.3. Each subsection provides a focused view of that particular area or function. Section 14.1 covers design tradeoffs within the constraints of cost reduction. Section 14.2 presents techniques and challenges experienced in delivering the optimum power consumption with a focus on two key contributors: the application engine and the cellular power amplifier. Many other key aspects encountered in modern terminal designs deserve attention, but cannot be covered within the scope of this short overview.

This chapter aims to show that, in this highly competitive and complex ecosystem, the key to winning designs is to deliver easily reconfigurable, highly integrated hardware solutions. In this respect, LTE is included in most subsections, since it becomes less and less cost-effective to deliver dual-mode-only chipsets.

14.1 Cost Reduction in Multimode Multiband Terminals

14.1.1 Evolution of Silicon Area and Component Count

The trend in mobile phone hardware cost metrics is illustrated in Figure 14.4, by plotting component footprint area and total component count vs. year of introduction. The graph is generated with data extracted from the teardown reports of 69 mono-mode, 48 dual-mode, and 7 triple-mode handsets [3]. While the total number of components in EGPRS has remained virtually constant from 1996 to 2005, the required area has been reduced considerably over that time – from 14 cm^2 (1998) to reach a minimum at approximately 5 cm^2 (2006–present). This represents a reduction of about 65%. The maturity plateau has been reached thanks to highly integrated single chip solutions in which four key components are integrated into a single die: RF transceiver, baseband modem, power management unit, and multimedia processor. It took dual-mode handsets only four to five years to reach the level of complexity that took nearly eight years to reach in the GSM realm. Due to the low number of teardown reports available at the time of printing, it is difficult to make an accurate assessment of the trend in triple-mode devices. The metrics tend to show that the introduction of LTE has not significantly increased

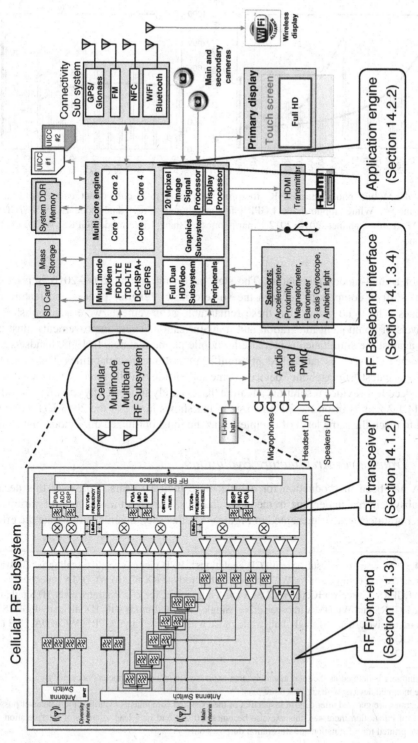

Figure 14.3 Generic multimode multiband terminal top level block diagram

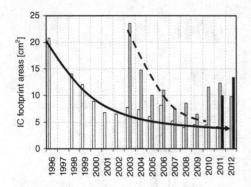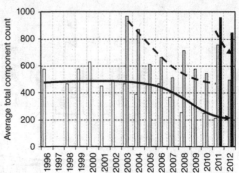

Figure 14.4 Mobile phone stacked IC footprint area[3] and total component count vs. year vs. supported modes. White: mono-mode EGPRS feature phones, Gray: dual-mode EGPRS-WCDMA (2003–2009 "feature" phones, 2010–2012 "smartphones"), Black: triple-mode EGPRS, WCDMA, LTE "smartphones"

the cost of a dual-mode smartphone. The step occurring around 2009–2010 is primarily due to the shift in complexity between the voice centric "feature" phones and feature rich smartphones. The trend shows that these remarkable achievements were accomplished by higher levels of component integration and miniaturization. These improvements must also be offset against the simultaneous increase in mobile phone features. In 1998, handsets used monochrome LCD screens and lacked any significant multimedia capabilities. This is in stark contrast to the latest "super-phone" devices, some supporting full HD screens, which will soon be able to decode 4 K video resolution, a feature that is barely supported by cable TV decoders. Section 14.1.2 zooms in on the RF subsystem and shows how fast the 2013 LTE solutions managed to reach the same level of complexity as the most optimized dual-mode platforms.

14.1.2   Transceiver Architecture Evolutions

WCDMA has been in production for a little more than 10 years. But, over this decade, the RF subsystem has undergone tremendous changes. Through the selection of various RF transceiver architectures, as presented in Figure 14.5, this section aims to illustrate the efforts

Figure 14.5   10 years of RF subsystem PCB evolutions[4]. PCB pictures relative scaling is adjusted to illustrate relative size comparison (a) WCDMA super-heterodyne RX IC, (b) WCDMA super heterodyne TX IC, (c) EGPRS transceiver IC, (d) WCDMA DCA transceiver, (e) EGPRS transceiver, (f) Single-chip dual-mode EGPRS and WCDMA transceiver, (g) Single-chip dual-mode with RX diversity EGPRS and WCDMA transceiver, (h and i) single-chip triple-mode with RX diversity EGPRS, WCMDA, LTE FDD (and GPS RX)

---

[3] IC footprint area is defined as the total assembly area expressed in square centimeters occupied by the packaged and surface mounted devices with eight or more pins.
[4] Block diagrams are guessed after careful inspection of the PCB and components identification whenever possible. As the level of integration increases, this exercise becomes difficult and may lead to errors of interpretation. The diagrams are printed for information and illustration purposes only.

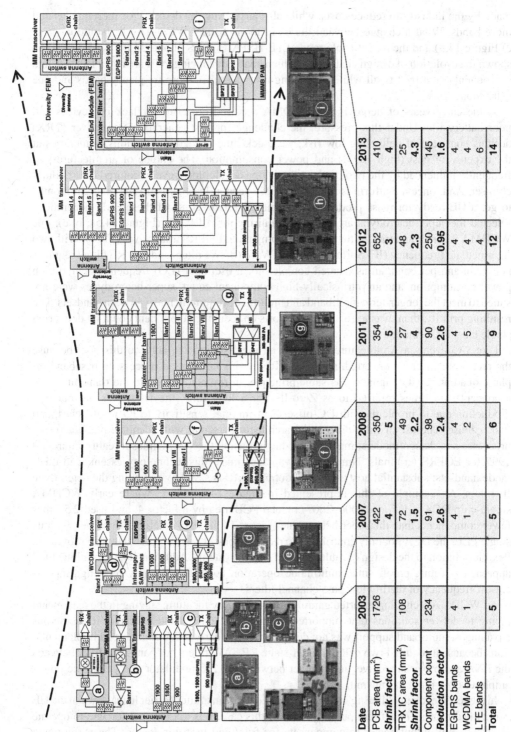

Date	2003	2007	2008	2011	2012	2013
PCB area (mm²)	1726	422	350	354	652	410
Shrink factor		4	5	5	3	4
TRX IC area (mm²)	108	72	49	27	48	25
Shrink factor		1.5	2.2	4	2.3	4.3
Component count	234	91	98	90	250	145
Reduction factor		2.6	2.4	2.6	0.95	1.6
EGPRS bands	4	4	4	4	4	4
WCDMA bands	1	1	2	5	4	4
LTE bands					4	6
Total	5	5	6	9	12	14

Figure 14.5

made by the industry to reduce costs, while also answering the demand for more modes and more bands. What a change between the early mono band I WCDMA UE (solution (a, b, c) Figure 14.5) and the recent triple mode LTE-HSPA-EGPRS phone. In this section will be shown the evolution of design challenges, which enabled reducing PCB area by a factor 4, and component count by 1.6, all while supporting nearly 3 times more frequency bands and one extra mode: LTE.

In the early years of deployment, WCDMA received criticism for lack of devices and poor battery life. Under this pressure, the challenge for first-generation transceiver (TRX) designs consisted of delivering low risk, yet quick time-to-market, solutions, sometimes at the expense of cost (high BOM) and power consumption. The choice of architecture was frequently influenced by the two- to three-year-long development cycle required for a cellular chip set. And, once a platform is released, OEMs need at least an extra five to eight months to get a UE ready for mass production. To minimize risk, some of these early generations selected the super-heterodyne architecture, requiring three RF ICs (Figure 14.5a, b, c): one WCDMA receiver and transmitter IC, and a companion EGPRS transceiver (TRX). With two Intermediate Frequency (IF) Surface Acoustic Wave (SAW) filters and two RF filters required per chain and per band, an associated complex local oscillator (LO) frequency plan, a high power consumption, and an intrinsically-high component count, super-heterodynes were not suited to meet the longer-term multiband and multimode requirements of modern handsets. The resulting priority, then, was to focus R&D efforts on reducing component count by delivering single chip TRX solutions.

Direct Conversion Architecture (DCA) is an ideal solution to achieve this goal because the received carrier is converted, in a single RF mixing conversion process, to baseband in-phase/quadrature (IQ) signals. The same principle, in reverse, applies to the transmit chain, a concept sometimes referred to as Zero-IF. For a long time, the DCA was not put into production due to problems with DC-offsets/self-mixing, sensitivity to IQ mismatches, and flicker noise. The use of fully-differential architectures, along with adequate manufacturing processes and design models, made IC production using these principles a reality in around 1995 for EGPRS terminals. Numerous advantages make this technology essential in dual-mode handsets: substantial cost and PCB footprint area is saved by removing the heterodyne IF filters. An example of this is presented in Figure 14.5d and e, where early WCDMA single-chip DCA solutions (2006–2007) reduce PCB area by a factor 4, and used 2.5 times fewer components than the super-heterodyne (Figure 14.5a, b, c) solution. Moreover, with its single PLL, the DCA not only contributes to reducing power consumption, but also provides a flexible solution to the design of multiple band RF subsystems. Additionally, this architectural approach is future-proof, since multimode operation consists in simply reconfiguring the cut-off frequency of the I/Q low pass channel filters.

As WCDMA technology started gaining commercial momentum, staying in the lead meant racing to deliver solutions that supported multiple WCDMA bands. Despite its numerous advantages, multiband support was not cost-efficient for early DCA commercial implementations because it required two external inter-stage RF SAW filters per band: one filter between the LNA and mixer in receive, and another between the TX modulator output and the power amplifier (PA) input in transmit.

In the receiver chain, filtering is required to reject the transmitter carrier leakage when the UE is operated near its maximum output power. In this case, it is not unusual for PAs to operate at 26-27 dBm output power to compensate for front-end insertion losses. The requirement to support more bands, and in the case of LTE the need to support more complex carrier aggregation scenarios, results in increasing RF front-end insertion losses. This, in turn, means that the PA maximum output power requirements may exceed 27 dBm in certain platforms. Because duplexer isolation is finite, TX carrier leakage can reach the LNA input port with a power as high as −25 dBm. In the presence of this "jammer," the mixer generates enough baseband products through second order distortion[5] to degrade sensitivity, thereby partially contributing to a phenomenon known as UE self-desensitization, or "desense" [4]. In this case, and depending on system dimensioning assumptions, RX SAW-less operation sets a stringent mixer IIP2 requirement on the order of 48 to 53 dBm (LNA input-referred) [4].

In the transmitter chain, further UE self-desense occurs due to TRX and PA noise emissions generated at the duplex distance. Depending on duplexer isolation, SAW-less operation sets TX noise requirements in the range of −160 dBc [4]. Meeting this requirement in bands with short duplex distances, such as band 13 or 17, constitutes a serious challenge to both TRX and duplexer designers as performance may become dominated by PA ACLR emissions rather than the far out of band noise level. The duplexer high-Q factor often pushes designers to trade isolation against insertion loss. In these bands, 3GPP has introduced relaxations to ease factors related to these design constraints [5].

As the number of bands to be supported was increasing rapidly, the race towards continuous cost reduction called for novel TRX architectures that would enable both TX and RX SAW-less operation. For example, removing the interstage filters in a penta-band RF subsystem may save the cost of a PA and the area of almost an entire RF transceiver[6]. SAW-less operation also means that RF transmitter ports can be shared between several RF bands. For example, a single low-band TX port could be used to support all UHF bands, for example, LTE bands 8, 13, 17, and 20. Similarly, a high band port could be shared to support either WCDMA or LTE transmissions in bands 1, 2, 3, 4. The SAW-less TX architecture is therefore the *sine qua non* to enable replacing discrete single-band PAs with a single multiband, multimode (MMMB) PA. The cost savings and design tradeoffs associated with the use of MMMB PAs are presented in Section 14.1.3.2 using a North American variant. Initial RX SAW-less commercial solutions appeared around 2008 (Figure 14.5f). Some of the techniques developed to meet the high IIP2 requirements range from careful design of fully differential structures, use of passive mixers or trap filters to reject TX leakage, as well as the integration of IIP2 calibration circuitry.

In contrast, TX SAW-less operation is more challenging and took a longer time to emerge since every block of the transmitter chain impacts the overall performance. The efforts required to meet the stringent −160 dBc requirement are illustrated in Figure 14.6a. This graph plots UE self-desense in a SAW-less application using the reported band I TRX noise performance[7]. With a desense less than a decibel under typical duplexer isolation, TX SAW removal started to become a reality around 2007–2008, and the first handsets using this technique made

---

[5] Through third order distortion, TX leakage mixes with CW tones defined by 3GPP narrowband blocker tests in the LNA. Cases where blockers are injected at half and double duplex distance drive the LNA IIP3 requirements.

[6] Cost saving assumptions: SAW unit price of approximately 0.10 to 0.15$, total cost savings for penta-band estimated at $1 to $1.50. PCB area assumptions: pi-matching network made up of three 0402 passives, SAW filter packages of 1.1×0.9 mm and 1.1×1.4 mm. Area savings per filter: 3 to 4 mm^2 based on [Figure 14.5h] PCB layout, total area saved in a penta-band application: 30 to 40 mm^2.

[7] Assumptions: 24 dBm UE output power, 52 dB duplexer TX to RX carrier rejection, 15 dB LNA gain, 4 dB NF referred to LNA input, 0.2 dB allowed desensitization, 2.8 dB PA to antenna insertion losses, IMD$_2$ rejection by analog and digital filters of 3 dB, and 9 dB DC to IMD$_2$ power ratio.

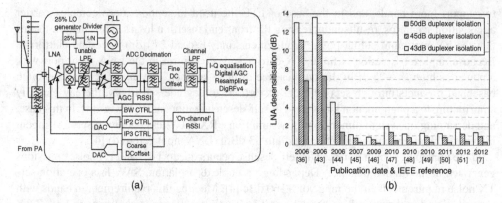

(a)   (b)

Figure 14.6 (a): Modern RF CMOS DCA receiver. Adapted from Xie *et al.* [8], (b): Evolution of LNA desensitization due to TX noise in a hypothetical TX SAW less application (IEEE survey). LNA input referred noise figure of transceiver is assumed to be 3 dB

their appearance a couple of years later (Figure 14.5g). Note that bands with low duplex gap and distance, such as bands 13,17, may still require inter-stage filtering for certain vendors (Figure 14.5i). The primary principle for achieving low noise consists in minimizing the number of cascaded gain stages at RF. One example [6] proposes to carefully balance the 74 dB gain control range between digital, analog IQ, and a bank of parallel RF mixers. In [7] a single pre-amplifier RF stage made up of multiple parallel amplifiers delivers the equivalent of an RF power DAC. Each solution provides a mixture of pros and cons. For example, controlling carrier leakage in an architecture based on a bank of mixers can be a challenge at minimum output power, due to mixer mismatches. The mixers must deliver approximately −65 dBm[8] output power with 15 to 20 dB LO rejection, so keeping LO leakage below −85 dBm while transmitting around 0 dBm sets, a non-trivial requirement on the LO-to-RF port isolation when dealing with multiple mixers. One key problem is not only to be able to implement an architecture which meets low noise requirements at high transmit powers, but also able to meet the high WCDMA dynamic range requirements with better than 0.5 dB gain step accuracy. The step size accuracy challenge can partially be solved using a high dynamic range IQ DAC. But for many years, these were either not available, or they consumed too much power.

Despite achieving an impressive level of integration, the BiCMOS solution in (Figure 14.5f) did not support RX diversity, a feature which was not absolutely necessary for early Cat-6 HSDPA terminals, but which quickly became mandatory for LTE operation. A series of factors pushed this technology from the forefront, in favor of CMOS. With digital BB production volumes moving rapidly from 130 nm to 28 nm, the lower cost of manufacturing made this technology process very attractive. Beyond the intrinsic cost advantage, using CMOS for the RF-IC implementation enabled novel architectures to deliver an unprecedented level of performance and power consumption. With low operating supply voltages, CMOS significantly lowers the power consumption of functions that were implemented in an analog fashion in

---

[8] 3GPP antenna referred minimum output power is −50 dBm. Assumes 15 dB PA gain in its low power mode.

BiCMOS designs. High dynamic range (HDR) ADCs [5] and DACs are key blocks which allow for a reduction in complexity for the analog low pass filters. For example, due to limited ADC resolution, the receiver in Figure 14.5f introduced rather complex auto-calibrated analog I/Q group delay equalizers to flatten the channel filter group delay distortions [9]. With high dynamic range ADCs, the role of the analog filter can be changed from a channel filtering to a simple anti-aliasing function. Not only is the filter complexity reduced as the filter order is relaxed, but also EVM performance is improved since the level of in-band distortion is kept to a minimum. Moreover, with most of the channel filtering implemented in digital filters, the difficulties in supporting multimode operation are eased, since reconfigurability then becomes a matter of reprogramming an FIR or an IIR filter. The emergence of a standardized digital baseband interface such as DigRF^{SM}v4 (Section 14.1.3.4) has further facilitated the transition towards a more digital-oriented CMOS transceiver design. With microcontrollers now available at lower cost impact, traditional digital BB functions can be moved into the RF transceiver. An illustration of a modern DCA receiver is shown in Figure 14.6a, where the use of a Digital Signal Processor (DSP) allows automatic calibration of LNA IIP3 and IIP2 mixer, digital DC offset compensation, and I/Q image rejection correction to deliver a level of performance that would have been nearly impossible to achieve in analog BiCMOS designs. Comparing receiver [8] with solution (f) in Figure 14.5, 70 dBm IIP2 (vs. 58 dBm) provides ample margin for SAW-less operation and, with an EVM performance below 3% (vs. 6%), near zero impact on LTE/HSPA+ demodulation performance is expected. And all of this is also achieved at half the power consumption (43 mW vs. 92 mW). In Figure 14.5, transceivers (g), (h), and (i) are all designed in CMOS technology. Finally, designing in CMOS paves the way for a low-cost, highly integrated single-chip solution in which the RF TRX, the digital BB, and power management ICs may be manufactured in a single die. Initially developed for EGPRS low-cost terminals, monolithic solutions are now in production for HSPA and have recently been announced for triple-mode LTE platforms [10, 11]. With such a high level of integration, only a handful of ICs is required to build the platform.

With the removal-of-SAW-filters challenge solved for the majority of commercial chip sets and frequency bands, what are the new challenges the industry faces today?

In single chip devices (also sometimes denoted as System-On-Chip, or "SoC"), the collocation of multiple noisy circuits next to the sensitive RF TX/RX chains create numerous co-existence challenges. Solving these myriad of co-existence issues is a fertile field of innovations, and is briefly introduced in Section 14.1.3.5.

Another interesting challenge lies in TRX IC area reduction. With the large number of bands required for a WW LTE solution, the number of LNAs to be integrated increases. The problem is exacerbated due to requirements for RX diversity for each band of operation. The resulting increase in number of RF ports is shown in Figure 14.7b. For example, a NA-EU triple mode LTE, HSPA+, EGPRS variant may require up to 32 RF ports[9]. This is a 50% increase compared to a typical dual-mode WW phone. Unfortunately, RF blocks, such as LNAs and RF PGAs, do not typically shrink with decreasing CMOS nodes. As Figure 14.5 shows, the area occupied by TRX packages has shrunk over the years, reaching, in early 2013, the ultimate Wafer Level

---

[9] Assumptions: LTE NA-EU device is able to support: 7 high-bands and 4 low-bands in receive, 6 high-bands and 4 low-bands in transmit, i.e., a total of 11 RF ports for each receiver branch, and 10 ports for transmit. It is assumed that each RX port is single-ended.

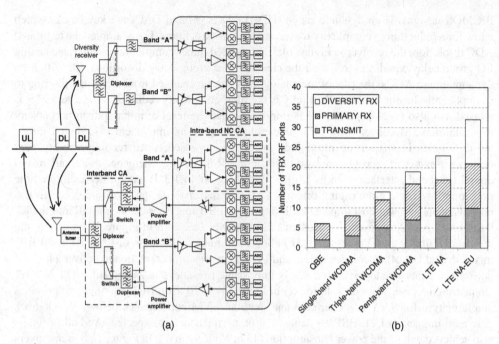

(a)   (b)

Figure 14.7 (a): Example RF front end complexity in supporting triple-mode and multiple downlink only carrier aggregation scenarios: contiguous and non-contiguous, intra and inter band. Only a pair of bands is shown. (b): RF transceiver number of RF ports required per application

Packages (WLP), in which the package is basically the size of the active die (25 mm^2). If pin count continues to grow, the modern RF CMOS transceiver package size may be dominated by pin count. In solutions using differential LNAs, the RF ports of the NA_EU variant would require a total of 54 RF pins. With such a high number of ports, ensuring acceptable pin-to-pin isolation is also further complicated. Lastly, PCB routing becomes more complex because high pin density packages often require additional PCB layers, which further raises the total cost to manufacture.

Finally, the introduction of Carrier Aggregation (CA) adds complexity to the entire RF subsystem as illustrated in Figure 14.7a. There are three categories of CA scenarios from 3GPP: inter-band, intra-band contiguous (IB-C-CA), and intra-band non-contiguous (IB-NC-CA). The immediate impact on TRX design is the need to support greater modulation bandwidth. The initial commercial deployments started in August 2013 with 10+10 MHz. The next phase, 20+20, will start during 2014, thereby requesting UE receivers to support a total of 40 MHz BW. There are numerous interesting challenges associated with CA, but a detailed description would go far beyond the scope of this introduction. While most challenges impact the RF-FE linearity, filter rejection, and topology, the special case IB-NC-CA is problematic for the design of a single-chip TRX solution. In this CA category, the wanted downlink carriers may be located at, or near, the edges of a given band as depicted in Figure 14.8. This situation creates at least two types of new challenges.

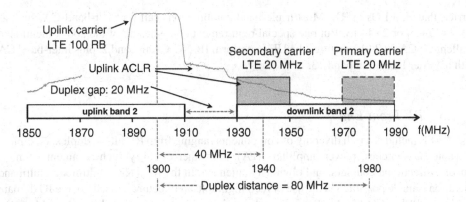

Figure 14.8 Example of worst case uplink–downlink carrier frequency spacing in the case of band 2 intra-band CA

The first problem results from the fact that, while the primary carrier is located at the duplex distance, the secondary carrier could be located at a distance slightly greater than the duplex gap ('DG' – cf. Figure 14.8). In bands with large DG, such as band 4, this is not an issue. But, for example, in the case of band 2, where the DG is only 20 MHz, the secondary carrier can be desensitized by the PA ACLR emissions. Thus even if inter-stage saw-filtering is applied, for certain combinations of uplink and downlink resource block allocations, the secondary downlink carrier sensitivity performance is degraded. This case is similar in nature to the desensitization occurring in band 13/17, where only a 3GPP performance relaxation can solve the issue. At the time of writing, these relaxations are being debated.

Secondly, it becomes apparent that opting for a single LNA/mixer chain to demodulate two carriers that could be located at the opposite edges of the band does not appear feasible for linearity and bandwidth reasons. As a consequence, a dual receiver and dual LO architecture is required to down-convert each carrier, as shown in Figure 14.7a. Contribution [12] summarizes in great detail the implications, and shows that due to limited isolation in a single-chip solution, cross-LO leakage occurs. In practice, this means that each mixer will not only be driven by its intended LO, but also by the leakage of the second LO, and by the multiple IMD products generated by non-linear mixing of the two LOs. The analysis indicates that requirements on cross-LO leakage and LO IMD product rejection might be difficult to achieve in a single chip. VCO-pulling is another phenomenon which is difficult to solve in single-chip devices. Pulling occurs when VCOs, operating either at close, or at harmonic, frequencies of another VCO, couple to one another, which may push the PLL to go out-of-lock. In the case of IB-NC-CA, [13] since carrier frequencies are signaled by the network, the reprogramming of VCO divide ratios has to be done "on the fly." During this time, PLL synthesizers must be disabled and consequently IQ samples are interrupted. Only a 3GPP relaxation can help solve this issue. It is suggested to introduce a 2-ms interruption, a penalty equivalent to 4 LTE time-slots. Note that this issue is also applicable to inter-band CA. Finally, the problem is not expected to abate as the recent introduction of triple-band downlink carrier aggregation requires the UE to activate simultaneously four local oscillators (LO): one uplink and one for each downlink carrier, that is, 3 LOs in RX. Most triple-band combinations call for inter-band-CA, such as CA 4+17+30 or 2+4+13. But one special requirement combines all of the above mentioned challenges: CA 2+2+13 requires the UE to perform IB-NC-CA in band 2, plus inter-band CA with a carrier located in band 13.

14.1.3   RF Front End

The RF-FE includes a rich diversity of components ranging from RF filters, duplexers, diplexers, antenna switches, power amplifiers (PA), and other circuitry such as antenna tuners, sensors, directional couplers, and obviously antennae. In the RF TRX, multimode multiband operation is made possible via the use of reconfigurable architectures, mostly in the IQ domain and in the PLL/VCO blocks. Reconfigurability in the front-end is much more problematic, since it is difficult to implement tuneable RF filters without incurring either insertion loss or attenuation penalties. This section aims to illustrate the solutions found by the industry to address cost reduction in the front-end, while also meeting the needs posed by greater complexity.

14.1.3.1   Trends in Filters and Switches

In the receiver path, filters are used to eliminate the unwanted RF signals. In the transmit direction, they attenuate out-of-band noise and spurious emissions to ease coexistence effects between bands and terminals. For each supported FDD band of operation, Figure 14.10 shows that three filters and an antenna switch are required: two band-pass filters (BPF) in the duplexer and one BPF in the diversity receiver branch. The number of throws in antenna switches is driven by the number of bands that must be supported. For example, Figure 14.5 shows that switches have evolved from single pole 6 throw (SP6T) devices in 2003, to SP12T in 2013. This number is obviously bound to grow in coming years with increasing band count, support for LTE-TDD and Carrier Aggregation. Carrier Aggregation also raises additional RF performance requirements – improved linearity for example (Section 14.1.3.5) – and the need for a diplexer to further separate the aggregated bands.

For switches, only an improvement in technology can lead to significant cost reductions. Legacy GaAs PHEMT solutions are being replaced with cheaper single die SOI (Silicon on Insulator) semiconductor technology. In the world of RF filtering, the ideal solution would make use of tuneable RF filters to address cost reduction. Research is ongoing in this area and initial tuneable duplexers [14] show promising results, but these are far from meeting the stringent FDD operation isolation/insertion loss requirements. In practice, the approaches taken to reduce cost/PCB footprint area essentially fall into three categories: component miniaturization, co-banding, and module integration.

The main challenge in filter miniaturization has been and still remains to maintain low insertion loss while delivering high attenuation. Filter technology rapidly evolved from ceramic to SAW filters, a technology based on ceramic piezoelectric materials. It is complemented by silicon BAW/FBAR[10] technology for the most demanding requirements related to short duplex

---

[10] BAW: Bulk Acoustic Wave, FBAR = Film Bulk Acoustic Resonators.

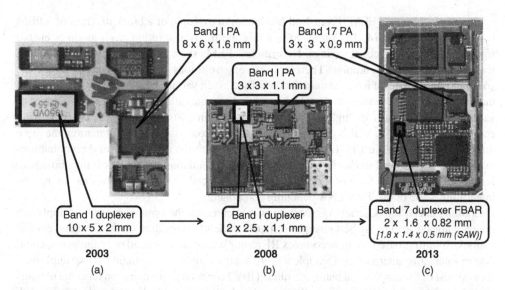

Figure 14.9 Example of duplexer miniaturization. (a): Ceramic 2003, (b): SAW 2008, (c): FBAR 2013 – PCB pictures are extracted from Figure 14.5 and scaled to reflect exact relative sizes

distance and for frequency of operation beyond 2 GHz. An illustration of the efforts made for duplexers is shown in Figure 14.9: the size of the 2003 ceramic duplexer was $10 \times 5 \times 2$ mm. SAW-based duplexers introduced around 2005–2008 shrank to $2 \times 2.5 \times 1.1$ mm (Figure 14.9b), whereas the latest mass-produced models come in packages of $1.8 \times 1.4 \times 0.5$ mm for SAW and $2.0 \times 1.6 \times 0.9$ mm for FBAR (Figure 14.9c). Compared to the initial ceramic devices, the size is remarkably reduced to 1/15 in area ratio and 1/60 in volume ratio for SAW devices.

Co-banding is a technique which consists of sharing RF filters in bands for which multimode operation must be supported. For example, in band II (PCS 1900), rather than implementing an EGPRS RX chain with a dedicated PCS RX BPF, it is tempting to reuse the antenna to RX path of the band II/band 2 FDD duplexer required for LTE and HSPA operation. By routing EGPRS through the duplexer, one RF BPF is saved. The drawback of this approach is that the duplexer insertion losses are usually higher than that of a standalone BPF, thereby leading to a slight degradation in EGPRS sensitivity. An example of co-banding is shown in Figure 14.5g (2011). In practical terms, if co-banding was applied to UE [15], the 17 frequency bands supported by this terminal could be reduced to only 8 unique RF receiver paths and 4 transmitter line-ups: bands 1, 2, 5, and 8 could be common to LTE, WCDMA, and EGPRS, band 4 common to LTE and WCDMA, band 3 common to LTE and EGPRS, leaving band 20 and band 7 as specific bands to LTE. In reality, the tradeoff between cost savings and performance degradation depends on OEMs/operator targets. For example, in many instances, the performance loss of co-banding EGPRS with LTE in band 3 is not acceptable and leads to separate RF paths for each air interface.

For a higher level of integration, and therefore in a move to achieve further PCB area savings, the switching and filtering functionalities may be integrated into a module also commonly known as a FEM (Front-End Module). An example of a block diagram of a FEM is presented in Figure 14.10. Compared to a discrete filtering solution, such modules enable significant component count savings. Nevertheless FEMs show less flexibility when adapting the band support configuration. Therefore, existing solutions are customized to match the specific requirements of a phone manufacturer for a given variant. In this respect, FEMs make the most economical sense when applied to a band combination needed in a large number of variants where volumes are high and therefore production costs are lower. The 3G WW band combination I, II, IV, V, VIII is one example. The more exotic the band combination, the more difficult it becomes to justify a custom FEM design. With the number of band combinations increasing, OEMs need to carefully study which band combinations are worth integrating for future products. At the time of writing, the most advanced FEMs deliver further savings by integrating some of the TRX LNA matching components.

Looking ahead into Carrier Aggregation (CA) solutions, the optimized bank of duplexers may not always lead to the best cost/performance tradeoff, depending on the CA band combinations. Solutions based on more complex RF multiplexers, such as quad or hexaplexers, could deliver interesting alternatives. Quadplexers are not the simple concatenation of two duplexers: they consist of four individual bandpass filters (BPF) connected to a common terminal antenna port. As such, each BPF differs from those used in duplexer designs. For example, the TX BPF of a duplexer is designed with the main constraints of delivering high TX noise rejection at the RX band (cf. Section 14.1.2). In the example of a quadplexer, each TX BPF has to reject TX noise into two distinct RX frequency bands. This increases the number of constraints (zeros in the stopband) under which the BPF must be designed. Depending on band combination, the designer may have to tradeoff isolation in return for insertion losses. One additional factor leading to higher losses comes from the matching networks required to ensure that each filter does not load each other. The main advantage of multiplexers is that they simplify the RF front-end architecture. An example using two high and two low hypothetical frequency bands is shown in Figure 14.11 below and illustrates the tradeoff between cost and complexity.

Figure 14.11a uses a common diplexer to all bands and supports only high–low CA, for example band 1–band 8, or band 2–band 5. The main drawback of this approach is that bands and air interfaces which are not meant to be used in Carrier Aggregation pay an insertion-loss penalty in both uplink and downlink. The use of a quadplexer in Figure 14.11b considerably simplifies the front-end circuitry, since only one antenna switch is required, and each band of operation benefits from a removal of the diplexer. In this example, only a high–high CA is supported, for example band 2–band 4. Figure 14.11c supports both high–high and high–low CA, but the price that must be paid for this extra capability is that of adding a diplexer. Note that this latter configuration also supports triple-band CA at no or little extra cost compared to the dual-CA solution (a). For example, solution (c) could support inter-band triple band CA of bands 2–4–13. It can be seen that with CA, a variety of front-end architectures must be assessed to deliver the optimum cost/performance tradeoff. For each combination, a tradeoff must be found between complexity and performance. These tradeoffs have generated many discussions at 3GPP on ways to determine the insertion loss associated with each configuration, and how these extra losses should be taken into consideration to define reference sensitivity and maximum output power relaxations. As a consequence, for each dual downlink CA scenario, many filter manufacturers' data had to be collected to agree upon an average insertion loss per type of multiplexer. Further complication occurs since insertion losses are heavily design

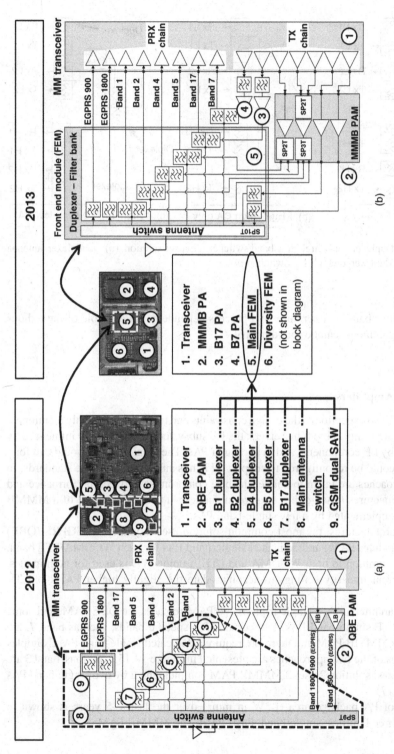

Figure 14.10 Enlargement of the RF front end of (block diagram[11] and PCB pictures) of 2012 and 2013 terminals from Figure 14.5. (a): Discrete solution, (b): Integrated front end module example which supports quad-band EGPRS, WCDMA bands I, II, IV, V, and LTE bands 1, 2, 4, 5, 7, 17

---

[11] Block diagrams are guessed after careful inspection of the PCB and components identification whenever possible. As the level of integration increases, this exercise becomes difficult and may lead to errors of interpretation. The diagrams are printed for information and illustration purposes only.

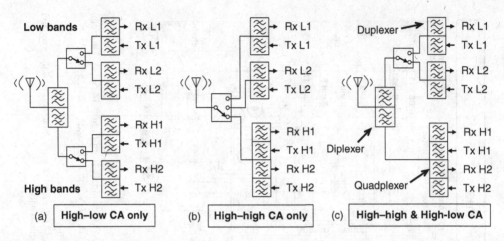

(a) **High–low CA only**   (b) **High–high CA only**   (c) **High–high & High-low CA**

Figure 14.11 (a): Duplexer bank, high–low band switches, diplexer solution, (b): quadplexer, antenna switch solution, (c): duplexer, quadplexer, switch, diplexer solution

dependent, as well as band, temperature, and process dependent. An example of the tedious and hard-fought agreements can be found in [16].

14.1.3.2 Power Amplifiers

Power amplifiers are no exception to the design constraints and tradeoffs previously mentioned. With highly integrated single chip transceivers, the RF subsystem area is heavily influenced by the area occupied by FE components and in particular PAs. The ideal solution would call for a unique PA which could be reconfigured to support all bands and all air interface standards. In practice, two approaches are taken to deliver the best compromise between performance and cost: discrete architectures based either on several single-band PAs, or the use of a MMMB PA architecture complemented by a few single-band PAs.

In discrete PA architectures, the most common solution uses a quad-band EGPRS (QBE) power amplifier module (PAM) and dedicated single-band PAs for each WCDMA/LTE band. When the band is common to both WCDMA and LTE, a unique PA is used for both modes, usually with different linearity settings to accommodate the slightly higher PAPR of LTE transmissions.

In MMMB PA architectures, a single triple-mode PAM replaces the QBE PAM, and some of the WCDMA/LTE single-band PAs. Yet, certain LTE specific bands, such as band 7, are not yet covered by MMMB PAM and therefore require a dedicated single-band PA. Examples of implementations of the two approaches are presented in Figure 14.15: solutions 1 and 3 are discrete architectures; solution 2 uses a MMMB PAM complemented by two single-band PAs (band 4 and band 17).

The evolution of PA package area (L*W in mm^2) over the last 4–5 years is shown in Figure 14.12 for three families: single-band, QBE PAM, and MMMB PAMs.

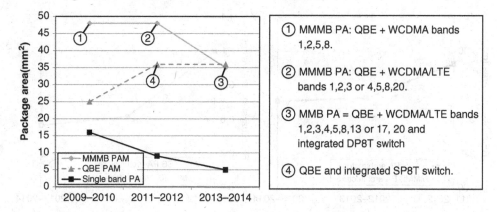

Figure 14.12 PA package size evolution and number of integrated bands – compiled from major suppliers' public domain data

Single-band PAs (squares) have reduced their footprint by a factor 3, from about 16 mm^2 to about 5 mm^2. This category has reached a maturity plateau and future packaging technology improvements are unlikely to significantly impact the overall radio area. With 25 mm^2 package area, and their long history of production and optimization, QBE PAM (triangles) have also reached a level of maturity and will most likely remain at this level for the coming years. The recent increase in area to 36 mm^2 is due to the integration of an SP8T antenna switch. The initial 2009–2010 MMMB PAMs (diamonds) combined a QBE and support for quad-band WCDMA operation. They rapidly evolved to support dual-mode penta-band, and the triple-mode operation octa-band operation. Remarkably, the 2013–2014 MMMB PAM, which covers QBE, octa-band WCDMA/LTE PAM, and a DP8T band distribution switch comes in a package size comparable (35 mm^2) to that of a QBE–SP8T module. It is expected that PA suppliers will invest further efforts into these solutions as triple-mode will soon become a *de facto* requirement.

The choice of architecture is primarily driven by the tradeoff between cost and size for a given band-set. The vast majority of dual-mode EGPRS-WCDMA terminals support two or three WCDMA bands. In these solutions, the discrete PA architecture is often the most preferred solution as there is little or no cost advantage to using a MMMB PAM. An MMMB PAM becomes advantageous when there are four or more WCDMA/LTE bands to be supported. An example of cost tradeoff selection metrics is shown in Figure 14.13 in the specific case of a North America (NA), triple-mode product[12]. Over time, the MMMB PA architecture is expected to provide both size and cost benefits over the discrete PA architecture. From an RF subsystem PCB area perspective, it requires an octa-band MMMB PAM to better the discrete PA solution. From a cost perspective, the dynamics are less trivial as the introduction

---

[12] The terminal supports quad-band EGPRS, penta-band WCDMA (I, II, IV, V, VIII), and LTE over at least band 4 and 17. Assumptions on MMMB PA band support: 2011–2012 supports bands 1, 2, 5, 8; 2012–2013 supports bands 1, 2, 4, 5, 8; 2013–2014 supports bands 1, 2, 4, 5, 8, 17.

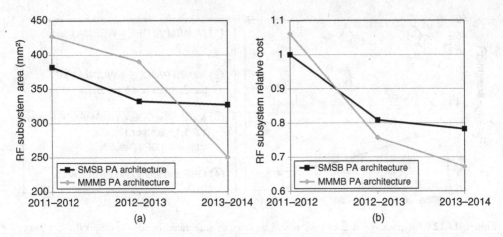

Figure 14.13 Single band vs. MM-MB PAs: RF subsystem area (a) and cost (b) tradeoffs in a NA triple-mode application

of such highly integrated products often benefits from a positive spiral effect: the introduction of a smaller package with increased band coverage induces a cost reduction, which induces a production volume increase, which in itself induces a reduction in sales price.

Several other factors come into play when making a selection of architecture:

– Power consumption: single-band PAs exhibit better power consumption than MMMB PAMs. This is mainly because discrete PA matching networks can be optimized for a narrow frequency range while MMMB PAMs must be matched over a wider band. In addition, MMMB PAM efficiency is lower as they absorb the insertion losses of the band distribution switch,
– Design effort: MMMB PAMs considerably simplify the PCB place and route task, especially in variants covering a large number of bands. An example of complexity in attempting to cover a future NA-EU variant using discrete PAs is shown for illustration purposes in Figure 14.15.
– Multiple vendor sourcing management: MMMB PAMs bring a significant advantage over discrete solutions as they reduce the number of references for a terminal model/variant,
– Ease of reconfigurability: MMMB PAMs, complemented with one or two single-band PAs, provide the best tradeoff today between cost, PCB place and route complexity, area, and ease of reconfiguration to support multiple ecosystems. An illustration of the reconfigurability concept is shown in Figure 14.14 [17], where a MMMB PAM is used to cover QBE and quad-band WCDMA (I, II, V, VIII) and LTE band 20. This example shows that a single reference design may be quickly adapted to serve three different regions of the world with minimal changes, and yet can achieve rather aggressive, and nearly identical, PCB area and component count metrics.

Finally, Figure 14.15 summarizes all of the previous discussions in one place. RF solutions 1 and 2 address the same NA, triple-mode telecom operator variant and illustrate both

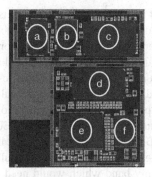 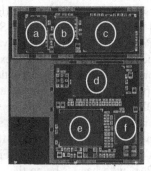 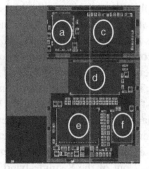

Triple-mode EU variant	Triple-mode NA variant	Dual-mode global variant
(a) Band 7	(a) Band 17	(a) Band 4
(b) Band 3	(b) Band 4	(b)
(c) MMMB PAM	(c) MMMB PAM	(c) MMMB PAM
(d) Main front-end module	(d) Main front-end module	(d) Main front-end module
(e) Transceiver	(e) Transceiver	(e) Transceiver
(f) Diversity front-end module	(f) Diversity front-end module	(f) Diversity front-end module

Figure 14.14 Reconfigurability concept with MMMB PA architecture. MMMB PAM covers quadband EGPRS, penta band 1, 2, 5, 8, 20 in WCDMA and LTE modes [17]. Triple-mode NA variant: QBE, WCDMA I, II, IV, V, VIII, LTE 1, 2, 4, 5, 8, 17, triple-mode European (EU) variant: QBE, WCDMA I, II, IV, V, VIII, LTE 1, 2, 3, 5, 7, 8, 20, WW dual-mode variant: QBE, WCDMA I, II, IV, V, VIII. Renesas 2012 [17]. Reproduced with permission of Renesas Mobile Corporation

	RF Solution 1 Discrete 3G/LTE PAs NA variant 2012 (non SAW-less)	RF Solution 2 MMMB PA NA variant 2012 (SAW-less architecture)	RF Solution 3 Discrete 3G/LTE PAs NA+EU variant prototype (SAW-less architecture)
PCB area **Shrink factor**	650 mm²	427 mm² **1.5**	633 mm² **1.03**
Comp. count **Reduction factor**	250	119 **2.1**	280 0.9
EGPRS bands	850/900/1800/1900	850/900/1800/1900	850/900/1800/1900
WCDMA bands	I, II, IV, V	I, II, IV, V, VIII	I, II, IV, V, VIII
LTE bands	1, 2, 4, 5, 17	1, 2, 4, 5, 8, 17	1, 2, 3, 4, 5, 7, 17, 20

Figure 14.15 Examples of triple-mode RF solutions (discrete and MMMB) – (a) Single-Band PAs (3 × 3 mm), (b) MMMB PAM (5 × 7 mm), (c) Triple-mode transceiver IC, (d) 2.5G Quad-band amplifier module (5 × 5 mm), (e) Duplexers (2.0 × 2.5 mm and 2.0 × 1.6 mm) + antenna switch

graphically and quantitatively the significant gains that MMMB PAMs bring to such products: the area is reduced by 150%, and uses half the components. Solution 3 is a prototype which uses single-band PAs and a QBE PAM to target a NA and EU variant in triple-mode operation. In comparison with solution 1, one can see that this PCB is extremely well-optimized, and it covers three additional LTE and one extra WCDMA band with nearly identical PCB metrics. Yet, it is easy to note that the complexity and the cost associated with such architecture would most likely be unacceptable for OEMs. It is evident from this prototype that next-generation variants which attempt to target more than one ecosystem, such as this effort for NA-EU coverage, can only be cost effective through the use of MMMB PAMs. It is estimated that in this example, the use of the latest octa-band MMMB PAM (Figure 14.12, 2013–2014) could save 7 out of the 8 single-band PAs, with band 7 being the only band which would need a dedicated PA. It thus comes as no surprise that the focus for future PA design will almost surely target the use of MMMB modules.

14.1.3.3 Over-the-Air (OTA) Performance

While chip set suppliers are benchmarked by OEMs to deliver the best RF performance in conducted test conditions, and often have to iterate a design to improve performance by a fraction of a dB, telecom operators, on the other hand, are primarily interested in OTA performance. Between these, the OEMs are more concerned with ensuring their products will pass the impressive set of 3GPP/PTCRB/GCF tests. Typically, the test plan of a triple-mode phone involves approximately 1500 tests, the majority of which are conducted tests, and which typically lead to several months of costly testing. When it comes to OTA performance, the dual-mode EGPRS/HSPA UE is measured against two simple Figures of Merit (FOM)[13]: Total Radiated Power (TRP) and Total Isotropic Sensitivity (TIS). Each FOM is then re-measured in different user interaction (UI) scenarios: UE in Free Space (FS), held in a hand phantom[14], placed against a head phantom or held in hand and placed against the head phantom [18]. Figure 14.16 plots the WCDMA TRP vs. TIS performance, in band V and band II, across a sample of 30 class 3 recently PTCRB-certified smartphones. Looking at the conducted results (circles), UEs generally perform within 1 or 2 dB from each other in both axes. In output power, performance ranges across the class 3 tolerances (24 dBm +1/−3 dB), the majority being calibrated to deliver 23 dBm. In sensitivity, all UEs pass with a comfortable margin, but because this metric is used by OEMs to benchmark chip set platforms, the performance spread is much tighter than in TX power. With OTA TRP and TIS, the spread across UI increases dramatically between bands.

For example, in band V the spread between conducted and BHHR varies between 10 and 18 dB. But in band II, the spread is nearly half that of band V, ranging from 5 to 9 dB in both TRP/TIS. In a given band, the difference between conducted and FS gives a good indication of antenna gain, with superior performance in band II as compared to that in band V. This illustrates that designing an antenna with desired performance in low bands is a significant challenge in modern smartphones. Figure 14.17b, shows that the task of the antenna designer

---

[13] CTIA acronyms are used throughout this section, where TRP and TIS are the average of several 3D measurements performed across different spatial angles of the radiation sphere and different polarizations.
[14] Both left and right hand phantoms are defined in CTIA OTA test plan.

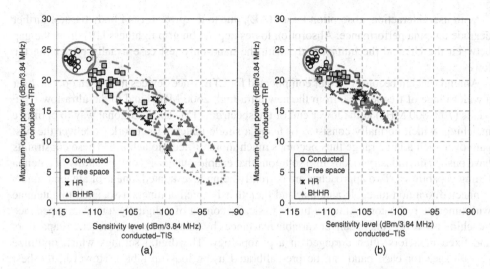

Figure 14.16 Example of modern WCDMA OTA TRP vs. TIS performance of 30 recent smartphones extracted from PTCRB reports based on CTIA OTA test plan. (a): band V mid-channel, (b): band II mid-channel. (HR: Phantom Hand Right only, BHHR: Beside Head and Hand Right Side, that is, head and hand). These graphs compiled with permission of PTCRB

is not becoming any easier, since the volume made available is continually reduced. Because of these constraints, it is not unusual for antennae to reach alarming VSWR values as high 10 : 1, corresponding to mismatch losses (ML) of 4.8 dB. This loss impacts TRP and TIS. One can see that with chip sets no longer being loaded/sourced with the ideal 50 Ohm of the conducted tests, the 0 dBi antenna gain assumption made in 3GPP is far from being met in actual situations.

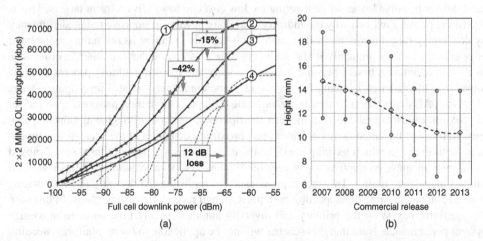

Figure 14.17 (a): example of MIMO throughput loss vs. antenna correlation: ① Static bypassed RF channel, ② Low correlation EPA model, ③ Medium correlation EPA model, ④ High correlation EPA model, (b): trends in smartphones heights. Reproduced with permission of Videotron Ltd.

With user interaction, absorption losses (AL), due to the proximity of body tissues, further degrade antenna performance. Absorption losses depend on grip tightness [19] or on the gap between the UE and the human body, with the hand-only test case usually presenting the lowest loss.

Antenna design becomes further complicated for LTE, since devices must then operate over a wider range of frequencies, from the new ultra-high 2600 MHz band to the ultra-low UHF bands (700–800 MHz) at the lower end of the spectrum [15]. The traditional way to optimize matching, which normally consisted of using a single matching network to deliver the best gain over such a large range thus becomes much more difficult to achieve. These constraints have pushed the industry to look into solutions that enable antenna reconfigurability or antenna tuning capabilities. Two approaches are currently being investigated: one is of antenna-tuners connected to a non-tuneable antenna, and the other is antenna tuners associated to antennae with tuneable resonators. Antenna tuners basically consist of a digitally tuneable impedance matching network. Tuners deliver variable reactances by using an array of switched capacitors and fixed inductors often arranged in a pi topology. The digital settings which optimize performance for each band can be pre-calibrated using look-up tables. However, the best agility is accomplished by using a coupler at the antenna port, so real-time VSWR can be monitored and used in a closed-loop operation to continuously deliver optimal performance, independent of UI. In this respect, the advantage of standardizing an RF-FE digital control bus, such as MIPI® RFFE, allows multiple vendors to be easily interchanged (cf. Section 14.1.3.4). In theory, an antenna tuner should aim at delivering close to 50 Ohm impedance. In practice, depending on the band and antenna structure, the tuner will only be able to partially correct mismatch losses. For example, the study on the impact of grip style in [19] indicates that in most configurations performance is largely dominated by absorption losses, with values ranging from 1.5 to 7 dB loss, while impedance MLs only account for 0.5 to 2.5 dB. In this example an antenna tuner connected to a non-tuneable antenna would not bring significant gains. These moderate gains must be weighed against design challenges associated with front-end circuitry. The linearity requirements for components are high[15] [20], and yet components must deliver minimal losses while meeting the low cost and low PCB footprint targets. This is one of the primary reasons why the industry is looking at ways to improve radiated efficiency by developing tuneable antennae. If the antenna is capable of tuning its resonators to the band of operation, then theoretically it should be possible to deliver enhanced antenna efficiency. In this case, antenna tuners become mandatory since the resonator tuning process induces impedance mismatches. Looking further ahead, some companies are also investigating ways of solving AL degradation by designing antennae with preset radiation patterns. With the set of proximity sensors embedded in modern smartphones, one could imagine a phone capable enough to adapt the antenna tuners based upon proximity detection of the user. At this time it is difficult to assess which technique will be the most promising. Some of the most advanced techniques in mass production today rely on antenna selection switching, with or without limited impedance tuning. This technique consists of replacing the diversity receive antenna with an antenna capable of supporting both transmit and receive frequency bands. In this case the platform may swap the primary and diversity antennae at any time so as to maximize system performance. Note that this scheme will not be applicable to future platforms needing to support transmit diversity.

---

[15] For band I, [20] estimations based on double and half duplex 3GPP CW blockers in a band I WCDMA applications set an IIP3 requirement of 65 dBm and 110 dBm IIP2 for the antenna switch.

With MIMO LTE, downlink throughput is the new FOM. MIMO throughput depends on the end-to-end correlation between the two data streams. Consequently, the UE demodulator experiences the product of three matrices: eNodeB TX antennae, RF propagation channel, and UE RX antennae correlation matrix.

Figure 14.17a, shows the impact of the RF channel correlation matrix on a recent triple-mode UE in a conducted test setup. The measurement is performed in AWS band, 10 MHz cell BW, using a 2×2 MIMO RF multipath fader. The graph plots UE DL throughput measured by an Anritsu 8820c eNodeB emulator vs. DL full cell BW power under four RF propagation conditions: static bypass mode (dots), low (squares), mid (triangles), and high (crosses) correlation matrices of the 3GPP EPA fading model. For each RF fading model, the DL MCS index is varied to capture the UE behavior across its entire range of demodulation capabilities. To avoid overloading the figure, only a few waterfall curves are shown in light gray dotted lines. The resulting envelope of each waterfall is plotted in plain lines.

This graph provides insights into the susceptibility of LTE MIMO performance to correlation. There are two ways to read this experimental data:

- Case of a user located under an eNodeB, experiencing a high DL RF power of, say, −65 dBm. The difference between low vs. mid, and low vs. high correlation results in a 19% and 42% throughput loss, respectively. In the latter case, the throughput drops from 70 Mbit/s to about 40 Mbit/s. Under high correlation conditions, the impairment is so bad that the BB is unable to deliver more than 54 Mbit/s even at maximum input power (not shown on graph). From a user experience perspective, it is unlikely that this lower-than-expected performance would be noticed, since absolute throughput is not a sufficient metric to guarantee a good user experience (see Chapter 13 for more details). From a telecom operator's perspective, the situation is quite different as this loss prevents the cell from delivering the maximum theoretical performance.
- At a given target throughput, increasing correlation leads to a degraded UE RF sensitivity. For example, at a target 40 Mbit/s, a user experiencing high correlation suffers from a 12 dB penalty in link budget. This penalty increases with increasing target throughput because of the floor in performance of that particular test case. This graph can also be interpreted as a fair illustration of the differences one can expect between operation in high bands (e.g., 2600 MHz) where the antenna should perform well, and low band operation (e.g. 700 MHz). Note that at cell edge, in low SNR conditions, there is little difference in performance.

This graph shows how crucial it is for telecom operators to jointly define with 3GPP a standardized OTA test method. In an ideal world, OTA testing should be able to reproduce and predict field performance. In practice, predicting OTA MIMO user experience appears to be a nearly impossible task because so many variables can impact performance. At worst, OTA testing will deliver a reliable tool to benchmark terminals against one other. Here are a few examples of the challenges – based on recent contributions:

- Contribution [21] has shown that antenna correlation in a smartphone is so sensitive to its close surroundings that the presence of tiny co-axial semi-rigid cables impacts the measurement accuracy. Using optical fiber feeds, FS correlation reached 0.8, a high value, while measurements with semi-rigid feeds indicated a correlation on the order of 0.4. If tiny semi-rigid cables can change the correlation by a factor of two, it is reasonable to assume that the presence of larger objects could induce greater impairments. Unfortunately, the nature of interactions is so complex that no general rule governing the degradation of antenna correlation can be drawn. For example, the study in [22] shows that a smartphone with good FS performance can turn into a poor terminal in the presence of a human body – but the exact opposite is also true. The study in [23] goes a step further by measuring the impact of a real user on antenna correlation vs. FS and phantom head. The study uses prototypes specifically designed to deliver high and low correlation performance in the 700 and 2600 MHz bands. Measurements demonstrate that the difference between high and low correlation decreases with UI, so much so that at 700 MHz, it becomes difficult to distinguish the two prototypes in the presence of a real person. This suggests that any ranking between good and bad devices based on FS and phantom head measurements might not be in agreement with the more subjective rankings resulting from actual use (see also [24]).

- In addition, antenna correlation varies vs. carrier frequency in a given band.
- From the above results, it may be intuitively understood that the angle of arrival also plays an important role. For example, one may expect that the use of phantom head only might not block as many incident waves as would be blocked by real person. This implies that measuring user experience requires a 3D isotropic test setup. 3GPP is assessing three test methods that could fulfill this requirement: anechoic, reverberation chambers, and a two-stage method. With their intrinsic 3D isotropic properties, reverberation chambers are an attractive solution to this problem.
- Another difference between lab measurements and practical user experience is that 3GPP test cases do not include closed-loop interactions between a NodeB and UE. Good MIMO performance is about making the best of the instantaneous radio conditions, either by using frequency selective scheduling or by adapting rank. These closed-loop algorithms are proprietary to each network equipment vendor and cannot be replicated using eNodeB emulators.

These are some of the complex challenges that OEMs, telecom operators, and chip set vendors are currently facing when it comes to defining OTA performance test cases. Not only must 3GPP assess and recommend test methods (AC, RC, or two-stage) but it must also define test conditions and pass/fail criteria to provide operators with a reliable and realistic tool to benchmark devices. Considering the complexity of the interactions between UE antennae and the surroundings, this task is far from trivial.

14.1.3.4   RF Subsystem Control and IQ Interfaces

The conventional RF subsystem architecture relies on analog IQ RF-BB interfaces and a myriad of control interfaces specific to each of the RF-FE components. The primary disadvantages of this approach are twofold:

- The associated number of pins tends to be larger, which increases package size, cost, and further complicates PCB routing.
- Dealing with a wide variety of proprietary interfaces restricts the ease with which handset makers can "mix and match" RF-FE components, and often leads to software segmentation.

The example in Figure 14.18 shows a hypothetical downlink-only carrier aggregation RF subsystem focusing on control and IQ interfaces only. In total, 47 pins are required. This

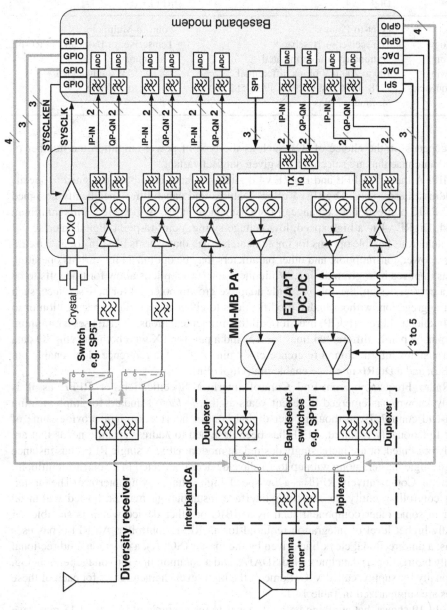

Figure 14.18 LTE inter-band carrier aggregation hypothetical block diagram using conventional buses. Pin count: 14 (switches) +20 (IQ) + 8 (PA-CTL) + 3 (SPI) + 2 (sysclk) = 47 pins. Interconnect for MM-MB PA temperature sensor (*), and antenna directional coupler (**) not included. For illustration purposes only. Pin count required for RF switch control could be reduced by sharing one or several sets of GPIO lines

Table 14.1   Selected DigRF v4 and RFFE characteristics. DigRF clock frequencies (*) assume a 26 MHz reference crystal oscillator. DigRFv4 has defined a set of alternate clock frequencies associated with 19.2MHz crystal reference oscillators

	DigRF v4	RFFE
	Point-to-Point	Point-to-Multipoint
Bus topology	Transceiver ↔ Baseband	Transceiver or Baseband → RF-FE
Termination	Terminated or unterminated	Unterminated
Voltage swing	100 to 200 mV pk-pk differential	1.8 or 1.2 V single-ended
Clock frequencies	1248, 1456, 2496, 2912, 4992, 5824* MHz	32 kHz to 26 MHz
Pin count	Minimum 7 pins	3 pins

number depends on the choice of ICs, such as the type of power amplifier, the number of bands and antennae that are selected for a given handset variant.

The MIPI Alliance[16] RFFE and DigRF v4 digital interfaces have been designed to specifically address these issues. DigRF v4 offers features and capabilities specifically designed for BBIC-RFIC bidirectional exchange of both data and control. With 1.248 GBit/s minimum bus speed, DigRF v4 is a high speed, low voltage swing, point-to-point digital interface. It provides numerous flexible options for implementers, such that aspects like pin count, spectral emissions, power consumption, and other parameters may be optimized for various operating conditions imposed by a given design. While these interface options allow for tradeoffs to be made in a current design, they also provide adaptive growth potential for new services, such as carrier aggregation, without undue modifications to either the interface specification or to existing Intellectual Property (IP) used for building implementations. DigRF uses a minimum of seven pins: a pair of differential lines for RX and a pair for TX with both carrying IQ data and control/status information, a reference clock pin (RefClk), a reference clock enable pin (RefClkEn), and a DigRF interface enable pin (DigRFEn).

The Radio Frequency Front-End Control Interface Specification, or RFFE as it is commonly known, has emerged in recent years as the *de facto* standard for implementing RF front-end control. Even though other digital control interfaces might provide some of the basic functionality required, RFFE has been designed to address specific needs that are frequently presented, or are increasingly desired, in modern UEs. A single RFFE bus instance supports a single master, along with up to 15 slave devices connected in a point-to-multipoint configuration. Comparatively, RFFE is a low speed, high voltage-swing interface. The master, or major controlling entity for an RFFE interface instantiation, may be hosted within an RFIC, or in some other component such as a BBIC or other device which is suitable for the slightly higher level of integration required for an RFFE controller. An RFFE bus uses three pins: a unidirectional clock line driven by the master (SCLK), a common unidirectional (optionally bidirectional) data line called SDATA, and a common line for voltage referencing, and optionally for supply, called VIO. Some of the high-level characteristics for each of these interfaces are summarized in Table 14.1.

Figure 14.19 shows that applying both interfaces to the example of Figure 14.18 may save up to 32 pins. Note that the system partitioning is implementation-specific and Figure 14.19 is

---

[16] MIPI® is a registered Trademark and DigRF^SM is a registered Service Mark of the MIPI® Alliance, Inc. All rights reserved.

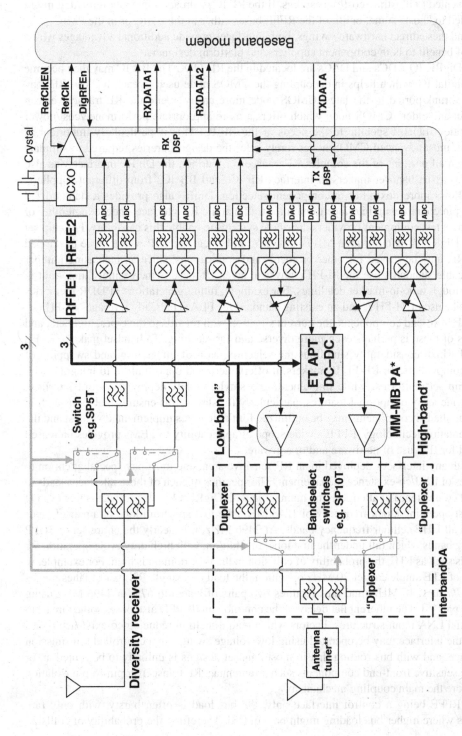

Figure 14.19 LTE inter-band carrier aggregation hypothetical block diagram using DigRF and RFFE. Assumes DigRF v4 is clocked at 2496 MHz, assumes a separate RFFE bus for switches and for PA control. Interconnect for MM-MB PA temperature sensor (*), and antenna directional coupler (**) not included.9 (DigRF) + 6 (RFFE) = 15 pins

only presented to illustrate cost/pin savings. If the RF IC is already pin count limited, it might be decided to "host" some, or all, of the RFFE buses either in the BBIC or in the PMIC.

Beyond these direct hardware savings, both interfaces provide additional advantages which can be of benefit to both component suppliers and platform designers.

With DigRF, IQ ADCs and DACs are located in the RF TRX. The BB IC may then become a pure digital IC, which helps in decoupling the CMOS node used in each IC. The BB may then be shrunk/ported to the latest CMOS node more easily, while the RF transceiver can remain in an "older" CMOS node which offers a more appropriate performance/cost/power consumption tradeoff specific to the needs of the RF IC. This is particularly important as the latest, most advanced CMOS nodes rarely offer the design libraries required to simulate and design all portions of the complex RF transceiver. Further, the DigRF interface was also defined to assist handset makers to interface RF ICs and BB ICs from different suppliers, which allows more flexibility in component selection. DigRF also provides a standardized and low pin count debug interface to monitor traffic at the IQ interface. Indeed, a number of well-known test equipment providers distribute systems to enable this capability. For chip set makers, DigRF v4 is built upon MIPI's M-PHY, a physical layer IP which, once developed for DigRF, may be reused for other protocols and applications, since an increasing number of applicable standards use the M-PHY physical layer. IP reuse is increasingly important in meeting tough time-to-market deadlines. For example, future generations of DDR memories could make use of M-PHY, and an existing standard for FLASH already leverages M-PHY.

For RF front-end components, the current situation with regards to the types, numbers, and suppliers of these is perhaps even more diverse and far-reaching. To handset makers, RFFE is a tool which considerably simplifies the selection, ease of integration, and swapping of RF-FE components. To RF-FE IC makers, it offers increased opportunities to integrate with many chip set platforms, while implementing a single, widely-deployed control interface. With a wide array of optional features, particularly for slaves, this ensures that a bus and, in particular, the FE components may be optimized for the features implemented vs. cost and the implementation technology. RFFE's wide scope of applicability to FEMs provides increased potential for IP reuse of hardware and/or software.

As with any interfaces deployed in an RF-sensitive area, one must pay special attention to the issues of EMI/co-existence management. The specifics of each of these interfaces leads to different co-existence challenges. The main issue with DigRF v4 in this respect is that even at its lowest operating clock frequency of 1248 MHz, the first spectrum lobe of a pair of lanes overlaps all UHF cellular frequency bands. At 2496 MHz, it is nearly the entire set of 3GPP operating bands which fall under the first lobe. In addition, with high duty cycle operation for standards such as LTE, the probability of collision in the time domain is high. For example, in the case of LTE single carrier 20 MHz operation, the load of a single RX pair of lanes ranges from 137% at 1248 MHz (and thus requires two pairs of lanes) to 67% at 2496 MHz using one lane pair [5]. The inherent flat decay of the common mode PSD also poses some threats to differential LNA input structures. Factors which tend to help in reducing co-existence issues are that the interface may be operated using low voltage swings, in a controlled transmission impedance, and with bus routing that in most handset designs is unlikely to be routed at, or near, the sensitive front-end components such as antennae. RF transceiver pin-to-pin isolation is therefore the main coupling mechanism.

With RFFE being a control interface only, the bus load is often bursty, with only rare instances where higher bus loading might be required. Therefore, the probability of collision in the time domain is lowered. In addition, its relatively low frequency range of operation (32 kHz to 26 MHz) is such that even the lowest cellular frequency bands, such as the LTE 700 MHz bands, are located at the 28th and 22 500th harmonic of the extreme interface rates of 26 MHz and 32 kHz, respectively. This provides plenty of "room" to implement spectrally efficient pulse shaping techniques. Yet, in contrast to DigRF, RFFE voltage swing is higher, and its point-to-multipoint topology can lead to longer PCB traces, some of which may be routed close to sensitive front-end components. Further, since the bus is not designed to be terminated, its impedance is more difficult to control, and any reflections will tend to impair the decay of spectral lobes. Because PSD decays slowly at high harmonic numbers, and pin-to-pin isolation decreases as operating frequency increases, RFFE co-existence issues may dominate above 1 GHz of operation. The most sensitive victim of this may be GPS, which requires aggressor noise PSD < -180 dBm/Hz to prevent less than 0.2 dB desense.

To help both chip set vendors and handset makers foresee co-existence issues in the early phase of product development, and to prepare for them, both DigRF and RFFE working groups anticipated the need to include a set of EMI mitigation tools. These unique features provide a rich, efficient, and yet relatively simple-to-implement set of techniques for reducing the associated impacts in the frequency and the power domains.

> **Pulse shaping/slew rate control**: Perhaps the most efficient of all EMI tools made available to handset makers. This feature allows shaping of the rising and falling edges of each transmitted bit so as to reduce the power spectral density (PSD) of the digital bit stream. This is available in both DigRF [5] and in RFFE.

> **Amplitude control**: A simple and yet efficient way of reducing the aggressor PSD. This option is available in DigRF, and also to a certain extent in RFFE, where 1.2 V is also specified in addition to 1.8 V.

> **Clock dithering**: A useful feature in cases when repetitive and frequent trans-missions of a given pulse pattern generate discrete spurs colliding with a cellular victim. This can be the case, for example, for the DigRF training sequence trans-mitted at the beginning of each burst to ease clock recovery. This tool is available as a part of the DigRF specification. It is also inherent in RFFE, where the master clock may be dithered between messages, since the clock used for control infor-mation need not be related to RF data. Also in RFFE, the timing between messages may be randomized, offering some ability to affect the "signature" for both clock and data streams.

> **Alternate clock frequency**: This feature is available in both interfaces and can be used to either place the victim in the vicinity of a spectral "zero" or "null," or by placing the victim under a higher order spectral lobe. For example, in DigRF v4, if GPS RX is desensitized when the bus is operated at 1248 MHz, the alternate clock rate of 1496 MHz places the GPS RX in close proximity of the first DigRF spectral null. In this instance, the aggressor PSD is significantly reduced with minimal impact on DigRF performance. In RFFE the extent of any harmonics is relatively narrow, and a slight change in the fundamental may often be utilized to move a resultant spike away from a specific band frequency of interest.

With the increasing need for more reconfigurable RF subsystems, and the mounting complexity in RF-FEs, these interfaces offer handset designers effective means for the solution of complex challenges. The ability to tailor them to specific needs provides component suppliers with an efficient method to achieve faster design cycles and to foster reuse. And, because EMI effects have been taken into consideration from the outset, these interfaces provide features well-suited to the goals and situations of all those involved in UE design. The DigRF interface will be adding 4998 MHz operation as a means to extend applicability to the future higher bandwidth requirements of carrier aggregation. And work is underway on RFFE to further enhance interface throughput, which will help to improve critical timing, and provide even more bandwidth for increasingly complex RF-FEs. Thus both of these interfaces offer continued promise for the future of handset design through cost minimization and complexity reduction, while also maximizing business opportunities.

14.1.3.5   Co-Existence Challenges

The principles of co-existence have been introduced in Section 14.1.2, where due to the full duplex nature of WCDMA, a UE located at cell edge may be desensitized due to its own transmitter being operated at maximum output power. In this relatively simple example, solutions range from the use of external SAW filters, to novel RF TRX architectures, and in the most challenging cases to 3GPP relaxations.

Co-existence issues become significantly more complex when multiple radios start to interact with each other, or when a radio is operated in the presence of internal jammers that are not enabled during conformance tests. For example, WCDMA receiver sensitivity may be impacted by USB high speed harmonics during a UE to external PC media transfer. In SoCs, aggressors range from simple clock harmonics, to noisy high speed digital buses associated with DSP, CPUs, camera and display interfaces, high current rating DC–DC converters, USB and HDMi ports. As for victims, nearly all RF subsystems are susceptible to EMI interference. Further complexity arises from the fact that RF transmitter chains are not exempt from becoming victims of these digital aggressors. Desensitization occurs when aggressors collide with victims in time, frequency, and power domains. In SoCs, both conducted and electromagnetic coupling may occur. The WCDMA UE RF self-desense is one example of conducted coupling. Coupling via power supply rails and/or ground currents is another. Electromagnetic coupling may occur via pin-to-pin interactions, bonding wires, PCB track-to-track, and even antenna-to-antenna coupling. Problems are exacerbated when collisions occur between the cellular and the connectivity radios. For example, the third harmonics of band 5 fall into WLAN 2.4 GHz, while the WLAN 5 GHz receiver might become victim of either band 2, or band 4 third harmonics, or band 5 seventh harmonics. The number of scenarios may become so great that co-existence studies often have to be performed using a multidimensional systems analysis. This topic constitutes an excellent playing field for innovative mitigation techniques, which commonly fall into two categories:

- Improve victim's immunity by increasing isolation in all possible domains: floor plan optimization, careful PCB layout, the use of RF and power supply filters as well as RF shielding. Victims may also be protected by ensuring non-concurrent operations in time.

- Reduce the aggressor signal level. Slew rate and voltage swing control are two of the most efficient techniques to reduce the level of interference at RF. Slew rate control consists in shaping the rising and falling edges of digital signals. This results in an equivalent low-pass filtering of the digital signal high-frequency content, and therefore helps to reduce the aggressor power spectral density in the victim's bands. For example, DigRF^{SM}v4 line drivers are equipped with slew rate control as a tool to reduce EMI. Frequency evasion is another commonly used technique [25]. For example, if the harmonic of a digital clock is identified as a source of desensitization, it is tempting to slightly alter the clock frequency only when the victim is tuned to the blocked channel. Frequency avoidance can be implemented by either simple frequency offset, or by dithering (e.g., DigRF^{SM}v4, Section 14.1.3.4), frequency hopping, or even by using direct sequence spreading of the clock.

The well-documented example of Band 4 (B4) – Band 17 (B17) carrier aggregation is a good illustration of co-existence issues within the cellular RF subsystem. B17 transmitter chain third harmonics (H3) can entirely overlap the B4 receiver frequency band. Nearly all components in the RF-FE generate B17 H3, with the PA being, of course, the dominant source. It has been shown in [26] that the sole contribution of the PA would cause 36 to 43 dB B4 RX desensitization, therefore calling for the insertion of a harmonic rejection filter. This might not solve the problem entirely since other sources of leakage, such as PCB coupling and TRX pin-to-pin isolation may dominate system performance [27]. Given reasonable PCB isolation and typical component contributions, B4 RX desense should not exceed 7 to 9 dB for 10 MHz and 5 MHz bandwidth operation respectively [28].

14.2   Power Consumption Reduction in Terminals

14.2.1   Smartphone Power Consumption

Poor battery life is probably at the top of the list of criticisms that smartphones face today. Users commonly have to recharge their terminals on a daily basis, and perhaps more often with heavy usage. Such a need clearly affects the user's experience. Figure 14.20 provides some of the factors to help illustrate the current experience. The bar graph on the left plots the evolution of the WCDMA subsystem power consumption for feature phones, compared to a selection of recent smartphones. The measurements are performed with the LCD screen "switched-off," and at minimum transmit output power, so as to minimize both the impact of the power amplifier and screen-related contributions. Not surprisingly, the trend in power consumption reduction is similar in both families of terminals. It is interesting to note the latest generation of chip sets used in smartphones outperforms even the best cost-optimized solution of the recent feature phones. For example, phone F consumes nearly twice as much as smartphone J.

The pie charts in Figure 14.20 show the impact of display-related contributors, such as screen, backlighting, and application engine, on the overall power consumption over the duration of a WCDMA voice call. The voice call consumption is measured according to the GSMA TS09 guidelines. In feature phone F, 70% of the total power consumption is related to the cellular subsystem, while screen-related activities only account for the remaining 30%. In smartphone J, the situation is completely reversed: the "cellular"-"screen-related" contribution split is 30–70%. Despite its state-of-the-art cellular power consumption, the smartphone J total

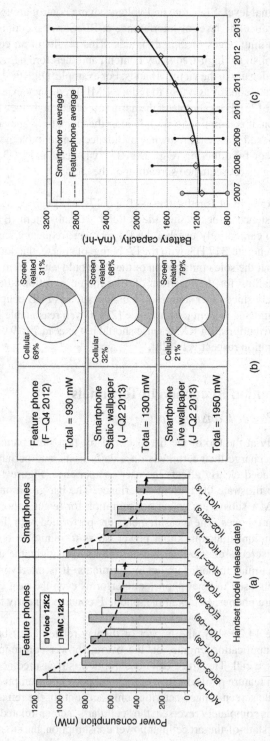

Figure 14.20 (a): screen "OFF", WCDMA power consumption at minimum output power (−50 dBm), (b): Screen "ON" brightness at 50% level, WCDMA power consumption integrated over TS09 profile, (c): Battery capacity trends vs. year: feature phone average (dashed-line), smartphone average (plain line)

power consumption, particularly when used with a live (or animated) wallpaper, is double that of feature phone F. Feature phones are almost entirely designed with an eye towards communication functions, such as voice calls, SMSs, and MMSs. However, with smartphones, recent user activity statistics [29] show that users spend only 25% of their time making voice calls. The remaining 75% of the user's time is spent on activities such as video streaming, web surfing, gaming, or activities related to entertainment or social networking. Each of these use cases implies heavier screen activity. This serves to explain why the power consumption is so high for smartphones. It also explains why, with nearly 2000 mAh battery capacity (Figure 14.20b), the modern smartphone, with its high resolution displays and more powerful application engine, requires a battery nearly twice as large, and therefore heavier, than the average feature phone. It also helps to understand why, despite this fact, that battery life is still a factor for users. And finally, the split between cellular and screen-related activity also explains why advanced 3GPP battery saving features, such as continuous packet connectivity, with discontinuous reception and transmission (DRX, DTX), have a limited impact on user experience. Section 14.2.2 highlights screen related contributors with a focus on the application engine. Techniques in mass production to improve the cellular power amplifier efficiency are presented in Section 14.2.3. Challenges associated with the implementation of CPC DRX are presented in Section 14.2.4.

With this high dependency on screen content and user activity, the estimation of battery life has become a rather complex task as the UE power consumption should now be assessed over a 24-hour user activity profile. The UE power consumption should then be assessed for each subsystem (WiFi, cellular, BT, FM, GPS), for each activity, and then could be finally averaged for a given customer profile. An example of such a profile can be found in Chapter 13. It is worth noting that, at the time of writing, the CTIA battery *ad hoc* group has launched an initiative which aims at defining such profiles so that published battery life better reflects the end user experience.

14.2.2 Application Engines

Over the last 15 years, mobile phones have evolved from voice centric, low-resolution, and small display size "feature phones" to data-centric "smartphones." With their high screen resolution and large size displays, smartphones have considerably improved web browsing and gaming user experience. Figure 14.21 shows that display size has gained on average 1 inch over the last three years, with most devices now using between 4″ and 6″ displays. At the same time, display resolution has grown exponentially to reach the full HD resolution of TV screens (1080 p). The growth in screen resolution has been one of the key drivers in the race to deliver more processing power in graphic processor units (GPU) and application engines (APE).

As pointed out in Figure 14.20, screen-related activity dramatically increases the UE power consumption. Figure 14.22 provides a detailed breakdown of the "screen-related" contributions vs. use case in a low-end Android smartphone. Camcorder, gaming, and video playback estimations assume flight mode. HSPA+ data and WCDMA voice call assume the UE screen is in "off" state. The HSPA background emulates an email synchronization use case, using a single carrier HSDPA cat 14, HSUPA cat 6, and an animated wallpaper identical to that used in the pie chart in Figure 14.20. Summing all screen-related contributors associated with

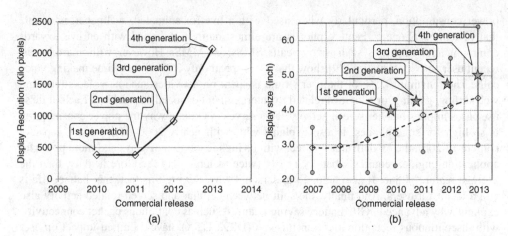

Figure 14.21 (a) Display resolution for four generations of an iconic tier-one smartphone family, (b) Display size trend

the animated wallpaper use case, the cellular/screen-power consumption ratio is 40–60% and 32–68% for HSPA background and voice call respectively. Zooming in on the screen-related contributors, the screen and its backlight circuitry consume nearly as much as the whole DC–HSPA+ cellular subsystem. The second biggest contributor is the center of attention in this section: the application engine (APE).

The APE CPU is in charge of two main tasks: general purpose processing and off-loading of computing intensive tasks to the GPU. Both CPU and GPU performance has constantly increased over the last decade. Figure 14.23a shows that in the last three years, the growth has risen exponentially, reaching near 35 000 Dhrystone Mega Instructions Per Second (DMIPS).

Relating user experience to APE processing power is not trivial, as performance experienced by the end user depends on the nature of the tasks required by the application, its mapping to either CPU, GPU, or both, and also the performance of external key components such as the dynamic random access memory (DRAM). For example, Figure 14.23b shows that doubling the CPU performance from 750 to 1500 MHz on an Android dual core CPU platform reduces the loading and initialization time of the Temple Run2 game by only a factor of 1.5. In other words, increasing the CPU/GPU clock frequency does not always result in a linear increase in performance across all use cases. In this simple example, the interface bus speed, Double Data Rate (DDR) DRAM, and IOs (eMMC, μSD) latencies may dominate the APE performance perceived by the end user. The ever-increasing CPU and GPU operating clock frequencies have pushed mobile phone DRAM vendors to reach an unprecedented level of performance. Recent DRAM can sustain transfer rates on the order of 17 GB/s. Despite this impressive level of performance, DDR may, under certain use cases, remain the performance bottleneck above a certain CPU/GPU clock frequency. One illustration is presented in Figure 14.24b which shows that layout Vellamo scores increase linearly with CPU clock frequency. Vellamo is a benchmark tool which measures HTML5 performance and user experience. However, the Ocean zoomer and scroller scores in Figure 14.24a exhibit a performance floor. In this use case, increasing CPU clock frequency above 1 GHz does not significantly improve performance.

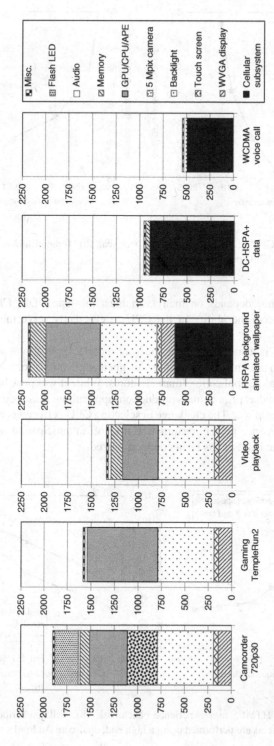

Figure 14.22 Estimated power consumption for a low-end Android 4.1 smartphone: Video playback assumes 1080p at 30 fps. All cellular activity is performed at 0 dBm transmit power. Dual-cell HSPA+ data session assumes 42 Mbit/s downlink, 11 Mbit/s uplink data, HSPA background assumes 21 Mbit/s downlink, 5.6 Mbit/s uplink data. Cellular subsystem= RF sub-system + digital baseband modem

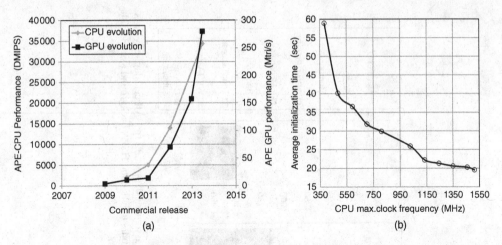

Figure 14.23 (a): CPU and GPU performance evolution over year. (b): TempleRun2 game application load and initialization time vs. CPU performance

In summary, user experience depends on many variables: intrinsic DDR, CPU and GPU performance matters, but also the ability of the CPU to efficiently serve the GPU comes into play.

Finally, user experience is also impacted by the way mobile applications are developed. The trend has evolved from pre-embedded, custom designed applications for a given terminal platform, to the world of online application (market) stores. While it was possible to optimize applications for a certain handset prior to mass production, applications must now be developed for a heterogeneous set of chip sets. The challenge in selecting CPU performance now consists in tailoring the best performance power consumption tradeoff in anticipation of applications that will be created after the handset has been commercialized.

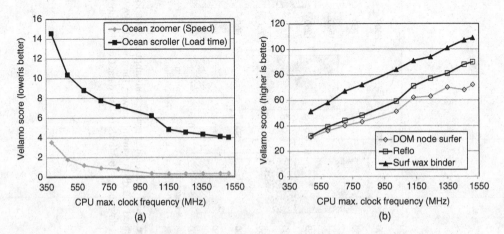

Figure 14.24 (a): Vellamo HTML5 user experience benchmark. (b): Vellamo scripting and layout performance. Both measurements are performed using a high end, dual-core Android smartphone

When it comes to increasing APE performance, three techniques are commonly used:

- Increase CPU clock speed. Increasing CPU clock to the 2 GHz mark while keeping power consumption reasonably low has been made possible thanks to the rapid CMOS technology shrink over the last decade. Two key tools have made this achievement feasible: Digital Voltage Frequency Scaling (DVFS) in hardware and Operating Processing Point (OPP) in software.
- Increase CPU pipeline length: this technique improved the DMIPS/MHz performance. In the example of the ARM[17]® CPU core family, performance has improved from 1.25 DMIPS/MHz for the ARM1176™ to 3.5DMIPS/MHz for the Cortex®-A15 processor.
- Increase the number of cores per CPU. In 2011, the introduction of Multi Processing (MP) capable cores such as the ARM Cortex-A9 MP capable core, together with symmetrical multiprocessing (SMP), enabled parallelizing of software processing on 2 to 4 Cortex®-A9 cores. These recent advances multiply the theoretical maximum performance of the CPU by a factor of 4. Nevertheless, increasing the number of cores only makes sense if the OS kernel and the applications are designed to provide a high level of parallel tasking. Applying Amdahl's law [30] to a quad-core CPU shows that performance gains become negligible if less than 75% of parallelism is effective.

However, this increase in CPU performance comes at the expense of both static and dynamic power consumption. ARM big.LITTLE™ processing aims to deliver the best combination of performance and power consumption by pairing two processors in a coherent system. In the first instance a "big" high-performance quad-core Cortex-A15 processor is paired with a "little" energy efficient, quad-core Cortex-A7 processor. The Dhrystone benchmark shows that "little" cores are 3.5 times more energy efficient than the "big" cores, while "big" cores are 1.9 times more powerful. Ideally, demanding tasks would be scheduled on the "big" cores while less-demanding and background tasks would be scheduled on the "little" cores to optimize power consumption.

There are currently three possible modes of operation for ARM big.LITTLE™ technology. The simplest approach called the "cluster migration" consists in creating two types of clusters: a cluster made up of a maximum of four "little" cores, and a cluster of up to four "big" cores. The software tasks are then either mapped to one cluster or another. This has the advantage of presenting a homogeneous set of processors to the programmers. The main disadvantage is that, at best, only half of the total number of cores is active.

The second approach, called CPU migration, is slightly more complex. It consists in creating logical pairs made up of one "big" and one "little" physical core. The special case of a quad-core APE is shown in Figure 14.25a. Compared to cluster migration, this solution remains relatively simple to operate as switching between physical cores can be done by the Linux governor using both OPP, DVFS, and CPU cores hot-plugging techniques. The main drawback of the CPU migration approach is that it also uses only half of the available cores. Further, a workload oscillating between low and high performance can lead to user experience degradation. To avoid these oscillations, a hysteresis scheme is often implemented using OPP sticky points.

---

[17] ARM and Cortex are registered trademarks of ARM Limited (or its subsidiaries) in the EU and/or elsewhere. big.LITTLE is a trademark ARM Limited (or its subsidiaries) in the EU and/or elsewhere. All rights reserved.

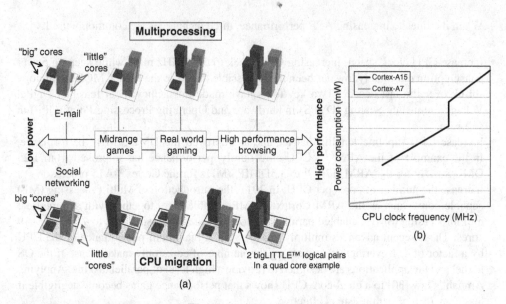

Figure 14.25 (a): ARM big.LITTLE™ CPU migration and multiprocessing concepts. (b): consumption vs. frequency vs. cores

Sticky points serve to maintain task scheduling on a given core type, and a task is migrated only if the required performance load is maintained over a certain period of time. This concept is similar to the scheme used in PA gain-switching, where hysteresis is sometimes used to prevent phase jumps from violating 3GPP requirements (see Section 14.2.3). In this sense, toggling between a "little" and a "big" core is analogous to switching the PA between its high and low power mode. Note that, in the future, nothing precludes pushing the analogy with PA control scheme further by assuming the concept of big.LITTLE™ could be extended to adopt a big-middle-little concept, in a fashion similar to the low-mid-high power modes of the cellular power amplifier.

The third approach, called "Multi-Processing" (MP), or also global task scheduling, allows all cores, or any combination of cores, to be active simultaneously (Figure 14.25a). As demand for processing power increases, load may be scheduled from one to all cores. The MP approach is more complex but it has the advantage of handling all cores. Its implementation is challenging because the kernel must be aware of task attributes in order to decide to which core the task should be scheduled. The kernel needs to have a good understanding if the task attribute is that of a foreground or background task, and of the amount of CPU load required.

A number of initiatives are currently being studied to make the best use of "big.LITTLE™" processing technologies. Among the more promising candidates is the big.LITTLE MP developed by ARM [31]. The solution allocates small tasks on "little" CPUs, without aggressively packing the tasks on as few CPUs as possible. The solution makes use of the task statistics to predict future task computation load. At the time of writing, big.LITTLE MP has just been released, and consequently early commercial implementations made use of the CPU migration and cluster migration approaches. In the near future, ARM big.LITTLE MP will most likely be the solution of choice since it delivers the best performance/power consumption tradeoff.

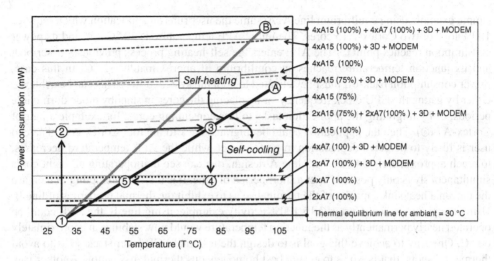

Figure 14.26 Power consumption vs. performance and thermal runaway in a 4 × 4 AMR® big.LITTLE™ processor configuration. Thick plain lines A and B are two examples of PCB thermal equilibrium lines: B corresponds to a greater heat sink capability than A

Finally, thermal dissipation is another critical aspect that must be considered when employing high performance APEs in the tiny enclosures of the modern smartphone. Transistor current leakage for process nodes such as CMOS 28 nm becomes an important factor at high operating temperatures. The higher the energy drawn by the SoC, the higher the thermal dissipation. In turn, the higher the junction temperature, the higher the current leakage, thereby leading to a positive feedback loop also sometimes referred as "thermal runaway." This phenomenon is shown in Figure 14.26 and is best illustrated when operating all eight cores at 100% load (four Cortex-A7 and four Cortex-A15). Note that the current increase with temperature tends to adopt a slight exponential curve which will be exacerbated in future CMOS shrinks. Under these circumstances, the system enters a positive feedback loop which eventually may lead to chip destruction. APE thermal monitoring then becomes necessary to stay within the safe thermal envelope of the chip. To give an idea of the power/thermal constraints, it is worth recalling that the APEs can draw up to 4 Amperes peak current.

Figure 14.26 plots the battery power consumption of an octa-core ARM® big.LITTLE™ APE for different core configurations and clock frequency vs. APE junction temperature. In this graph, it is assumed that each subsystem is active 100% of the time[18], and that the initial ambient temperature is +30 °C. Two thermal equilibrium lines (A and B) are plotted to illustrate the design tradeoff between the mobile phone heat sink capacity and APE performance under the thermal runaway constraints. Thermal stabilization occurs when one of the system configuration lines crosses the plain bold lines A or B. If the configuration line is above either A or B lines, the system is in runaway, that is, thermal self-heating. Otherwise natural cooling effect takes place due to PCB heat sink effects. Let's illustrate this through a

---

[18] This is a worst case assumption only valid when running stress benchmarking tests such as the Antutu benchmark. In real life, the APE is never running at 100% load permanently.

simple example using equilibrium line A. Assume the user runs an application which requires 100% load of two Cortex-A15 cores. The initial APE temperature is 30 °C (①), and its power consumption reaches point ②. The APE enters the self-heating process previously described, and its junction temperature reaches an equilibrium of approximately 82 °C. In this case, power consumption back-off must be activated to cool the system down. This can be achieved either by gating the CPU clock frequency, or by placing the cores in standby mode during idle periods, or by migrating the application tasks to less consuming cores, for example a pair of Cortex-A7 (④). Then the application could be migrated back to the four Cortex-A7s. The end user is likely to perceive degradation in performance while the APE temperature decreases, to reach approximately 54 °C. With line A design, one can see that operating all eight cores simultaneously is only possible over a short period of time. It is therefore tempting to increase the terminal heat sink capacity of the terminal so as to exhibit the thermal properties of line B. Using the previous dual Cortex-A15 100% activity example, using line B, the APE could be operated nearly permanently as the junction temperature would now stabilize at approximately 60 °C. One way to achieve this goal is to design the terminal PCB layer stack so as to avoid thermal hotspots, that is, so as to ensure best homogeneous thermal dissipation. Another way to increase terminal heat sink capacity is to equip smartphones with a "heat pipe" cooling system such as that implemented in [32]. But each of these solutions comes at the expense of cost, weight, or complexity.

From a user experience view, thermal back-off strategies are rather observable, since they result in a sudden slower response. In the case of bigLITTLE^{TM} processor configuration this is due to migration of foreground tasks running from the "big" cores to the "little" cores. In order to avoid such situations, next generation systems may apply a strategy similar to that of the power amplifier "average power tracking" (cf. Section 14.2.3), that is, a strategy which consists in delivering an average acceptable performance while maintaining a sufficient thermal headroom to absorb unexpected high-demand tasks initiated by the end user.

14.2.3   Power Amplifiers

Power amplifiers (PAs) must operate over a wide range of output power. For instance, an HSPA terminal must be able to adjust its output power from −50 dBm up to +23 dBm with ±0.5 dB step accuracy. The main challenge for PAs is to deliver the correct output power while fulfilling spectral emission requirements at high efficiency. This is a key facet to minimize the terminal thermal dissipation. Loss of efficiency, either induced by increased insertion losses due to the increasing complexity of the RF front-end, or by PA performance (including its supply strategy), directly translates into heat dissipation. For example, it is not uncommon for a WCDMA PA package when operated continuously at maximum output power to reach temperatures in the range of 85–90 °C. With the application engine performance being thermally limited (cf. Section 14.2.2), it is of prime importance that the PA, the second most important source of thermal dissipation, delivers the best possible efficiency. In that respect even a 3–5% gain in PA power added efficiency (PAE) can be the key differentiator to win OEM designs.

However operation at maximum power may not be representative of the most frequent power level generally in use in real networks. For instance, the GSMA TS09 statistical distribution states that the average output power is close to 0 dBm for a WCDMA voice call. In the

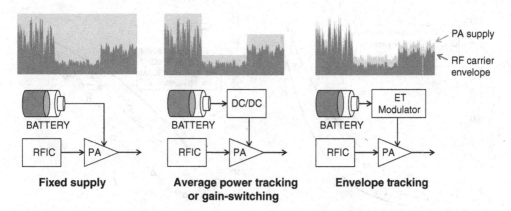

Figure 14.27 PA power supply control schemes

case of HSPA data sessions, an output power of +10 dBm is a common value used by the semiconductor industry. In LTE, the UE output power statistical distribution depends heavily on the network vendor scheduler algorithms. For example, some vendors tend to request the UE to transmit near maximum output power to save resources in the time domain. Because field performance depends on so many variables, there is no consensus when it comes to relating a telecom operator's deployment strategy to the terminal TX power distribution profile. Nevertheless, recent efforts have focused on developing power control strategies to deliver respectable PA efficiencies even at low power levels. Among these, two primary techniques are in production today: Gain Switching (GS) and Average Power Tracking (APT). Also the topic of High Efficiency PA (HEPA) recently brought up at 3GPP, a technique also known as Envelope Tracking (ET), is presented here as a third, promising alternative. At the time of writing, ET has only started appearing in a couple of high-end smartphones, and therefore only the principles and associated challenges are presented based on a survey of recently published articles in IEEE. Each control scheme is illustrated in Figure 14.27.

Gain-switching strategy relies on several PA gain modes, with each optimized for different output power ranges. Gain-switched PAs can be operated in either two or three modes. For example, a low power mode may be used from −50 dBm to around +6 dBm, a mid-power mode from +6 to around +16 dBm, and a high power mode for powers above +16 dBm. Each mode may be optimized separately to deliver the best compromise between spectral emissions and high efficiency. Since HSPA terminals spend most of their time transmitting at 0 dBm or below during voice calls, the low power mode of operation is one of the keys to delivering maximal battery life. In this mode, one common strategy consists of using RF switches to bypass an amplifier stage. Alternatively, PAs may use a dedicated secondary stage optimized for low current consumption. The benefits of toggling PA gain at a level of +17 dBm over a fixed power supply linear PA are illustrated in Figure 14.28 (diamonds vs. triangles). The relative savings remain advantageous even at low output powers. For example, at 17 dBm gain switching saves 40% current consumption (70 mA vs. 120 mA), while at 0 dBm battery current is reduced by 33% (14 mA vs. 21 mA). However, this technique presents several challenges: the toggling point must be selected carefully to ensure sufficient spectral emission margins to account for fabrication process spread, temperature, frequency, and battery voltage variations.

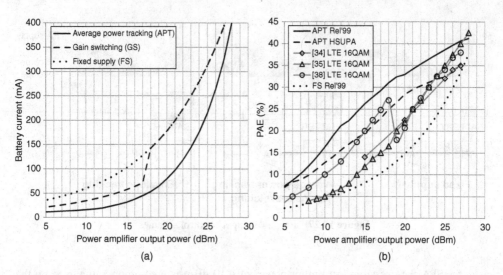

Figure 14.28 (a): APT, GS, FS PA battery current consumption for WCDMA rel'99 uplink transmissions. (b): Comparison of APT, FS, and ET power added efficiencies vs. output power. ET curves extracted from [34, 35, 38]

Additionally, mode toggling in most cases generates a sudden gain and phase jump. Both impairments are bounded by specific 3GPP test cases. Each gain step must be calibrated in mass production and compensated by the UE transmitter chain to meet the TPC accuracy of ± 0.5 dB. Phase jumps must not exceed the phase discontinuity requirements [33] and may be digitally corrected by application of I/Q constellation rotation. If the phase jump is below 60 degrees, another simple technique consists in using separate threshold points for up and down power control commands. The toggling points are separated by at least 6 dB, effectively resulting in a hysteresis scheme which provides compliance to 3GPP requirements.

Average Power Tracking (APT) is a technique in which either the PA supply is adjusted, or both supply and bias voltages are altered, according to the amplifier output power level. In this scheme, a DC–DC converter decreases the supply voltage as the UE output power decreases. APT can be seen as an extension of the GS strategy, where adjustments are made on a slot-by-slot basis to deliver the best performance tradeoff over the mid- to high-output power range. Implementing APT in an open-loop (OL) fashion delivers the best cost and simplicity of implementation, but calls for mass production calibration of gain steps associated to each bias and supply setting. This technique may also be implemented in a closed-loop (CL) fashion, wherein the PA RMS output power is measured, digitally sampled, and compared to the targeted TX power. This algorithm may apply one, or several, fine gain corrections during the first few microseconds of a timeslot to ensure the target value is met within 3GPP accuracy requirements. Early implementations of this approach can be traced back to the first few generations of UEs (cf. Figure 14.5a, b and c), where the loop uses an RF directional coupler, a logarithmic RF power detector, a low-pass filter, and auxiliary ADC [4]. Modern CL implementations may replace the discrete power detector with a dedicated measurement receiver embedded within the RF TRX. For low power levels, the loop may be disabled, leaving the PA to operate OL in its low power mode. The main challenge in CL APT consists of ensuring that repetitive gain corrections do not impair the carrier's modulation properties. One example of this is the case of HS-DPCCH burst transmissions which are not time-aligned to the slot boundaries of other uplink physical channels. Also, power detector linearity must be excellent across a wide range of environmental conditions. Yet, as of today, APT remains the most efficient technique. Figure 14.28a shows an example of current consumption savings with APT (crosses). At 0 dBm, the savings are around 60% vs. a fixed supply. Even at high power levels the current consumption is slightly decreased because some amplifiers may be able to operate at 3.2 V, which is a lower voltage than the nominal battery voltage of 3.8 V.

Finally, a third technique called Envelope Tracking (ET) has recently received much attention. The concept, illustrated in Figure 14.27, consists in applying the smallest needed supply voltage in real-time, following the instantaneous envelope of the RF carrier. This may be seen as a real-time extension of APT. The current consumption gains are expected to be higher than with APT because the supply voltage is continuously adjusted to its lowest possible level without compromising the spectral emissions requirements. Interestingly, it enables better savings for high Peak-to-Average Power Ratio (PAPR) modulations, whereas APT efficiency decreases as PAPR increases because it requires a higher supply voltage to avoid PA clipping of the envelope peaks. For example, Figure 14.28b shows that the impressive level of performance achieved by the APT scheme for WCDMA rel'99 transmissions where the peak-to-average power ratio is typically 3 dB (at 0.1% probability of presence) and is noticeably degraded as soon as HSUPA transmissions occur (dashed-lines, cubic metric = 2). In this case, ET PAE exceeds that of HSUPA APT in the upper 6 to 7 dB output range.

However, ET implementation requires high-performance, high-speed DC–DC converter cores that are able to track the envelope of high bandwidth modulated carriers, such as LTE. The envelope being roughly the square of the IQ waveform, the bandwidth over which the DC–DC converter must deliver high efficiency becomes quite large. For instance, the envelope of a 20 MHz LTE carrier occupies 40 MHz. Delivering high supply modulator efficiencies over a large bandwidth and a large range of output powers is the key to ET exceeding APT performance. ET also requires tight time synchronization between the PA supply control path and the transceiver data IQ-to-RF path. Furthermore, high-speed supply modulators can degrade out-of-band noise emissions [37].

In both APT and ET, the total system efficiency is the product of PA and DC–DC converter efficiencies. In comparison to APT, the higher PA efficiency of ET is counter-balanced by a poorer DC–DC efficiency. It is easier to achieve high converter efficiency with APT than with ET, especially when ET is applied to a 20 MHz or greater modulation BW. Typically APT DC–DC converter efficiency can reach 90%, as opposed to the 70–80% range typical for ET [38]. Some solutions use a dual-supply modulator architecture resulting in an overall performance profile similar to that of GS (Figure 14.28 [38]). At the time of writing, it is difficult to assess the exact gains of ET, since only a couple of terminals have enabled this feature. Generally, current consumption gains are expected for the highest 6 dB output power levels. Considering that ET is still in its infancy, and that great progress has been made in only a couple of years (Figure 14.28 [34, 35, 38]), there are good reasons to believe that ET will exceed APT performance levels. Since ET modulators may be entirely implemented in CMOS, this technology paves the way for commercial production of integrated linear CMOS PAs, provided that the cost of such products becomes attractive.

14.2.4   Continuous Packet Connectivity

CPC DRX / DTX is the most efficient technique to achieve significant power consumption savings as the UE entire cellular subsystem can be gated "On and Off." Under ideal conditions [4], the power consumption can be in theory halved. Yet, in most cases, the end user perception is dominated by screen-related activity power consumption which remains constant during CPC operation. For example, halving the consumption in the HSPA background use case of Figure 14.22 reduces the UE power consumption by approximately 14%, a figure close to that reported in Chapter 13. This section describes some implementation aspects of DRX with a focus on the cellular subsystem.

CPC operation is complex to analyze. The first challenge consists in deriving the UE "On/Off" activity ratio which depends on a complex timing relationship between the various uplink and downlink physical channels. An illustration of this timing complexity can be found in [4] chapter 20 (section 20.4.5), a typical list of configuration parameters in chapter 4, table 1. To simplify the UE behavior, this section uses experimental measurement plots performed in LTE DRX connected mode with no or very little TX activity. This simplifies the examination of the UE "ON" time with respect to the eNodeB DRX cycle "on" duration. This section does not pretend to present an exhaustive list of events simply because DRX operation is optimized and fine tuned for a given chip set architecture. A detailed analysis would therefore reflect choices tailored to a specific vendor implementation.

DRX is often considered as a simple ON/OFF gating with instantaneous UE wake-up "t_{wup}" (or rise time) and power down "t_{pd}" (or fall) transition times. In reality, it takes several milliseconds for the UE to transition from its deep sleep (DS) state to the active state and back again to DS. As a consequence, the real UE power consumption profile is better modelled as that of a wake-up, active "on" time "t_{on}" and a power down profile as shown in Figure 14.29a. If the DRX cycle is too short, the UE may not be able to enter the DS state. In this case, the UE maintains a certain number of HW blocks and SW tasks active to deliver the best compromise

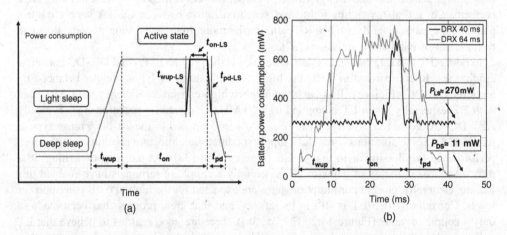

(a)   (b)

Figure 14.29   (a): UE power consumption model. (b): Power consumption traces measured on a recent Android based smartphone, in LTE DRX connected mode (no data transfer, no CQI transmission), in band 4. DRX cycle length 40 ms (black) is too short and UE remains in light-sleep state, with DRX 64 ms (gray) UE enters deep sleep state. eNodeB DRX "on" duration is set to 1 ms

between power consumption and system performance. The power consumption reaches a level that may be called "light sleep" (LS) as the power drained is somewhat at a level between the active and the DS state. Power consumption traces shown in Figure 14.29b have been measured on one recent triple mode UE and illustrate all above mentioned states and transitions. The minimum DRX cycle to enable DS on this UE is 64 ms. This phone would enter LS state for DRX cycle length between 64 and 32 ms, and would then stay at nominal power consumption for DRX cycle duration less than 32 ms. Assuming a similar wake-up/sleep processing time in WCDMA, this means that in numerous CPC configurations where the On/Off gap is only a few timeslot long, the UE cannot be gated to its complete "off" or DS state.

14.2.4.1 Deep Sleep State

This state is similar to that achieved in standby/idle cell state. The UE runs off a low power, low frequency clock. Most UEs rely on a low cost 32.768 kHz crystal to maintain the minimum required number of HW blocks active. This state allows timers such as real-time clock and system frame number (SFN) to be updated at the lowest possible power consumption. In low-cost solutions, a tradeoff is often found between crystal oscillator cost/power consumption and oscillator jitter performance. A high 32 kHz jitter may require the UE to remain in DS state for a time long enough to ensure accurate SFN tracking. This prevents the UE from entering DS for certain DRX cycle lengths. In DS state, the entire RF subsystem is switched off, the BB modem is partially supplied, but no SW code is executed. A limited number of low-drop out (LDO) and DC–DC converters are activated. All power supply regulators dedicated to peripherals are switched off whenever possible (GPS, SD card, USB bus, etc.). Figure 14.30a shows that continuous efforts are made to improve the deep sleep state power consumption: the performance has improved by a factor of 2 across three generations of chip set. Controlling leakage current is key to delivering best in class performance in this state. One common strategy in SoC devices relies on hierarchical power domain distribution schemes as a complement to a clock distribution tree [39, 40].

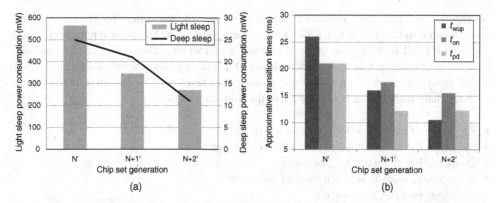

Figure 14.30 (a): Light (gray bars) and deep (black line) state power consumption vs. chipset generation. (b): State transition time evolution vs. chip set generations when UE transitions from DS to active state

14.2.4.2   Light Sleep State

The cellular subsystem runs off a low phase noise reference clock, which, depending on chip set vendors, is either a 26 MHz or a 19.2 MHz crystal. Most of the BB modem LDOs and DC–DC converters are active and the modem executes SW code. In most implementations, the RF subsystem is switched off. Architectures which rely on "smart RF transceivers" may maintain the TRX digital core active. The LS power consumption follows the general trends in reducing the active state power consumption optimization across generations of platforms [41]. Figure 14.30a shows power consumption has nearly improved by a factor of 2 over three generations.

14.2.4.3   DS to Active State Transition (wake-up "t_{wup}")

This transition is a complex cascade of HW and SW events to progressively bring all HW blocks to the active state, in the fastest (shortest rise time) and most power efficient way. Power supplies rails, or power domains, are usually activated in a hierarchical scheme [39]. When a large number of gates must be activated, the sequential power-up technique helps prevent current rush/surge issues [40] from occurring. Due to battery internal resistance, these surges may generate a drop in battery voltage which at low operating voltages could, for example, trigger a UE shutdown sequence if detected by the power management unit. The sequential power-up delay between successive domains depends on the nature of the activity of the block being supplied. Delays range from 50–100 μs for digital PLL, 80–150 μs for RF LO PLL synthesizers, to 500 μs DC–DC converter typical settling time, to 2 ms required to stabilize the low phase noise reference clock oscillator (26 or 19.2 MHz). As soon as HW blocks are supplied, numerous activities can be parallelized to minimize wake-up time (t_{wup}). The near factor of 2 reduction of t_{wup} achieved across three generations of triple mode UEs shown in Figure 14.30b shows that despite the complexity of the task, chip sets are able to continuously improve performance.

14.2.4.4   Active State ("t_{on}") – Transition from DS

Several SW and HW tasks must be executed to prepare the subsystem for physical channel demodulation/modulation. From a SW perspective, resuming the SW context requires fetching data in LP-DDR to fill up the processor local cache memory. Depending on the operating system, and modem protocol stack complexity, this task can take several milliseconds. From an RF subsystem point of view, multiple initialization tasks must be executed. Preparing the receiver chain includes but is not limited to the following steps:

- Receiver self calibration (VCOs, channel filters).
- Switch on local oscillator and tune RF PLL frequency.
- Configure RX chain transceiver filters.
- Set antenna switch to selected band of operation.
- Switching on LNA and mixer.
- Prepare AGC algorithm: initial fast gain acquisition or tracking mode?
- Set the initial RX chain gain (LNA, mixer, I/Q analog VGAs).
- Switch on I/Q ADC (in RF transceiver if digital IQ interface, in modem BB otherwise).

Once the RF IC is ready, a critical algorithm to deliver a power efficient performance is that of fast gain control, also sometimes referred to as fast gain acquisition (FGA). In WCDMA, this should ideally be completed within less than one timeslot (666.66 μs). FGA is common to compressed mode and DRX operation. Its goal is to ensure the I/Q ADC is operated at its optimal back-off. Because of multipath fading, a compromise must be found between RRSS (receiver radio signal strength) integration period to ensure good I_{or} power measurement accuracy, the number of gain corrections, gain step size boundaries, and initial gain settings to ensure convergence to the target ADC back-off across the entire UE dynamic range in a minimum number of steps. Most chip sets manage to achieve this task using three gain adjustments. One of the challenges with FGA is that ZIF receiver self-mixing due to uplink carrier leakage generates a time varying DC offset which, if not cancelled, could lead to erroneous I_{or} power measurements. DC offset compensation in WCDMA is often achieved via an equivalent high-pass filtering function, either in analog IQ, digital IQ, or a combination of coarse analog IQ and fine digital IQ DC offset compensation. The higher the equivalent HPF cut-off frequency, the higher the EVM. Since, during FGA, only RRSS measurement accuracy matters (i.e., EVM is not a concern), fast DC settling time can be achieved by selecting a high cut-off frequency. Once FGA is completed, the DC offset compensation circuitry must resume its low EVM, low cut-off frequency. The target ADC back-off can then be locked through a standard AGC loop. Finally, the UE must perform RF channel estimation. With the introduction of recent high-speed train fading profiles, the UE may have to adapt its channel estimation averaging time and update rate based on an estimation of its velocity. At high speed, a longer and more frequent channel estimation might be required. These are some of the key tasks the system must complete prior to demodulation.

14.2.4.5 LS to Active State Transition ("$t_{\text{on-LS}}$")

In most UEs, this task consists in activating the RF subsystem as the entire modem and SW code is active. The delays associated with this task are those of the previously listed sequence. The active time which follows a transition from LS state ("$t_{\text{on-LS}}$") should in theory be shorter since the only tasks required are: program default gain, and frequency values, perform FGA, and channel estimation. In the example of Figure 14.29b, $t_{\text{wup-LS}}$ lasts less than 1 ms and $t_{\text{on-LS}}$ is approximately equal to 2.6 ms for an eNodeB DRX cycle "on duration" of 1 ms.

14.2.4.6 Active State to LS to DS Transition

Switching off the RF subsystem is more straightforward and faster than the activation procedure. All blocks can be switched off nearly instantaneously. The remaining power-down delays are SW stack-related.

Finally, Figure 14.31 shows the CPC battery savings that are achieved in practice on a recent triple-mode UE operating in band 4. The measurements are performed using an Anritsu 8475a NodeB emulator during an FTP download sequence restricting the UE to operate in single cell mode, HSDPA cat 14, HSUPA cat 6. CPC parameters are similar to those of Chapter 4, Table 4.1. It can be seen that despite large inactive gaps, the UE power consumption profile is far from reaching the ideal DS power consumption. In this case, the UE power consumption

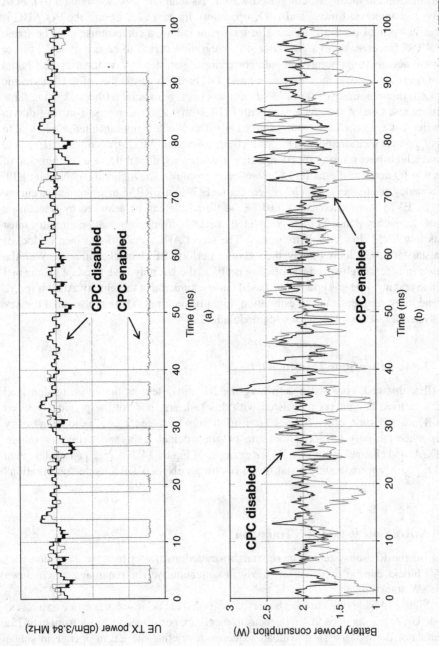

Figure 14.31 CPC battery consumption savings and UE TX power gating vs. time during an FTP file download in HSDPA cat 14. (a): UE TX power with CPC enabled vs. CPC disabled. (b): UE battery power consumption with screen "switched OFF"

savings are 18% with screen "off," and 7% when the screen is activated. The average downlink MAC layer throughput is approximately 12 Mbit/s.

Figure 14.32 has been captured by stalling the FTP download process so as to minimize the impact of background FTP server application SW tasks onto the cellular subsystem power consumption. The UE remains in CPC DTX DRX connected mode, with very few packets to receive (near zero downlink throughput), or transmit. The graph shows the UE power consumption profile closely matching the UE TX power mask when CPC is activated. The power consumption bumps between two transmissions are likely related to DRX activity. The screen is "off" but the FTP server application is still running in the background. The UE power consumption savings are close to 51%.

14.3 Conclusion

The key to successful designs in the mobile terminal industry has always been, and still remains, to deliver a timely solution that meets the desired price/performance/power consumption tradeoffs. However, it is important to note that the proper balance of these criteria is a goal that remains constantly in flux, due to the ever-changing demands and expectations of the consumer. With the wide adoption of smartphones, this tradeoff becomes ever more difficult, as the number of use cases has increased considerably. From an application engine standpoint the terminal must support a wide range of additional performance requirements, from the lower demands of texting/chatting applications, to the highly demanding 3D gaming, satellite navigation, augmented reality, or even video editing applications. The always-on requirement combined with the need to minimize product variants force transceiver designers to produce solutions that must be able to support as many of the 40 standardized frequency bands (29 FDD-LTE, 12 TD-LTE) as is possible. This, in turn, requires a multiplicity of implementation improvements, including antennae to deliver optimal efficiency across a wide range of frequency bands. From a power consumption point of view, and despite the availability of moderately higher capacity batteries, designers must deliver solutions that ensure at least a full day of use under these demanding usage conditions. In this context, the traditional design challenges of earlier generations of terminals are exacerbated as a consequence of the increasing device complexity and increased consumer expectations. Some of the resultant outcomes of this new usage paradigm for terminal designers include:

- OTA performance will have a major impact on MIMO user experience.
- Cellular transceivers must be able to support a continuously increasing number of bands, as well as the complex band combinations to support carrier-aggregation.
- As a consequence, co-existence scenarios are becoming more complicated to resolve as the number of radio-to-radio, as well as the number of digital-to-radio subsystems operating simultaneously increases.
- The limited thermal envelope of the device places severe constraints on how to best deliver expected user experience across a wide range of use cases in a small form factor.

Admittedly these challenges present opportunities for innovation, and the industry has a history of embracing challenges, so there is reason to believe it will do so again and thus will find innovative means to effectively solve these issues. This short introduction to this topic has

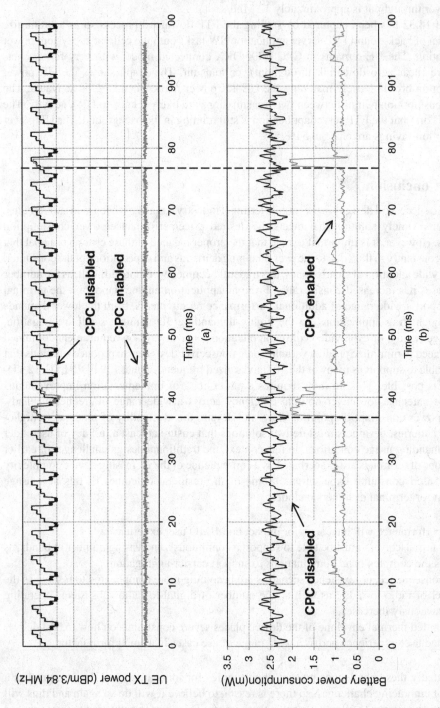

Figure 14.32 CPC battery consumption savings and UE TX power gating vs. time during a stalled FTP file download session. (a): UE TX power with CPC enabled vs. CPC disabled. (b): UE battery power consumption with screen "switched OFF"

attempted to highlight that one of the keys to delivering successful solutions has become to rely more heavily upon reconfigurable hardware architectures in as many blocks and functions as possible, such as:

- Use of hybrid CPUs in the application engine.
- Extensive use of digital signal processing in the RF transceiver, thanks to the availability of low power, high dynamic range ADCs and DACs, making the transceiver easily reconfigurable to support multiple air interfaces.
- Tunable antennae to optimize performance across a wide range of operating frequencies.
- Multimode, multiband power amplifiers.

The RF front end remains an area which still possesses less-than-optimal reconfigurability. Since today high performance dedicated RF filters and RF multiplexers are required per band or for selected band combinations, this makes the cellular RF-FE one of the primary areas where substantial challenges lie ahead for next-generation platforms. And looking ahead to carrier-aggregation, the RF front end will become even more complex, so there is a need to achieve greater reuse and reconfigurability in all areas.

Market segmentation does not help in this respect either, as platforms must be optimized for each market segment. A recent forecast indicates that low-end smartphones (<$200 price), and, to a certain extent mid-range devices ($200–300 retail price), are expected to dominate market growth, with 3x and 50% increase, respectively, by 2018. Therefore, most of the design tradeoffs which have been mentioned in this chapter must be re-assessed for each market segment and appropriate changes must be made. This increases the R&D work load for terminal designs – all in an industry where meeting an aggressive time-to-market is a primary key to success. The combination of all these constraints, as well as the pressure of huge production volumes, probably qualifies the design of the modern mobile terminal as one of the most challenging design environments in the electronics industry.

To put this in perspective – how many terminal devices or chip sets will have been shipped throughout the world by the time you read this sentence?

References

[1] Gartner press release (April 4, 2013), "Gartner Says Worldwide PC, Tablet and Mobile Phone Combined Shipments to Reach 2.4 Billion Units in 2013", http://www.gartner.com/newsroom/id/2408515 (last checked August 2013).
[2] R2-124396, "LS on extending E-UTRA band number and EARFCN numbering space", 3GPP TSG RAN WG2 Meeting #79bis, Bratislava, Slovakia, 8–12 October 2012.
[3] Graph compiled using data extracted off a total of 125 mobile phone teardown reports, mostly licensed from UBM Techinsights (www.ubmtechinsights.com) and partially from ABI research (http://www.abiresearch.com). Breakdown: 69 monomode EGPRS teardown reports, 48 dual-mode EGPRS-WCDMA terminals, 7 triple-mode EGPRS-WCDMA-LTE terminals.
[4] Holma, H. and Toskala, A. (2010) *WCDMA for UMTS: HSPA Evolution and LTE*, 5th edn, chapter 20, John Wiley & Sons, Ltd, Chichester.
[5] Holma, H. and Toskala, A. (2011) *LTE for UMTS: Evolution to LTE-Advanced*, 2nd edn, John Wiley & Sons, Ltd, Chichester.
[6] Jones, C., Tenbroek, B., Fowers, P. et al. (2007) Direct-Conversion WCDMA Transmitter with −163 dBc/Hz Noise at 190 MHz Offset. Solid-State Circuits Conference, 2007. ISSCC 2007. Digest of Technical Papers. IEEE International, pp. 336–607.

[7] Kihara, T., Sano, T., Mizokami, M. et al. (2012) A multiband LTE SAW-less CMOS transmitter with source-follower-drived passive mixers, envelope-tracked RF-PGAs, and Marchand baluns. Radio Frequency Integrated Circuits Symposium (RFIC), IEEE, pp. 399–402.

[8] Xie, H., Oliaei, O., Rakers, P. et al. (2012) Single-chip multiband EGPRS and SAW-less LTE WCDMA CMOS receiver with diversity. *IEEE Transactions on Microwave Theory and Techniques*, **60**(5), 1390–1396.

[9] Gaborieau, O., Mattisson, S., Klemmer, N. et al. (2009) A SAW-less multiband WEDGE receiver. Solid-State Circuits Conference – Digest of Technical Papers, 2009. ISSCC 2009. IEEE International, pp. 114–115, 115a.

[10] Intel XMM6140 dual-mode EGPRS-HSDPA single-chip modem, power management unit, RF transceiver solution, http://www.intel.com/content/www/us/en/wireless-products/mobile-communications/mobile-phone-platforms.html.

[11] Broadcom BCM21892, "4G LTE Advanced FDD and TDD; 3G HSPA +and TD-SCDMA; 2G-EDGE Modem with Integrated World-Band Radio", February 2013, http://www.broadcom.com/products/Cellular/4G-Baseband-Processors/BCM21892.

[12] R4-121361, "LO coupling issues for NC intra-band CA", 3GPP TSG-RAN WG4 Meeting #62bis, Jeju Island, Korea, 26–30 March, 2012.

[13] R4-131235, "Considerations for single-chip implementations of carrier aggregation", 3GPP TSG-RAN WG4 Meeting #66 Bis, Chicago, U.S.A, 15th April – 19th April, 2013.

[14] Abdelhalem, S.H., Gudem, P.S. and Larson, L.E. (2012) A tuneable differential duplexer in 90nm CMOS. *Radio Frequency Integrated Circuits Symposium (RFIC)*, 2012 IEEE, pp. 101–104.

[15] Sony Mobile C6903 (PM-0450-BV), supported bands LTE: 1,2,3,4,5,7,8,20; HSPA+/UMTS:I,II,IV,V,VIII; GSM-GPRS-EDGE:850/900/1800/1900 MHz http://www.gsmarena.com/sony_xperia_z_ultra-5540.php (last checked Aug. 2013).

[16] R4-133985, 'On the additional insertion-loss for CA_2A-4A', TSG-RAN Working Group 4 (Radio) meeting #68, Barcelona, Spain, 19–23 August 2013.

[17] 2012 CES Innovations Design and Engineering Awards in category 'Embedded Technologies': Renesas SP2531 (Pegastick) triple mode (EDGE,HSPA+,LTE) USB datacard, http://www.cesweb.org/cesweb/media/CESWeb/Innovation%20Awards/2012%20Innovations%20Honorees/Embedded%20Technologies/Pegastick.bmp.

[18] "Test Plan for Mobile Station Over the Air Performance, Method of Measurement for Radiated RF Power and Receiver Performance", revision 3.2.1, CTIA – The Wireless Association®, March 2013.

[19] Pelosi, M., Franek, O., Knudsen, M.B. et al. (2010) Antenna Proximity Effects for Talk and Data Modes in Mobile Phones. Antennas and Propagation Magazine, *IEEE*, pp. 15–27.

[20] Ranta, T., Ella, J. and Pohjonen, H. (2005) Antenna switch linearity requirements for GSM/WCDMA mobile phone front-ends. The European Conference on Wireless Technology 2005, pp. 23–26.

[21] Del Barrio, S.C. and Pedersen, G.F. (2012) Correlation Evaluation on Small LTE Handsets. Vehicular Technology Conference (VTC Fall), 2012 IEEE, Quebec City, pp. 1–4.

[22] Yanakiev, B., Nielsen, J.O., Christensen, M. and Pedersen, G.F. (2012) On Small Terminal Antenna Correlation and Impact on MIMO Channel Capacity. *IEEE Transactions on Antennas and Propagation*, 689–699.

[23] R4-66AH-0003, "Effect of user-presence on MIMO OTA using an anechoic chamber and a reverberation chamber", TSG-RAN Working Group 4 (Radio) Meeting #66 Ad hoc, Munich, Germany, 12–13 March, 2013.

[24] Boyle, K. and Leitner, M. (2011) Mobile phone antenna impedance variations with real users and phantoms. International Workshop on Antenna Technology (iWAT), 2011, pp. 420–423.

[25] Wu, T.-H., Chang, H.-H., Chen, S.-F. et al. (2013) A 65-nm GSM/GPRS/EDGE SoC with integrated BT/FM. *IEEE Journal of Solid-State Circuits*, **48**(5), 1161–1173.

[26] R4-124113, "REFSENS analysis using MSD methodology for Band 4 and Band 17 carrier aggregation", 3GPP TSG RAN WG4 Meeting #64, Qingdao, China, Aug., 13th – 17th, 2012.

[27] R4-121862, "Cross-coupling of Harmonics in case of band 17/4 Carrier Aggregation", 3GPP TSG-RAN WG4 Meeting #62bis, Jeju, Korea, March 26 – 30, 2012.

[28] R4-124359, "Interband CA Class A2 MSD", TSG-RAN Working Group 4 (Radio) meeting #64 Qingdao, P.R.China, Aug 13th – 17th, 2012.

[29] Experian marketing (May 2013), John Fetto, "Americans spend 58 minutes a day on their smartphones", http://www.experian.com/blogs/marketing-forward/2013/05/28/americans-spend-58-minutes-a-day-on-their-smartphones/ (last accessed August 2013).

[30] Cameron, K.W. and Ge, R. (2012) Generalizing Amdahl's law for power and energy. *IEEE Computer*, **45**(3).

[31] Rasmussen, M. (2013) Using Task Load Tracking to Improve Kernel Scheduler Load Balancing. The Linux Foundation Collaboration Summit, Apr 15–17 2013.

[32] NEC Casio X medias N-06E product launched by NTTdocomo in July 2013, http://www1.medias.net/jp/sp/n06e, and http://www.gsmarena.com/nec_medias_x_has_a_watercooled_snapdragon_s4_pro_chipset-news-6043.php.

[33] 3GPP TS 25.101 V12.0.0 (2013-07), www.3gpp.org.

[34] Honda, Y., Yokota, Y., Goto, N. et al. (2012) A wide supply voltage and low-rx noise envelope tracking supply modulator IC for LTE handset power amplifiers. 2012 42nd European Microwave Conference (EuMC), 2012, pp. 1253–1256.

[35] Kim, D., Kang, D. and Jooseung, K. (2012) Wideband envelope tracking power amplifier for LTE application. Radio Frequency Integrated Circuits Symposium (RFIC), 2012 IEEE, pp. 275–278.

[36] Kaczman, D.L., Shah, M., Godambe, N. et al. (2006) A single-chip tri-band (2100, 1900, 850/800 MHz) WCDMA/HSDPA cellular transceiver. *IEEE Journal of Solid-State Circuits*, **41**(5), 1122–1132. Noise at 12.5 MHz offset is the only value published.

[37] R4-132882, "Envelope tracking measurement results", 3GPP TSG RAN WG4 #67, 20–24 May 2013, Fukuoka, Japan.

[38] Kim, J., Dongsu, K., Yunsung, C. et al. (2013) Envelope-tracking two-stage power amplifier with dual-mode supply modulator for LTE applications. *IEEE Transactions on Microwave Theory and Techniques*, **61**, 543–552.

[39] Kim, G.S., Je, Y.H. and Kim, S. (2009) An adjustable power management for optimal power saving in LTE terminal baseband modem. *IEEE Transactions on Consumer Electronics*, **55**(4), 1847–1853.

[40] Kanno, Y., Mizuno, H., Yasu, Y. et al. (2007) Hierarchical power distribution with power tree in dozens of power domains for 90-nm low-power multi-CPU SoCs. *IEEE Journal of Solid-State Circuits*, **42**(1), 74–83.

[41] Lauridsen, M., Noël, L. and Mogensen, P.E. (2013) Empirical LTE Smartphone Power Model with DRX Operation for System Level Simulations. Vehicular Technology Conference (VTC Fall), 2013 IEEE.

[42] R4-120442, Way forward for inter-band CA Class A2. 3GPP TSG-RAN WG4 Meeting #62, Dresden, Germany, Feb 6th – 10th, 2012.

[43] Tomiyama, H., Nishi, C., Ozawa, N. et al. (2006) Low voltage (1.8 V) operation triple band WCDMA transceiver IC. Radio Frequency Integrated Circuits (RFIC) Symposium, 2006.

[44] Koller, R., Ruhlicke, T., Pimingsdorfer, D. and Adler, B. (2006) A single-chip 0.13 /spl mu/m CMOS UMTS W-CDMA multi-band transceiver. Radio Frequency Integrated Circuits (RFIC) Symposium, 2006.

[45] Jones, C., Tenbroek, B., Fowers, P. et al. (2007) Direct-Conversion WCDMA Transmitter with −163 dBc/Hz Noise at 190 MHz Offset. Solid-State Circuits Conference, 2007.

[46] Sowlati, T., Agarwal, B., Cho, J. et al. (2009) Single-chip multiband WCDMA/HSDPA/HSUPA/EGPRS transceiver with diversity receiver and 3G DigRF interface without SAW filters in transmitter / 3G receiver paths. Solid-State Circuits Conference, 2009, pp. 116–117, 117a.

[47] Huang, Q., Rogin, J., XinHua, C. et al. (2010) A tri-band SAW-less WCDMA/HSPA RF CMOS transceiver with on-chip DC-DC converter connectable to battery. Conference Digest of Solid-State Circuits (ISSCC), 2010, pp. 60–61.

[48] Tsukizawa, T., Nakamura, M., Do, G. et al. (2010) ISO-less, SAW-less open-loop polar modulation transceiver for 3G/GSM/EDGE multi-mode/multi-band handset. Microwave Symposium Digest (MTT), 2010 IEEE, pp. 252–255. Transceiver.

[49] Hausmann, K., Ganger, J., Kirschenmann, M. et al. (2010) A SAW-less CMOS TX for EGPRS and WCDMA. Radio Frequency Integrated Circuit Symposium (RFIC), 2010, pp. 25–28.

[50] Giannini, V., Ingels, M., Sano, T. et al. (2011) A multiband LTE SAW-less modulator with −160 dBc/Hz RX-band noise in 40 nm LP CMOS. Solid-State Circuits Conference Digest of Technical Papers (ISSCC), 2011 IEEE, pp. 374–376.

[51] Oliaei, O., Kirschenmann, M., Newman, D. et al. (2012) A Multiband Multimode Transmitter without Driver amplifier. Solid-State Circuits Conference Digest of Technical Papers (ISSCC), 2012 IEEE, pp. 164–166.

[52] ABI Research, 'Sub-$200 Smartphone Shipments to Exceed 750 Million in 2018', Oyster Bay, New York – 07 Aug 2013 press release, http://www.abiresearch.com/press/sub-200-smartphone-shipments-to-exceed-750-million.

[53] Apple A1475 supports a total of 22 bands. LTE Band 1, 2, 3, 4, 5, 7, 8, 13, 17, 18, 19, 20, 25; HSPA+/UMTS:I,II,IV,V,VIII; GSM-GPRS-EDGE:850/900/1800/1900 MHz.

15

LTE Interworking

Harri Holma and Hannu Raassina

15.1 Introduction

Long Term Evolution (LTE) networks are being rapidly rolled out but the deployment process will take some time. Therefore, there is a clear need for smooth interworking between 3G and LTE for coverage, capacity, and for service reasons. The interworking requires a number of different functionalities depending on the LTE use cases. We can differentiate the following phases of LTE network deployment:

- Data only LTE network for USB modems but not for smartphones. The interworking for the data connection is required from 3G to LTE when LTE coverage is available. Also the coverage trigger from LTE to 3G is needed to transfer the connection to 3G when running out of LTE coverage area.
- Data only LTE network, also for smartphones. The interworking solution for voice calls is required, Circuit Switched (CS) fallback, to move the UE from LTE to 3G when the voice call is initiated.
- Enhanced interworking for data, including redirection and handovers to minimize the connection break during the data transition. The first phase of data interworking may rely on simple functions based on reselections.
- Voice over LTE (VoLTE) to also carry voice in LTE requires interworking between Voice over IP (VoIP) and CS voice when running out of LTE coverage area. This is called Single Radio Voice Call Continuity (SRVCC).

The selection of the target system and target cell in the idle mode is done by the UE based on the network parameters, and the UE has no freedom in the system selection. The idle mode parameters are typically defined so that the UE will camp in LTE as long as the LTE signal level exceeds a predefined threshold. Also, network parameters typically allow the UE to move from 3G to LTE as soon as the LTE signal is strong enough. The network can move

HSPA+ Evolution to Release 12: Performance and Optimization, First Edition.
Edited by Harri Holma, Antti Toskala, and Pablo Tapia.
© 2014 John Wiley & Sons, Ltd. Published 2014 by John Wiley & Sons, Ltd.

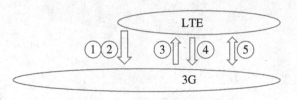

(1) = Coverage trigger from LTE to 3G for data
(2) = Coverage trigger from LTE to 3G for VoLTE (SRVCC)
(3) = Capability trigger from 3G back to LTE
(4) = CS fallback for voice service in 3G
(5) = Load and service based handover trigger in both directions

Figure 15.1   Inter-system mobility triggers

UEs between LTE and 3G also by redirection and handover procedures. The redirection is controlled by the network but the resources in the target cell are not reserved. In the case of handover, the resources in the target cell are reserved.

Several different network controlled triggers can be used for moving UEs. The different triggers for interworking are illustrated in Figure 15.1. The coverage trigger from LTE to 3G is needed when running out of LTE area. One such trigger is needed for the packet data connection to continue the data transfer in 3G, and a separate trigger is needed for the VoLTE connection to continue the voice call in the CS network in 3G. A trigger is also required to get the UE back from 3G to LTE when it is again in the LTE coverage area. It is also possible to utilize service-based or load-based interworking between LTE and 3G to balance the loading and the network utilization. CS fallback is one specific case of interworking where UE is moved from LTE to 3G for the voice call if VoLTE is not supported.

This chapter describes the interworking functionalities for data connections in Section 15.2 and for CS fallback in Section 15.3, discusses parameter optimizations in Section 15.4, presents SRVCC in Section 15.5, and concludes in Section 15.6. Interworking mobility methods are mainly described between LTE and 3G networks, but similar methods are also introduced for LTE and 2G networks. Depending on the mobile operator strategy and network rollouts, the mobility may be implemented only between LTE and 3G networks or between LTE and 2G networks as well.

The 3GPP radio protocols can be found in [1] for 3G and in [2] for LTE. The performance requirements for the interworking are defined in [3] for 3G part and in [4] for LTE part. The circuit-switched core network protocols are shown in [5] and the evolved packet core network protocols in [6]. The architecture for CS fallback is described in [7] and for SRVCC in [8].

15.2   Packet Data Interworking

The data interworking between LTE and 3G can utilize reselections, redirections, and handovers. In the early phase, reselection and redirection are used to move the UE from LTE to 3G while later handover can also be utilized to minimize the connection break. The return from 3G to LTE can use UE-based reselection or network controlled redirection and handover.

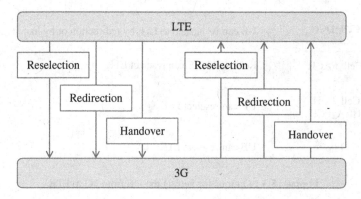

Figure 15.2   LTE to 3G packet data interworking features

The redirection and handover allow UE to get back faster to LTE. It may be hard to get UE quickly back to LTE if we only rely on reselection, since reselection is possible only if UE is in idle mode or a Cell/URA_PCH state long enough. The ongoing data transfer can delay the reselection-based mobility from 3G to LTE. These procedures are summarized in Figure 15.2.

The different interworking techniques are compared in Table 15.1. The reselection is based on the UE measurements without any network control, while redirection and handover are controlled by the network. The target cell measurements can be used with redirection, but it is also possible to make redirection blindly to a predefined target frequency without any measurements. Resource reservation in the target cell is done in the case of handover but not for redirection. The connection break is shortest with handover: typically below 1 s. The break with redirection and reselection from 3G to LTE is short, even below 2 s, while the break from LTE to 3G can be 5 to 10 s. The redirection from 3G to LTE needs RRC connection and Tracking Area Update to be performed on the LTE side, which is quick. The redirection from LTE to 3G needs more SIB reading, Location and Routing Area Updates, context modifications, and radio bearer setups. The reselection procedure takes more time than redirection because UE first needs to make measurements of the target cell and the UE needs to be in idle state, URA_PCH, or Cell_PCH before the actual decision to perform the reselection is made. Therefore, the UE can move faster from 3G to LTE with redirection than with reselection. The fast return to LTE is relevant after the CS fallback call.

Redirection from 3G to LTE can be used when the UE is in Cell_DCH state. The UE-based reselection is used when the UE is in Cell_PCH, URA_PCH, or in idle state. There used to

Table 15.1   Comparison of packet data interworking features

	Reselection	Redirection	Handover
Network controlled	No	Yes	Yes
Target cell measurements	Yes	Possible but not always used (blind)	Yes
Target cell resource reservation	No	No	Yes
Connection break	LTE to 3G: <10 s 3G to LTE: <2 s	LTE to 3G: <10 s 3G to LTE: <2 s	<1 s

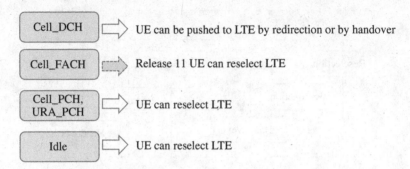

Figure 15.3   Options for getting the UE from 3G to LTE

be no possibility of going to LTE while in Cell_FACH state. The reselection capability from Cell_FACH to LTE was added only in 3GPP Release 11 as one of the sub-features of Further Enhanced FACH (FE-FACH), see Chapter 4. Cell_FACH state is normally very short and this limitation is not any major concern. But some of the 3G networks did not have Cell_PCH nor URA_PCH state active, and used instead long inactivity timers on Cell_FACH. That made it difficult for UEs to get back from 3G to LTE.

The redirection functionality can be used after the CS voice call has ended in 3G and the UE needs to be moved back to LTE. The redirection can be done in the RRC connection release in 3G when the voice call has ended. The redirection can also be used to push the UE back to LTE when the UE enters the LTE coverage area. The RNC can decide to use redirection when there is no data transfer happening in Cell_DCH state in order to minimize the impact of the redirection to the end user performance. The options for getting UE from 3G to LTE are illustrated in Figure 15.3.

When the CS fallback call is completed in 3G, UE will reselect back to LTE. The reselection procedure is shown in Figure 15.4. UE needs to run a Location Area Update (LAU) either

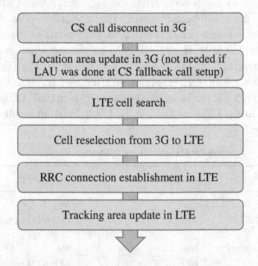

Figure 15.4   Reselection from 3G to LTE after a CS fallback call

during the CS fallback setup or after the CS call has been disconnected. The Location Area Update in 3G is required to make the incoming voice calls to route the paging to the right network. The Location Area Update takes a few seconds. It then takes some time for the UE to find LTE cells and make a Tracking Area Update to LTE. The total delay is typically 10 s to get back to LTE after the CS voice call is completed. The return to LTE may take more time even if the LTE coverage is good enough, if the UE remains in Cell_DCH or in Cell_FACH where cell reselection to LTE is not possible.

Reselection takes quite some time and also causes a noticeable break in the data transfer. Redirection or handover allow the UE to get quickly back to LTE after the voice call, improving the end user performance. The data interruption time with redirection is longer than with handover, but redirection has the benefit that it is relatively easy to create, maintain, and operate compared to the handover. The UE needs to know only the target frequency but no cell specific information in the case of redirection. With redirection, there is no need to maintain and update any cell specific neighbor information between LTE and 3G. In the case of handover, the neighbor information needs to be maintained between LTE and 3G. However, maintaining neighbor information linked to LTE and 3G relations could be more and more automatized in the future due to network and UE evolution. When the commercial Self Organizing Networks (SONs) evolve, UEs can report missing LTE to 3G neighbors to the network. Based on these UE reports, the network can create the necessary LTE-3G cell pair relations and store them in the network database.

The interruption time with handover is shorter than redirection since the resources in the target system are reserved in the case of handover. In the case of redirection, UE needs to perform registration to 3G and establish a data connection to the 3G network. This may increase the data interruption up to 12 s. The interruption time with redirection is shown in Figure 15.5. The solutions for speeding up redirection are considered in Section 15.3 together with CS fallback optimization.

Example packet handover measurements from LTE to 3G are shown in Figure 15.6. The break from the control plane point of view is just 0.2–0.3 s. The break from the user plane point of view is below 1 s, which is a major improvement compared to reselection and redirection.

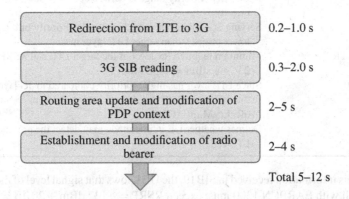

Figure 15.5   Interruption time with redirection from LTE to 3G

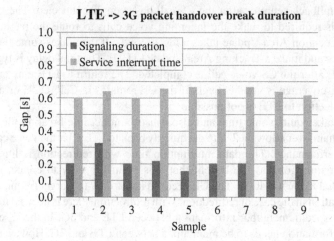

Figure 15.6   Handover measurements from LTE to 3G

15.2.1   Example Trace of 3G to LTE Cell Reselection

This example describes a typical scenario in 3G to LTE cell reselection. System information block type 19 (SIB19) is a broadcasted signaling message in 3G, which informs the UE of the criteria, how to perform 3G to LTE cell reselection. SIB19 information is shown in Figure 15.7 and explained below.

earfcn: 1300	Target LTE layer belongs to LTE band 3 meaning 1800 MHz downlink carrier frequency with E-UTRA Absolute Radio Frequency Channel Number 1300.
Measurement bandwidth:	Measurement bandwidth of target LTE layer here is 50 PRBs (mbw50). It means that target layer RSRP (Reference Signal Received Power) is measured over 10 MHz bandwidth.
Priority: 6	Target LTE layer has priority 6 in the eutra-Frequency And Priority Info List. Priority range is between 0…7. Value 7 means highest priority
Priority: 3	Serving 3G cell has priority 3 in the utra-Priority Info List, which is lower priority, than target LTE layer has.
qRxLevMinEUTRA: −66	Minimum required Rx level of the target LTE cell RSRP is 2×-66 dBm $= -132$ dBm
threshXhigh: 10	Since LTE layer has higher priority compared to 3G layer, 2×10 dB $= 20$ dB threshold is added on top of qRxLevMinEUTRA.
eutraDetection	Defines whether UE in idle mode may detect the presence of LTE cell.

Based on this information received in SIB19, the UE knows that signal level of the target LTE 1800 MHz cell with EARFCN 1300 must exceed RSRP $= -132$ dBm $+ 20$ dB $= -112$ dBm, before cell reselection from 3G to LTE can be done. The actual cell reselection decision is

Downlink

SYSTEM_INFORMATION_BLOCK_TYPE_19   (3GPP TS 25.331 ver 8.8.0 Rel 8)

SysInfoType19
  utra-PriorityInfoList
    utra-ServingCell
      priority   : 3
      s-PrioritySearch1   : 0
      s-PrioritySearch2   : 0
      threshServingLow   : 0
  eutra-FrequencyAndPriorityInfoList
    eutra-FrequencyAndPriorityInfoList value 1
      earfcn   : 1300
      measurementBandwidth   : mbw50
      priority   : 6
      qRxLevMinEUTRA   : -66
      threshXhigh   : 10
      threshXlow   : 0
      eutraDetection   : true

Figure 15.7   SIB19 in 3G

made by the UE itself based on its repetitive target LTE layer measurements. The requirement in 3GPP specification [1] is that the UE must measure LTE cells at least every 60 s in the case of one target LTE frequency and every 120 s in the case of two target LTE frequencies.

The UE measurement logs from Anite Nemo outdoor tool are shown in Figure 15.8. Based on log files, the test terminal measured the target LTE layer every 7th second. When the RSRP of the target LTE cell was strong enough, meaning the RSRP exceeding the threshold −112 dBm defined by network parameters in SIB19, cell reselection from 3G to LTE was performed. Cell reselection from 3G to LTE is possible only in the case that UE is in idle mode or Cell/URA_PCH-state.

When entering LTE, the UE performs a Tracking Area Update to indicate its location to Mobility Management Entity (MME). Authentication, ciphering, and integrity checks are done with the Authentication and Security Mode procedures. During these phases, the integrity and encryption algorithms for the EPS session are agreed. Along with these, the UE needs to establish the RRC connection which is needed to maintain the connection between UE and LTE eNodeB.

In the case that the UE QoS subscription parameters differ in LTE compared to 3G, a ModifyEPSBearerContextRequest-message may be sent by the network to the UE to modify the EPS bearer context proper for LTE after the Tracking Area Update has been done. The maximum downlink and uplink bitrates may be modified in this EPS bearer context modification procedure. The context modification does not typically take more than 0.5 s.

Event ID	System	Transf. dir.	Time	Message name
RRCSM	UMTS FDD	Downlink	7:41:14.504	SYSTEM_INFORMATION_BLOCK_TYPE_1
RRCSM	UMTS FDD	Downlink	7:41:14.504	SYSTEM_INFORMATION_BLOCK_TYPE_7
RRCSM	UMTS FDD	Downlink	7:41:14.524	SYSTEM_INFORMATION_BLOCK_TYPE_11
RRCSM	UMTS FDD	Downlink	7:41:14.524	SYSTEM_INFORMATION_BCH
RRCSM	UMTS FDD	Downlink	7:41:14.544	SYSTEM_INFORMATION_BCH
RRCSM	LTE FDD	Downlink	7:41:19.754	MasterInformationBlock
RRCSM	LTE FDD	Downlink	7:41:19.781	SystemInformationBlockType1
L3SM	LTE FDD	Uplink	7:41:19.802	TRACKING_AREA_UPDATE_REQUEST
RRCSM	LTE FDD	Uplink	7:41:19.804	RRCConnectionRequest
RRCSM	LTE FDD	Downlink	7:41:19.840	RRCConnectionSetup
RRCSM	LTE FDD	Uplink	7:41:19.896	RRCConnectionSetupComplete
RRCSM	LTE FDD	Downlink	7:41:19.940	DLInformationTransfer
L3SM	LTE FDD	Downlink	7:41:19.941	AUTHENTICATION_REQUEST
RRCSM	LTE FDD	Uplink	7:41:20.434	ULInformationTransfer
L3SM	LTE FDD	Uplink	7:41:20.434	AUTHENTICATION_RESPONSE
RRCSM	LTE FDD	Downlink	7:41:20.457	DLInformationTransfer
L3SM	LTE FDD	Downlink	7:41:20.457	SECURITY_MODE_COMMAND
L3SM	LTE FDD	Uplink	7:41:20.459	SECURITY_MODE_COMPLETE
RRCSM	LTE FDD	Uplink	7:41:20.460	ULInformationTransfer
RRCSM	LTE FDD	Downlink	7:41:21.485	DLInformationTransfer
L3SM	LTE FDD	Downlink	7:41:21.486	TRACKING_AREA_UPDATE_ACCEPT
L3SM	LTE FDD	Uplink	7:41:21.488	TRACKING_AREA_UPDATE_COMPLETE
RRCSM	LTE FDD	Uplink	7:41:21.489	ULInformationTransfer
RRCSM	LTE FDD	Downlink	7:41:21.500	SystemInformationBlockType1
RRCSM	LTE FDD	Downlink	7:41:21.508	DLInformationTransfer
L3SM	LTE FDD	Downlink	7:41:21.509	EMM_INFORMATION
RRCSM	LTE FDD	Downlink	7:41:21.855	RRCConnectionReconfiguration
RRCSM	LTE FDD	Uplink	7:41:21.856	RRCConnectionReconfigurationComplete
L3SM	LTE FDD	Downlink	7:41:21.857	MODIFY_EPS_BEARER_CONTEXT_REQUEST
L3SM	LTE FDD	Uplink	7:41:21.858	MODIFY_EPS_BEARER_CONTEXT_ACCEPT

Figure 15.8   3G to LTE cell reselection with tracking area update

15.2.2   Example Trace of LTE to 3G Redirection

This example shows blind redirection from LTE to 3G with an active packet connection. The UE asks for an RRC connection with RRCConnectionRequest with cause value Mobile Originated data in Figure 15.9. The UE already has the default EPS bearer context activated. Ciphering and integrity protection algorithms are agreed in the SecurityModeCommand for the active RRC connection. Since the UE informs in UECapabilityInformation that it supports InterRAT-mobility from LTE to 3G and 2G, eNodeB informs InterRAT-mobility criteria and related triggers in RRCConnectionReconfiguration-message.

Event ID	System	Transf. dir.	Time	Message name
RRCSM	LTE FDD	Downlink	15:39:35.150	DLInformationTransfer
RRCSM	LTE FDD	Downlink	15:39:35.151	RRCConnectionRelease
L3SM	LTE FDD	Downlink	15:39:35.159	EMM_INFORMATION
RRCSM	LTE FDD	Uplink	15:39:35.262	RRCConnectionRequest
RRCSM	LTE FDD	Downlink	15:39:35.292	RRCConnectionSetup
RRCSM	LTE FDD	Uplink	15:39:35.294	RRCConnectionSetupComplete
L3SM	LTE FDD	Uplink	15:39:35.294	SERVICE_REQUEST
RRCSM	LTE FDD	Downlink	15:39:35.319	SecurityModeCommand
RRCSM	LTE FDD	Uplink	15:39:35.321	SecurityModeComplete
RRCSM	LTE FDD	Downlink	15:39:35.321	UECapabilityEnquiry
RRCSM	LTE FDD	Uplink	15:39:35.322	UECapabilityInformation
RRCSM	LTE FDD	Downlink	15:39:35.342	RRCConnectionReconfiguration
RRCSM	LTE FDD	Uplink	15:39:35.347	RRCConnectionReconfigurationComplete
RRCSM	LTE FDD	Downlink	15:39:35.753	RRCConnectionReconfiguration
RRCSM	LTE FDD	Uplink	15:39:35.756	RRCConnectionReconfigurationComplete
RRCSM	LTE FDD	Downlink	15:39:36.725	Paging
RRCSM	LTE FDD	Downlink	15:39:44.406	Paging
RRCSM	LTE FDD	Downlink	15:39:45.045	Paging
RRCSM	LTE FDD	Uplink	15:39:51.185	MeasurementReport
RRCSM	LTE FDD	Downlink	15:39:51.212	RRCConnectionRelease
RRCSM	UMTS FDD	Downlink	15:39:51.641	SYSTEM_INFORMATION_BCH
RRCSM	UMTS FDD	Downlink	15:39:51.662	MASTER_INFORMATION_BLOCK
RRCSM	UMTS FDD	Downlink	15:39:51.662	SYSTEM_INFORMATION_BCH

Figure 15.9   LTE to 3G redirection

LTE to 3G or 2G redirect triggers are presented in Figure 15.10. InterRAT-mobility is based on event triggering. This means that the UE reports to the network if a predefined event happens. EventA2 means that the serving LTE cell has become worse than the threshold.

Hysteresis: 0	Hysteresis for A2 event
A2-Threshold RSRP: 18	UE must send a measurement report to the network if RSRP goes below -140 dBm + eventA2threshold + hysteresis = -140 dBm + 18 dB + 0 dB = -122 dBm
timeToTrigger: ms256	A2 criteria must be valid at least 256 ms, before measurement report A2 can be sent
triggerQuantity: rsrp	RSRP is the triggering quantity for LTE to 3G redirect

When LTE RSRP goes below the eventA2 threshold, the UE sends a measurement report to the network. In this example, the UE indicates in the measurement report that the measured RSRP has gone already to -140 dBm + 15 dB = -125 dBm, which is below the eventA2

```
reportConfigToAddModList value 3
 reportConfigId    : 3
 reportConfig
  reportConfigEUTRA
   triggerType
    event
     eventId
      eventA2
       a2-Threshold
        threshold-RSRP    : 18
      hysteresis   : 0
     timeToTrigger    : ms256
   triggerQuantity   : rsrp
   reportQuantity    : sameAsTriggerQuantity
   maxReportCells    : 8
   reportInterval    : min60
   reportAmount    : r1
```

Figure 15.10   LTE to 3G redirection threshold in RRC connection reconfiguration

threshold defined LTE to 3G redirect threshold of -140 dBm $+ 18$ dB $= -122$ dBm. The measurement report is shown in Figure 15.11.

Based on that measurement report and LTE to 3G redirect mobility relation defined in the network, eNodeB commands the UE to 3G network with the RRCConnectionRelease-message. The example in Figure 15.12 shows a blind redirect method where eNodeB informs the UE only about the target 3G UTRAN Absolute Radio Frequency Channel Number (UARFCN). Based on UARFCN utra-FDD: 10 837 the UE knows that the target 3G belongs to a 2100 MHz network. After this, the UE leaves the LTE network and starts seeking and camping on the 2100 MHz 3G cell with UARFCN 10 837.

UE camping to 3G is shown in Figure 15.13. When the UE has synchronized to the 3G cell, it starts reading 3G system information. The system information presents, for example,

```
Uplink

MeasurementReport    (3GPP TS 36.331 ver 9.5.0 Rel 9)

UL-DCCH-Message
 message
  c1
   measurementReport
    criticalExtensions
     c1
      measurementReport-r8
       measResults
        measId   : 4
        measResultServCell
        rsrpResult   : 15
        rsrqResult   : 21
```

Figure 15.11   Measurement report for eventA2

```
Downlink

RRCConnectionRelease     (3GPP TS 36.331 ver 9.5.0 Rel 9)

DL-DCCH-Message
  message
    c1
      rrcConnectionRelease
        rrc-TransactionIdentifier    : 3
        criticalExtensions
          c1
            rrcConnectionRelease-r8
              releaseCause    : other
              redirectedCarrierInfo
                utra-FDD    : 10837
```

Figure 15.12   Redirect from LTE to 3G with RRC connection release

the following information to the UE: Idle mode mobility criteria in 3G presented in SIB3, 3G to LTE mobility criteria in SIB19, and uplink interference information in SIB7. If the camped 3G cell is the same cell where the UE last time camped in 3G, the UE may have stored part of that specific 3G cell system information to its memory like the 3G cell neighboring cell information stored in SIB11. In this example, it can be seen that SIB11 is not read. It is stored to the UE's memory from the previous mobility session which was done to this same 3G cell.

After all the relevant system information is received, UE can start RRC connection establishment with cause value "registration." That is needed for the Location and Routing Area Updates, presented in Figure 15.14, which are needed for the network to know that the UE is now in the 3G network under the specific 3G Location and Routing areas defined with Location and Routing Area Codes (LAC, RAC). The UE identity is verified with IdentityRequests and indicated with International Mobile Subscriber Identity (IMSI) and International Mobile Station Equipment Identity and Software Version (IMEISV). Also, the authentication and ciphering are handled and agreed with the related procedures. 3G mobility criteria and connected mode neighbor lists are informed with MeasurementControl-messages. After the Routing Area Update is accepted, the UE context is modified with ModifyPDPContextRequest to meet the 3G QoS parameters, which are defined in the network based on customer subscription. For example, the maximum bitrates for downlink and uplink for 3G connection are set here.

After the PDP context is modified, UE may send a ServiceRequest to the network including information on active PDP context and pending uplink data. Pending data triggers the setup of radio bearer, which is modified with a RadioBearerReconfiguration to provide a packet data connection. In this example, the packet data connection was reconfigured to the HSPA connection with 21 Mbps downlink and 5.8 Mbps uplink capability. The data transfer can continue now. The 3G radio bearer setup is shown in Figure 15.15.

In this example, the data outage was around 10 s starting from LTE to 3G redirect triggered by RRC connection release in LTE and finally ending at RadioBearerReconfiguration in 3G.

Event ID	System	Transf. dir.	Time	Message name
RRCSM	LTE FDD	Downlink	15:39:44.406	Paging
RRCSM	LTE FDD	Downlink	15:39:45.045	Paging
RRCSM	LTE FDD	Uplink	15:39:51.185	MeasurementReport
RRCSM	LTE FDD	Downlink	15:39:51.212	RRCConnectionRelease
RRCSM	UMTS FDD	Downlink	15:39:51.641	SYSTEM_INFORMATION_BCH
RRCSM	UMTS FDD	Downlink	15:39:51.662	MASTER_INFORMATION_BLOCK
RRCSM	UMTS FDD	Downlink	15:39:51.662	SYSTEM_INFORMATION_BCH
RRCSM	UMTS FDD	Downlink	15:39:51.682	SYSTEM_INFORMATION_BLOCK_TYPE_1
RRCSM	UMTS FDD	Downlink	15:39:51.682	SYSTEM_INFORMATION_BCH
RRCSM	UMTS FDD	Downlink	15:39:51.682	SYSTEM_INFORMATION_BLOCK_TYPE_7
RRCSM	UMTS FDD	Downlink	15:39:51.701	SYSTEM_INFORMATION_BCH
RRCSM	UMTS FDD	Downlink	15:39:51.721	SYSTEM_INFORMATION_BLOCK_TYPE_19
RRCSM	UMTS FDD	Downlink	15:39:51.721	SYSTEM_INFORMATION_BLOCK_TYPE_5
RRCSM	UMTS FDD	Downlink	15:39:51.721	SYSTEM_INFORMATION_BCH
RRCSM	UMTS FDD	Downlink	15:39:51.742	SYSTEM_INFORMATION_BCH
RRCSM	UMTS FDD	Downlink	15:39:51.742	MASTER_INFORMATION_BLOCK
RRCSM	UMTS FDD	Downlink	15:39:51.761	SYSTEM_INFORMATION_BLOCK_TYPE_7
RRCSM	UMTS FDD	Downlink	15:39:51.761	SYSTEM_INFORMATION_BLOCK_TYPE_1
RRCSM	UMTS FDD	Downlink	15:39:51.761	SYSTEM_INFORMATION_BCH
RRCSM	UMTS FDD	Downlink	15:39:51.781	SYSTEM_INFORMATION_BLOCK_TYPE_3
RRCSM	UMTS FDD	Downlink	15:39:51.781	SYSTEM_INFORMATION_BCH
RRCSM	UMTS FDD	Downlink	15:39:51.816	SYSTEM_INFORMATION_BCH
RRCSM	UMTS FDD	Downlink	15:39:51.821	MASTER_INFORMATION_BLOCK
RRCSM	UMTS FDD	Downlink	15:39:51.821	SYSTEM_INFORMATION_BCH
RRCSM	UMTS FDD	Downlink	15:39:51.856	SYSTEM_INFORMATION_BLOCK_TYPE_1
RRCSM	UMTS FDD	Downlink	15:39:51.856	SYSTEM_INFORMATION_BCH
RRCSM	UMTS FDD	Downlink	15:39:51.856	SYSTEM_INFORMATION_BLOCK_TYPE_7
RRCSM	UMTS FDD	Downlink	15:39:52.041	SYSTEM_INFORMATION_BCH
RRCSM	UMTS FDD	Downlink	15:39:52.041	SYSTEM_INFORMATION_BLOCK_TYPE_19
RRCSM	UMTS FDD	Downlink	15:39:52.061	MASTER_INFORMATION_BLOCK
RRCSM	UMTS FDD	Downlink	15:39:52.061	SYSTEM_INFORMATION_BCH
RRCSM	UMTS FDD	Downlink	15:39:52.081	SYSTEM_INFORMATION_BLOCK_TYPE_1
RRCSM	UMTS FDD	Downlink	15:39:52.081	SYSTEM_INFORMATION_BLOCK_TYPE_7
RRCSM	UMTS FDD	Downlink	15:39:52.081	SYSTEM_INFORMATION_BCH
RRCSM	UMTS FDD	Uplink	15:39:52.154	RRC_CONNECTION_REQUEST
RRCSM	UMTS FDD	Downlink	15:39:52.514	RRC_CONNECTION_SETUP
RRCSM	UMTS FDD	Uplink	15:39:52.801	RRC_CONNECTION_SETUP_COMPLETE
RRCSM	UMTS FDD	Uplink	15:39:52.816	INITIAL_DIRECT_TRANSFER
L3SM	UMTS FDD	Uplink	15:39:52.816	LOCATION_UPDATING_REQUEST
RRCSM	UMTS FDD	Uplink	15:39:52.817	INITIAL_DIRECT_TRANSFER
L3SM	UMTS FDD	Uplink	15:39:52.817	ROUTING_AREA_UPDATE_REQUEST
RRCSM	UMTS FDD	Downlink	15:39:53.253	DOWNLINK_DIRECT_TRANSFER

Figure 15.13 Camping to 3G after LTE to 3G redirection

Event ID	System	Transf. dir.	Time	Message name
RRCSM	UMTS FDD	Uplink	15:39:52.801	RRC_CONNECTION_SETUP_COMPLETE
RRCSM	UMTS FDD	Uplink	15:39:52.816	INITIAL_DIRECT_TRANSFER
L3SM	UMTS FDD	Uplink	15:39:52.816	LOCATION_UPDATING_REQUEST
RRCSM	UMTS FDD	Uplink	15:39:52.817	INITIAL_DIRECT_TRANSFER
L3SM	UMTS FDD	Uplink	15:39:52.817	ROUTING_AREA_UPDATE_REQUEST
RRCSM	UMTS FDD	Downlink	15:39:53.253	DOWNLINK_DIRECT_TRANSFER
L3SM	UMTS FDD	Downlink	15:39:53.253	IDENTITY_REQUEST
L3SM	UMTS FDD	Uplink	15:39:53.259	IDENTITY_RESPONSE
RRCSM	UMTS FDD	Uplink	15:39:53.259	UPLINK_DIRECT_TRANSFER
RRCSM	UMTS FDD	Downlink	15:39:53.533	MEASUREMENT_CONTROL
RRCSM	UMTS FDD	Downlink	15:39:53.653	MEASUREMENT_CONTROL
RRCSM	UMTS FDD	Uplink	15:39:53.685	MEASUREMENT_REPORT
L3SM	UMTS FDD	Downlink	15:39:53.973	AUTHENTICATION_AND_CIPHERING_REQUEST
RRCSM	UMTS FDD	Downlink	15:39:53.973	DOWNLINK_DIRECT_TRANSFER
RRCSM	UMTS FDD	Downlink	15:39:54.053	ACTIVE_SET_UPDATE
RRCSM	UMTS FDD	Uplink	15:39:54.073	ACTIVE_SET_UPDATE_COMPLETE
L3SM	UMTS FDD	Uplink	15:39:54.161	AUTHENTICATION_AND_CIPHERING_RESPONSE
RRCSM	UMTS FDD	Uplink	15:39:54.161	UPLINK_DIRECT_TRANSFER
L3SM	UMTS FDD	Downlink	15:39:54.173	AUTHENTICATION_REQUEST
RRCSM	UMTS FDD	Downlink	15:39:54.173	DOWNLINK_DIRECT_TRANSFER
RRCSM	UMTS FDD	Uplink	15:39:54.388	UPLINK_DIRECT_TRANSFER
L3SM	UMTS FDD	Uplink	15:39:54.388	AUTHENTICATION_RESPONSE
RRCSM	UMTS FDD	Downlink	15:39:54.413	MEASUREMENT_CONTROL
RRCSM	UMTS FDD	Downlink	15:39:54.493	MEASUREMENT_CONTROL
RRCSM	UMTS FDD	Downlink	15:39:54.573	SECURITY_MODE_COMMAND
RRCSM	UMTS FDD	Uplink	15:39:54.577	SECURITY_MODE_COMPLETE
RRCSM	UMTS FDD	Downlink	15:39:54.973	SECURITY_MODE_COMMAND
RRCSM	UMTS FDD	Uplink	15:39:54.975	SECURITY_MODE_COMPLETE
RRCSM	UMTS FDD	Downlink	15:39:55.253	DOWNLINK_DIRECT_TRANSFER
L3SM	UMTS FDD	Downlink	15:39:55.253	IDENTITY_REQUEST
L3SM	UMTS FDD	Uplink	15:39:55.255	IDENTITY_RESPONSE
RRCSM	UMTS FDD	Uplink	15:39:55.255	UPLINK_DIRECT_TRANSFER
RRCSM	UMTS FDD	Downlink	15:39:56.613	DOWNLINK_DIRECT_TRANSFER
L3SM	UMTS FDD	Downlink	15:39:56.613	ROUTING_AREA_UPDATE_ACCEPT
RRCSM	UMTS FDD	Uplink	15:39:56.627	UPLINK_DIRECT_TRANSFER
L3SM	UMTS FDD	Uplink	15:39:56.627	ROUTING_AREA_UPDATE_COMPLETE
RRCSM	UMTS FDD	Downlink	15:39:56.931	DOWNLINK_DIRECT_TRANSFER
L3SM	UMTS FDD	Downlink	15:39:56.931	MODIFY_PDP_CONTEXT_REQUEST
RRCSM	UMTS FDD	Uplink	15:39:56.940	UPLINK_DIRECT_TRANSFER
L3SM	UMTS FDD	Uplink	15:39:56.940	MODIFY_PDP_CONTEXT_ACCEPT
RRCSM	UMTS FDD	Downlink	15:39:57.850	DOWNLINK_DIRECT_TRANSFER
L3SM	UMTS FDD	Downlink	15:39:57.850	LOCATION_UPDATING_ACCEPT

Figure 15.14 Mobility procedures in 3G after redirection from LTE

Event ID	System	Transf. dir.	Time	Message name
L3SM	UMTS FDD	Uplink	15:39:56.940	MODIFY_PDP_CONTEXT_ACCEPT
RRCSM	UMTS FDD	Downlink	15:39:57.850	DOWNLINK_DIRECT_TRANSFER
L3SM	UMTS FDD	Downlink	15:39:57.850	LOCATION_UPDATING_ACCEPT
L3SM	UMTS FDD	Uplink	15:39:57.853	TMSI_REALLOCATION_COMPLETE
RRCSM	UMTS FDD	Uplink	15:39:57.853	UPLINK_DIRECT_TRANSFER
L3SM	UMTS FDD	Uplink	15:39:57.902	SERVICE_REQUEST
RRCSM	UMTS FDD	Uplink	15:39:57.902	UPLINK_DIRECT_TRANSFER
RRCSM	UMTS FDD	Downlink	15:39:57.930	DOWNLINK_DIRECT_TRANSFER
L3SM	UMTS FDD	Downlink	15:39:57.930	MM_INFORMATION
L3SM	UMTS FDD	Downlink	15:39:57.931	SERVICE_ACCEPT
RRCSM	UMTS FDD	Downlink	15:39:57.931	DOWNLINK_DIRECT_TRANSFER
RRCSM	UMTS FDD	Uplink	15:39:57.931	RADIO_BEARER_SETUP_COMPLETE
RRCSM	UMTS FDD	Downlink	15:39:57.931	RADIO_BEARER_SETUP
RRCSM	UMTS FDD	Downlink	15:39:57.931	SIGNALLING_CONNECTION_RELEASE
RRCSM	UMTS FDD	Downlink	15:39:58.071	MEASUREMENT_CONTROL
RRCSM	UMTS FDD	Downlink	15:39:58.191	MEASUREMENT_CONTROL
RRCSM	UMTS FDD	Downlink	15:39:58.271	MEASUREMENT_CONTROL
RRCSM	UMTS FDD	Downlink	15:39:58.311	MEASUREMENT_CONTROL
RRCSM	UMTS FDD	Downlink	15:39:58.391	MEASUREMENT_CONTROL
RRCSM	UMTS FDD	Uplink	15:39:58.406	MEASUREMENT_REPORT
RRCSM	UMTS FDD	Uplink	15:39:58.631	MEASUREMENT_REPORT
RRCSM	UMTS FDD	Downlink	15:40:00.010	RADIO_BEARER_RECONFIGURATION
RRCSM	UMTS FDD	Uplink	15:40:01.062	RADIO_BEARER_RECONFIGURATION_COMPLETE

Figure 15.15   3G Radio bearer setup in 3G after LTE to 3G redirection

15.3   Circuit-Switched Fallback

The target of the CS fallback is to use the existing 3G or 2G network to carry voice calls while the LTE network is used for data connections. CS fallback allows provision of voice services for LTE smartphones without implementing VoIP in LTE. CS fallback is the main voice solution in LTE smartphones and networks currently. The CS fallback can be targeted to 3G, GSM, or CDMA-1X. The most common target system is 3G WCDMA.

The CS fallback procedure is illustrated in Figure 15.16 and in Figure 15.17. The CS fallback UE makes a combined attachment to the LTE network and the MME executes a Location Area Update towards a serving MSS (Mobile Switching Center Server) to announce the presence of the UE to the CS core network. This procedure uses the SG's interface between MME and MSS. MSS then knows the UE's Location Area and serving MME, and can thus forward the paging message to the Mobility Management Entity (MME) which sends paging to the correct Tracking Area. When MME has sent the paging, the UE establishes an RRC connection in LTE and responds to the paging with an Extended Service Request to MME, which forwards it to MSS. The MME indicates eNodeB to move the UE to 3G or 2G by sending an Initial Context Setup, and the eNodeB redirects the UE to the 3G network. The UE finds the 3G target cell and establishes a RRC connection. The UE may need to read the System Information Block (SIB) in the target cell. The UE will then establish a RRC connection to 3G. The UE may also

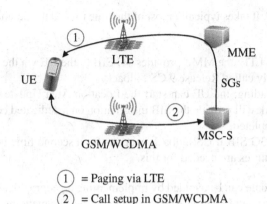

Figure 15.16 CS fallback functionality

need to perform a Location Area Update. If the UE does not perform a Location Area Update, it sends a Paging Response to the MSC, which continues the call setup. The Routing Area Update in 3G runs in parallel to the CS call setup and does not impact the call setup time.

The main factors impacting the CS fallback setup time are SIB reading in WCDMA and the Location Area Update in WCDMA. The setup time can be minimized by avoiding SIB reading and by avoiding Location Area Update. There are multiple solutions to avoid SIB reading or

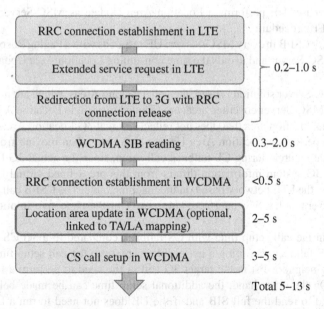

Figure 15.17 CS fallback call setup flow

part of the SIB reading. It takes typically most of the time to read the neighbor list information stored in SIB11/12:

- 3G SIB delivery via LTE: the MME provides 3G SIB to the UE via the LTE network. This procedure is normally called Release 9 CS fallback.
- Deferred 3G SIB reading: the UE can start the Location Area Update before all SIBs are read. The 3G network will provide the SIB information on a dedicated control channel after the call setup is completed.
- UEs can remember 3G SIB if CS fallback happens for a second time in a row to the same cell. No network features are needed for this.

The Location Area Update can be avoided by implementing a geographically correct mapping between Tracking Areas and Location Areas to the network. The mapping refers to a database in MME managing the relationship between geographically overlapping Tracking Areas and Location Areas. The corresponding Location Area Code is informed to the UE in the LTE network Tracking Area Update procedure with the Tracking Area Update Accept-message or already in the LTE attach phase with the Attach Accept-message. Location Area Update during CS fallback can be avoided if SGs interface is enabled to MSC-Server and Tracking Area/Location Area mapping matches with the Tracking Area and Location Area combination observed by the UE during mobility between LTE and 3G. The overlay solution refers to the case where at least one, but not all, of the MSC-Servers have the SGs interface. However, typically only one CSFB MSC-Server has SGs interface in this type of solution. In the terminating call case, CSFB MSC-Server pages the UE in LTE via MME in the same way as in 2G/3G. After the paging, the UE performs CSFB to 2G/3G MSC-Server and proceeds as follows in the two cases:

- CSFB overlay case: UE performs a Location Area Update as MSC-Server has changed during the CSFB procedure.
- SGs interface for CSFB in every MSC-Server: UE responds with a Paging Response message to the same MSC-Server and avoids the time consuming Location Area Update procedure.

If an overlay MSC-Server solution is used, then all Tracking Areas, which are not overlapping with the overlay MSC-Server coverage area, are mapped to a single Location Area, which can be termed a virtual location area that does not reflect any real 3G Location Area. This method forces the UE to perform a Location Area Update each time when moving from LTE to 3G to the virtual location area during CS fallback call setup, since the actual 3G Location Area Code read from 3G system information differs from this pre-defined virtual Location Area Code informed by the LTE network based on the mapping table. The CSFB call is then routed via the MSC-Server that has SGs interface. These two CSFB architecture solutions are shown in Figure 15.18.

The increase in the call setup time with CS fallback compared to a 3G CS call is shown in Table 15.2. CS fallback can bring a noticeable increase to the call setup time in the case that there is long neighbor list in the target 3G cell or the need to perform a Location Area Update in 3G. On the other hand, the additional setup time can be made below 1 s if the UE does not need to read the full SIB and if the UE does not need to run a Location Area Update. Figure 15.19 shows example setup time measurements in a well-planned live network

Overlay CSFB solution

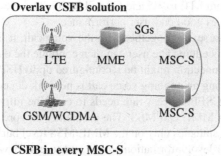

CSFB in every MSC-S

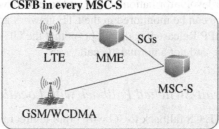

Figure 15.18 Overlay CSFB solution and CSFB in every MSC-S

Table 15.2 Call setup time increase with CS fallback

	Location area update avoided	With location area update
With 3G SIB reading	0.5–3.0 s	2.0–8.0 s
Skip 3G SIB reading (Deferred SIB or Release 9 CSFB)	<1.0 s	1.5–6.0 s

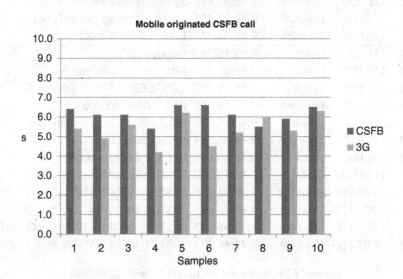

Figure 15.19 CS fallback setup time measurements

comparing CS fallback from LTE to 3G and CS call in 3G. The average increase in the setup time is 0.7 s, which provides good end-user performance.

If there is a parallel data session together with the voice call, it is handed over from LTE to 3G during the CSFB procedure. The user can then continue the data connection during the CS voice call. The data connection might be reconfigured up to HSPA data rates, but support of HSPA functionality during an ongoing voice call is network dependent.

The monitoring of the CSFB success rate needs to combine information from LTE radio, WCDMA radio, and from MME and MSS. The problem with core network counters is that they give the performance statistics only on the MME/MSS level but not on the cell level. The radio counters provide cell-level information but the end-to-end view is missing since the first part of the CSFB procedure can be monitored in the LTE network and the actual call setup in the WCDMA network. 3GPP Release 9 adds an option that the UE can indicate in WCDMA that the call is a CSFB case and not a normal 3G call.

15.3.1   Example Circuit-Switched Fallback with Location Area Update

This example describes LTE CS fallback to 3G with blind redirect for the incoming call with an overlay MSC architecture. In this case, the UE must typically perform a Location Area Update before the actual 3G call setup can start. The UE is initially in LTE in idle mode and receives a Paging-message from the MME due to the incoming voice call in Figure 15.20. The UE establishes RRC Connection with a cause value: mt-Access and responds to paging with ExtendedServiceRequest with a service type value: mobile terminating CS fallback. UE InterRAT capabilities are checked with UECapabilityEnquiry. If 3G support exists, the UE is redirected to 3G with a RRCConnectionRelease-message containing information on the

Event ID	System	Transf. dir.	Time	Message name
RRCSM	LTE FDD	Downlink	9:46:37.473	SystemInformation
RRCSM	LTE FDD	Downlink	9:46:37.473	SystemInformationBlockType1
L3SM	LTE FDD	Downlink	9:46:37.473	EMM_INFORMATION
RRCSM	LTE FDD	Downlink	9:46:45.750	Paging
RRCSM	LTE FDD	Uplink	9:46:45.750	RRCConnectionRequest
L3SM	LTE FDD	Uplink	9:46:45.750	EXTENDED_SERVICE_REQUEST
RRCSM	LTE FDD	Downlink	9:46:45.793	RRCConnectionSetup
RRCSM	LTE FDD	Uplink	9:46:45.795	RRCConnectionSetupComplete
RRCSM	LTE FDD	Downlink	9:46:45.824	UECapabilityEnquiry
RRCSM	LTE FDD	Uplink	9:46:45.825	UECapabilityInformation
RRCSM	LTE FDD	Downlink	9:46:45.855	RRCConnectionRelease
RRCSM	UMTS FDD	Downlink	9:46:46.233	SYSTEM_INFORMATION_BCH
RRCSM	UMTS FDD	Downlink	9:46:46.233	SYSTEM_INFORMATION_BLOCK_TYPE_1
RRCSM	UMTS FDD	Downlink	9:46:46.233	SYSTEM_INFORMATION_BLOCK_TYPE_7
RRCSM	UMTS FDD	Downlink	9:46:46.253	SYSTEM_INFORMATION_BLOCK_TYPE_3

Figure 15.20   Redirect from LTE to 3G with CS fallback

Downlink

RRCConnectionRelease   (3GPP TS 36.331 ver 9.5.0 Rel 9)

```
DL-DCCH-Message
  message
    c1
      rrcConnectionRelease
        rrc-TransactionIdentifier    : 3
        criticalExtensions
          c1
            rrcConnectionRelease-r8
              releaseCause    : other
              redirectedCarrierInfo
                utra-FDD    : 10837
```

Figure 15.21   RRC connection release in LTE

redirect carrier frequency channel number, which in this case belongs to UMTS 2100 MHz band. The RRCConnectionRelease message is shown in Figure 15.21.

After all the relevant broadcasted SIB information is read by the UE, it can start establishing a RRC connection with a cause value "registration" in Figure 15.22. The RRC connection is established for performing Location and Routing Area Updates, including agreement of ciphering with Security Mode procedures and identity and authentication requests. After the Location Area Update is accepted, the signaling connection is released with SignalingConnectionRelease towards the CS-domain in the case that the setup of the actual call has not started. The release of the signaling connection can be avoided by optimizing the core network settings.

Since the signaling connection towards the core network domain was already released after the UE registered to 3G network, the UE must be paged again in 3G to start the actual call setup in Figure 15.23. This is done with a PagingType2-message with cause: TerminatingConversationalCall. PagingType2 is used in the case that the UE has an active RRC connection established. The UE responds to the paging with PagingResponse. The ciphering is handled by SecurityModeCommand. The setup message includes, for example, information on the calling number and it is informed by the core network. With the CallConfirmed-message the UE informs its 3G and 2G speech codec capabilities. The radio resources for speech call are set with the RadioBearerSetup-message. In this case, SF128 is reserved, meaning capacity suitable for 12.2 kbps 3G speech codec. In the Alerting-phase, the UE starts to play the UE ring tone for subscriber B and in the Connect-phase subscriber B answers the call and finally the call is connected successfully.

In this example, the total setup time of the incoming CSFB call with blind redirect to 3G took around 7 s.

Event ID	System	Transf. dir.	Time	Message name
RRCSM	UMTS FDD	Downlink	9:46:46.393	SYSTEM_INFORMATION_BCH
RRCSM	UMTS FDD	Downlink	9:46:46.413	SYSTEM_INFORMATION_BLOCK_TYPE_3
RRCSM	UMTS FDD	Downlink	9:46:46.413	SYSTEM_INFORMATION_BCH
RRCSM	UMTS FDD	Uplink	9:46:46.519	RRC_CONNECTION_REQUEST
RRCSM	UMTS FDD	Downlink	9:46:46.755	RRC_CONNECTION_SETUP
RRCSM	UMTS FDD	Uplink	9:46:46.887	DCCH_RRC_CONNECTION_SETUP_COMPL
RRCSM	UMTS FDD	Uplink	9:46:46.893	INITIAL_DIRECT_TRANSFER
L3SM	UMTS FDD	Uplink	9:46:46.893	LOCATION_UPDATING_REQUEST
L3SM	UMTS FDD	Uplink	9:46:46.893	ROUTING_AREA_UPDATE_REQUEST
RRCSM	UMTS FDD	Uplink	9:46:46.893	INITIAL_DIRECT_TRANSFER
L3SM	UMTS FDD	Downlink	9:46:47.110	IDENTITY_REQUEST
L3SM	UMTS FDD	Uplink	9:46:47.110	IDENTITY_RESPONSE
RRCSM	UMTS FDD	Downlink	9:46:47.110	DOWNLINK_DIRECT_TRANSFER
RRCSM	UMTS FDD	Uplink	9:46:47.110	UPLINK_DIRECT_TRANSFER
RRCSM	UMTS FDD	Downlink	9:46:47.190	MEASUREMENT_CONTROL
RRCSM	UMTS FDD	Downlink	9:46:47.230	MEASUREMENT_CONTROL
L3SM	UMTS FDD	Downlink	9:46:47.728	AUTHENTICATION_REQUEST
RRCSM	UMTS FDD	Downlink	9:46:47.728	DOWNLINK_DIRECT_TRANSFER
L3SM	UMTS FDD	Uplink	9:46:47.773	AUTHENTICATION_RESPONSE
RRCSM	UMTS FDD	Uplink	9:46:47.773	UPLINK_DIRECT_TRANSFER
RRCSM	UMTS FDD	Uplink	9:46:47.888	SECURITY_MODE_COMPLETE
RRCSM	UMTS FDD	Downlink	9:46:47.888	SECURITY_MODE_COMMAND
RRCSM	UMTS FDD	Downlink	9:46:47.998	DOWNLINK_DIRECT_TRANSFER
L3SM	UMTS FDD	Downlink	9:46:47.998	IDENTITY_REQUEST
L3SM	UMTS FDD	Uplink	9:46:48.000	IDENTITY_RESPONSE
RRCSM	UMTS FDD	Uplink	9:46:48.000	UPLINK_DIRECT_TRANSFER
L3SM	UMTS FDD	Downlink	9:46:48.938	LOCATION_UPDATING_ACCEPT
RRCSM	UMTS FDD	Downlink	9:46:48.938	DOWNLINK_DIRECT_TRANSFER
L3SM	UMTS FDD	Uplink	9:46:48.941	TMSI_REALLOCATION_COMPLETE
RRCSM	UMTS FDD	Uplink	9:46:48.941	UPLINK_DIRECT_TRANSFER
RRCSM	UMTS FDD	Downlink	9:46:48.958	DOWNLINK_DIRECT_TRANSFER
L3SM	UMTS FDD	Downlink	9:46:48.958	MM_INFORMATION
RRCSM	UMTS FDD	Downlink	9:46:49.048	SIGNALLING_CONNECTION_RELEASE

Figure 15.22 Registration to 3G

Event ID	System	Transf. dir.	Time	Message name
RRCSM	UMTS FDD	Downlink	9:46:48.958	DOWNLINK_DIRECT_TRANSFER
L3SM	UMTS FDD	Downlink	9:46:48.958	MM_INFORMATION
RRCSM	UMTS FDD	Downlink	9:46:49.048	SIGNALLING_CONNECTION_RELEASE
RRCSM	UMTS FDD	Downlink	9:46:50.621	PAGING_TYPE_2
L3SM	UMTS FDD	Uplink	9:46:50.623	PAGING_RESPONSE
RRCSM	UMTS FDD	Uplink	9:46:50.623	INITIAL_DIRECT_TRANSFER
RRCSM	UMTS FDD	Downlink	9:46:50.853	SECURITY_MODE_COMMAND
RRCSM	UMTS FDD	Uplink	9:46:50.855	SECURITY_MODE_COMPLETE
RRCSM	UMTS FDD	Uplink	9:46:50.977	MEASUREMENT_REPORT
L3SM	UMTS FDD	Downlink	9:46:51.013	SETUP
RRCSM	UMTS FDD	Downlink	9:46:51.013	DOWNLINK_DIRECT_TRANSFER
L3SM	UMTS FDD	Uplink	9:46:51.018	CALL_CONFIRMED
RRCSM	UMTS FDD	Uplink	9:46:51.018	UPLINK_DIRECT_TRANSFER
RRCSM	UMTS FDD	Downlink	9:46:51.523	RADIO_BEARER_SETUP
RRCSM	UMTS FDD	Uplink	9:46:51.523	MEASUREMENT_REPORT
RRCSM	UMTS FDD	Uplink	9:46:51.777	MEASUREMENT_REPORT
RRCSM	UMTS FDD	Uplink	9:46:52.035	RADIO_BEARER_SETUP_COMPLETE
RRCSM	UMTS FDD	Uplink	9:46:52.040	UPLINK_DIRECT_TRANSFER
L3SM	UMTS FDD	Uplink	9:46:52.040	ALERTING
RRCSM	UMTS FDD	Uplink	9:46:52.194	UPLINK_DIRECT_TRANSFER
L3SM	UMTS FDD	Uplink	9:46:52.194	CONNECT

Figure 15.23 CS call setup in 3G

15.3.2 Example Circuit-Switched Fallback without Location Area Update

This section shows an example CS fallback setup where no Location Area Update is required. When the UE moves from a 3G to a LTE network, it must perform a Tracking Area Update. In the TrackingAreaUpdateAccept-message, the LTE network informs on which Location Area in 3G mapped by operator should overlap with the serving Tracking Area in LTE. The message is shown Figure 15.24.

In this example, the UE receives information that Location Area Code (LAC) 5008 should have geographical overlap with Tracking Area Code (TAC) 4080. The UE stores this information in its memory. When the CS fallback call attempt in LTE is made, the UE again sends an ExtendedServiceRequest to the network in Figure 15.25, establishes a RRC connection, and is redirected to 3G with a similar RRCConnectionRelease-message that was already presented in Figure 15.21.

The UE camps to the 3G cell that has the UARFCN informed in the RRCConnectionRelease-message and starts reading 3G system information. SystemInformationBlockType1 in Figure 15.26 shows the LAC of the 3G cell. In this example, mapping of LAC geographically

O TAI list
    Partial tracking area identity list 1:
        Type of list: list of TACs belonging to one PLMN, with non-consecutive TAC values
        Number of elements: 1
        MCC: 944
        MNC: 99
        TAC 1: 4080
O EPS bearer context status
    EBI(5): ACTIVE
    EBI(6): INACTIVE
    EBI(7): INACTIVE
    EBI(8): INACTIVE
    EBI(9): INACTIVE
    EBI(10): INACTIVE
    EBI(11): INACTIVE
    EBI(12): INACTIVE
    EBI(13): INACTIVE
    EBI(14): INACTIVE
    EBI(15): INACTIVE
O Location Area Identification
    MCC digits: 944
    MNC digits: 99
    LAC: 5008

Figure 15.24   Tracking area/location area-mapping information in tracking area update accept-message

Event ID	System	Transf. dir.	Time	Message name
L3SM	LTE FDD	Uplink	16:10:10.588	EXTENDED_SERVICE_REQUEST
RRCSM	LTE FDD	Uplink	16:10:10.589	RRCConnectionRequest
RRCSM	LTE FDD	Downlink	16:10:10.677	RRCConnectionSetup
RRCSM	LTE FDD	Uplink	16:10:10.679	RRCConnectionSetupComplete
RRCSM	LTE FDD	Downlink	16:10:10.713	RRCConnectionRelease
RRCSM	UMTS FDD	Downlink	16:10:11.068	SYSTEM_INFORMATION_BLOCK_TYPE_12
RRCSM	UMTS FDD	Downlink	16:10:11.068	SYSTEM_INFORMATION_BCH
RRCSM	UMTS FDD	Downlink	16:10:11.088	SYSTEM_INFORMATION_BCH
RRCSM	UMTS FDD	Downlink	16:10:11.108	SYSTEM_INFORMATION_BCH
RRCSM	UMTS FDD	Downlink	16:10:11.108	MASTER_INFORMATION_BLOCK
RRCSM	UMTS FDD	Downlink	16:10:11.108	SYSTEM_INFORMATION_BLOCK_TYPE_7
RRCSM	UMTS FDD	Downlink	16:10:11.128	SYSTEM_INFORMATION_BCH
RRCSM	UMTS FDD	Downlink	16:10:11.148	SYSTEM_INFORMATION_BCH
RRCSM	UMTS FDD	Downlink	16:10:11.168	SYSTEM_INFORMATION_BCH
RRCSM	UMTS FDD	Downlink	16:10:11.188	SYSTEM_INFORMATION_BCH
RRCSM	UMTS FDD	Downlink	16:10:11.188	MASTER_INFORMATION_BLOCK
RRCSM	UMTS FDD	Downlink	16:10:11.188	SYSTEM_INFORMATION_BLOCK_TYPE_7
RRCSM	UMTS FDD	Downlink	16:10:11.208	SYSTEM_INFORMATION_BLOCK_TYPE_4
RRCSM	UMTS FDD	Downlink	16:10:11.208	SYSTEM_INFORMATION_BLOCK_TYPE_1
RRCSM	UMTS FDD	Downlink	16:10:11.208	SYSTEM_INFORMATION_BCH
RRCSM	UMTS FDD	Downlink	16:10:11.228	SYSTEM_INFORMATION_BLOCK_TYPE_12
RRCSM	UMTS FDD	Downlink	16:10:11.228	SYSTEM_INFORMATION_BCH
RRCSM	UMTS FDD	Downlink	16:10:11.248	SYSTEM_INFORMATION_BLOCK_TYPE_19

Figure 15.25   CS fallback redirect to 3G

Downlink

SYSTEM_INFORMATION_BLOCK_TYPE_1   (3GPP TS 25.331 ver 8.8.0 Rel 8)

SysInfoType1
 cn-CommonGSM-MAP-NAS-SysInfo
   lac   : 5008
 cn-DomainSysInfoList
   cn-DomainSysInfoList value 1
     cn-DomainIdentity   : cs-domain
     cn-Type
       gsm-MAP
         T3212 timeout value   : 40 (4.0 hours)
         ATT   : MSs shall apply IMSI attach and detach procedure

Figure 15.26   Location area code in 3G system information

to TAC matches, since LAC is 5008 also in the 3G system information. Since the 3G LAC informed in SIB equals the information given in LTE with the TrackingAreaUpdateAccept-message, the UE does not need to perform Location Area Updates anymore, which speeds up the call setup time in Figure 15.27. The UE establishes a RRC connection with cause: OriginatingConversationalCall, requests the Routing Area to be updated, which is done in parallel with the actual call setup. The call setup is initiated by the UE with a CMServiceRequest-message with service type info: Mobile originating call establishment. Finally, the Routing Area Update is accepted, the context is modified to meet the 3G QoS requirements, and the radio bearer setup is done to reserve resources from the radio network for speech call. Setup and call proceeding messages are run in parallel to the call setup. The call setup is finalized with the Connect-message in Figure 15.28. The CS fallback call setup took in this example 5.4 s, which is shorter than the example presented in Section 15.3.1, where a Location Area Update was required.

15.4   Matching of LTE and 3G Coverage Areas

The 3G system should preferably have a larger coverage area than LTE when transferring the UE from LTE to 3G. This aspect is especially true with CSFB functionality. When VoLTE is introduced to the LTE network, then the LTE coverage area can be larger than 3G.

The interworking between LTE and 3G provides smoother data and voice performance and simpler operability compared to the interworking between LTE and 2G. If the UE cannot find the target 3G system, it may drop out of the network coverage for short period of time. Such a case may happen if the LTE is deployed on low frequency (LTE800) while 3G uses high frequency (UMTS2100). The preferred solution would be to also have low band 3G together

Event ID	System	Transf. dir.	Time	Message name
RRCSM	UMTS FDD	Downlink	16:10:11.587	SYSTEM_INFORMATION_BLOCK_TYPE_7
RRCSM	UMTS FDD	Downlink	16:10:11.587	SYSTEM_INFORMATION_BCH
RRCSM	UMTS FDD	Downlink	16:10:11.607	SYSTEM_INFORMATION_BCH
RRCSM	UMTS FDD	Downlink	16:10:11.607	SCHEDULING_BLOCK_1
RRCSM	UMTS FDD	Downlink	16:10:11.627	SYSTEM_INFORMATION_BLOCK_TYPE_3
RRCSM	UMTS FDD	Downlink	16:10:11.627	SYSTEM_INFORMATION_BCH
RRCSM	UMTS FDD	Uplink	16:10:11.732	RRC_CONNECTION_REQUEST
RRCSM	UMTS FDD	Downlink	16:10:11.969	RRC_CONNECTION_SETUP
RRCSM	UMTS FDD	Uplink	16:10:12.086	RRC_CONNECTION_SETUP_COMPLETE
RRCSM	UMTS FDD	Uplink	16:10:12.094	INITIAL_DIRECT_TRANSFER
RRCSM	UMTS FDD	Uplink	16:10:12.094	INITIAL_DIRECT_TRANSFER
L3SM	UMTS FDD	Uplink	16:10:12.094	CM_SERVICE_REQUEST
L3SM	UMTS FDD	Uplink	16:10:12.094	ROUTING_AREA_UPDATE_REQUEST
RRCSM	UMTS FDD	Downlink	16:10:12.389	MEASUREMENT_CONTROL
RRCSM	UMTS FDD	Downlink	16:10:12.419	SECURITY_MODE_COMMAND
RRCSM	UMTS FDD	Uplink	16:10:12.420	SECURITY_MODE_COMPLETE
L3SM	UMTS FDD	Uplink	16:10:12.521	SETUP
RRCSM	UMTS FDD	Uplink	16:10:12.521	UPLINK_DIRECT_TRANSFER
RRCSM	UMTS FDD	Downlink	16:10:12.529	MEASUREMENT_CONTROL
RRCSM	UMTS FDD	Downlink	16:10:12.649	SECURITY_MODE_COMMAND
RRCSM	UMTS FDD	Uplink	16:10:12.650	SECURITY_MODE_COMPLETE
RRCSM	UMTS FDD	Downlink	16:10:12.757	DOWNLINK_DIRECT_TRANSFER
L3SM	UMTS FDD	Downlink	16:10:12.757	CALL_PROCEEDING
RRCSM	UMTS FDD	Downlink	16:10:12.967	DOWNLINK_DIRECT_TRANSFER
L3SM	UMTS FDD	Downlink	16:10:12.967	ROUTING_AREA_UPDATE_ACCEPT
RRCSM	UMTS FDD	Uplink	16:10:12.971	UPLINK_DIRECT_TRANSFER
L3SM	UMTS FDD	Uplink	16:10:12.971	ROUTING_AREA_UPDATE_COMPLETE
L3SM	UMTS FDD	Downlink	16:10:13.137	MODIFY_PDP_CONTEXT_REQUEST
RRCSM	UMTS FDD	Downlink	16:10:13.137	DOWNLINK_DIRECT_TRANSFER
L3SM	UMTS FDD	Uplink	16:10:13.138	MODIFY_PDP_CONTEXT_ACCEPT
RRCSM	UMTS FDD	Uplink	16:10:13.138	UPLINK_DIRECT_TRANSFER
RRCSM	UMTS FDD	Downlink	16:10:13.517	RADIO_BEARER_SETUP
RRCSM	UMTS FDD	Uplink	16:10:14.005	RADIO_BEARER_SETUP_COMPLETE

Figure 15.27   Call setup with no location area update

Event ID	System	Transf. dir.	Time	Message name
RRCSM	UMTS FDD	Downlink	16:10:14.407	ACTIVE_SET_UPDATE
RRCSM	UMTS FDD	Uplink	16:10:14.447	ACTIVE_SET_UPDATE_COMPLETE
RRCSM	UMTS FDD	Uplink	16:10:14.527	MEASUREMENT_REPORT
RRCSM	UMTS FDD	Uplink	16:10:14.687	MEASUREMENT_REPORT
RRCSM	UMTS FDD	Downlink	16:10:14.721	DOWNLINK_DIRECT_TRANSFER
L3SM	UMTS FDD	Downlink	16:10:14.816	PROGRESS
RRCSM	UMTS FDD	Uplink	16:10:15.073	MEASUREMENT_REPORT
RRCSM	UMTS FDD	Downlink	16:10:15.088	DOWNLINK_DIRECT_TRANSFER
L3SM	UMTS FDD	Downlink	16:10:15.167	ALERTING
RRCSM	UMTS FDD	Uplink	16:10:15.488	MEASUREMENT_REPORT
RRCSM	UMTS FDD	Uplink	16:10:15.521	MEASUREMENT_REPORT
L3SM	UMTS FDD	Downlink	16:10:15.967	CONNECT

Figure 15.28   Call setup finalized

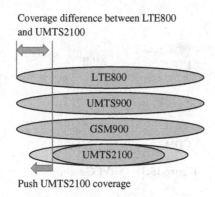

Figure 15.29   Inter-system mobility challenge if 3G coverage is smaller than LTE coverage

with low band LTE to provide similar coverage. This case is illustrated in Figure 15.29. If 3G coverage is smaller than LTE coverage, the following solutions may be needed

- Reduce the LTE coverage area by a lower transmit power or by a higher trigger level for the inter-RAT mobility.
- Increase the 3G coverage area by a higher transmit power, by lower inter-RAT mobility thresholds from 3G to 2G, or by new 3G base stations.
- Implement mobility from LTE also to 2G. In that case, inter-RAT measurements are needed to be activated to the network side to identify whether the target system providing sufficient coverage during Inter-RAT mobility is 3G or 2G. The inter-RAT measurements are done by the UE anyway in case of cell reselection. Blind redirection cannot be used because that feature must blindly use a predefined target system for each cell. The solution in that case should be redirection with measurements or handover with measurements. The mobility from LTE to 2G requires also that the LTE neighbor list includes 2G target frequencies.

15.5   Single Radio Voice Call Continuity (SRVCC)

Voice over LTE (VoLTE) began commercially in 2012, but without any interworking to 3G CS voice. Such a solution is possible only with excellent LTE coverage. A handover from VoLTE to CS voice is required in typical cases. The handover is called Single Radio Voice Call Continuity (SRVCC) and is illustrated in Figure 15.30.

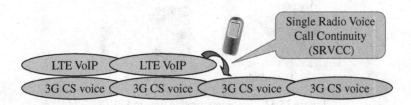

Figure 15.30   SRVCC handover from VoIP to CS voice

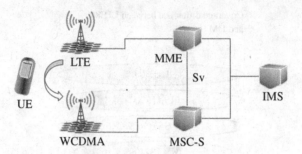

Figure 15.31   SRVCC architecture

The main reason for SRVCC handover occurs when a UE with a VoLTE call runs out of the LTE coverage area. SRVCC consists of two phases: inter-Radio Access Technology (inter-RAT) handover and session transfer. Inter-RAT handover is similar to the packet handover but it is done from Packet-Switched VoIP to Circuit-Switched voice. Session transfer is a mechanism to move voice media anchoring from the Evolved Packet Core (EPC) to the CS core network. The MSS initiates the session transfer and the IMS keeps control of the user during the process. The eNodeB requests the MME to make a PS-to-CS handover for the UE and the MSS requests the 3G network to prepare for the incoming handover. The SRVCC related architecture and interfaces are shown in Figure 15.31. SRVCC architecture includes two radio networks, evolved packet core, CS core, and IMS. The large number of network elements increases the complexity of SRVCC.

The target is to keep the user plane connection break below 0.3 s [9] which makes SRVCC practically unnoticeable to the end user. Such a short voice interruption time is possible to reach as all CS side radio and voice core (MSC-Server/MGW) resources are reserved beforehand. In the case of SRVCC, there is no need for the time consuming Location Update procedure in any circumstances, as the SRVCC MSC-Server is doing that on behalf of the UE. Figure 15.32 shows voice interruption time measurements: the downlink interruption was 150 ms and uplink 250 ms. These values are similar to the interruption times in circuit-switching inter-RAT handover from 3G WCDMA to 2G GSM.

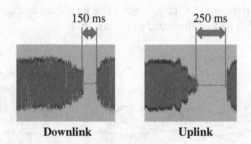

Figure 15.32   Voice interruption time in SRVCC

SRVCC was defined in Release 8 and further enhanced in the following releases:

- Release 8 defined SRVCC from LTE to WCDMA, GSM, and CDMA. SRVCC is also defined from HSPA to WCDMA and from HSPA to GSM for the case where VoIP is implemented in HSPA.
- Release 9 introduced SRVCC support for emergency calls that are anchored in the IMS core. It is not enough to have emergency call support for VoLTE, SRVCC functionality must also support emergency calls if there is a need to move from VoLTE to CS voice during emergency calls.
- Release 10 brings procedures for enhanced SRVCC (eSRVCC). Release 8 functionality requires that the home network is involved in SRVCC even if the procedure happens in the visited network. In order to avoid any unnecessary delays, eSRVCC was introduced to allow anchoring VoLTE within the visited network. In practice, LTE roaming even for data is currently in its very early phase (2014). When LTE data roaming agreements become more common, roaming cases may still use CS fallback even if VoLTE is used in the home network. eSRVCC also has the benefit that it hides SRVCC from the other operators in the case of inter-operator VoLTE – VoLTE call. Additionally, Release 9 also enables SRVCC in the alerting phase (aSRVCC).
- Release 11 enables two new capabilities: reverse SRVCC from CS voice back to VoLTE (rSRVCC) and SRVCC for video calls from LTE to CS video (vSRVCC). In the case of pre-Release 11, SRVCC was possible only from VoIP to CS voice, but not vice versa. Therefore, the voice call remained in the CS domain even if the UE returned to LTE coverage area. Release 11 allows movement of the call back to VoIP when LTE coverage becomes available again. The reverse SRVCC from 3G to LTE is controlled by RNC. Release 11 brings the relevant UE capability information to RNC to make RNC aware of UE VoIP capability.

15.6 Summary

3GPP has defined a comprehensive set of inter-working functions between LTE and 3G, both for data and for voice connections. The data connection can be transferred between LTE and 3G with reselection, redirection, and handovers. These features can be used for coverage, capacity, or service reason mobility between LTE and 3G. Most of the inter-working features need corresponding upgrades both in LTE and in 3G radio networks and core networks.

LTE and 3G voice inter-working relies initially on CS fallback where an LTE capable UE is moved to a 3G network for CS voice call. The simultaneous data connection can be handed over from LTE to 3G during a CS fallback procedure. 3GPP specifications in Release 8 also support the handover from VoLTE to CS voice in the 3G network called SRVCC with simultaneous packet data session continuity. SRVCC functionality is further enhanced in Releases 9, 10, and 11.

References

[1] 3GPP Technical Specifications 25.331 "Radio Resource Control (RRC); Protocol specification (UMTS)", v. 8.7.0, 2009.

[2] 3GPP Technical Specifications 36.331 "Radio Resource Control (RRC); Protocol specification (LTE)", v. 8.6.0, 2009.

[3] 3GPP Technical Specifications 25.133 "Requirements for support of radio resource management (UMTS)", v. 8.8.0, 2009.

[4] 3GPP Technical Specifications 36.133 "Requirements for support of radio resource management (LTE)", v. 8.8.0, 2009.

[5] 3GPP Technical Specifications 24.008 "Mobile radio interface Layer 3 specification", v. 9.7.0, 2011.

[6] 3GPP Technical Specifications 24.301 "Non-Access-Stratum (NAS) protocol for Evolved Packet System (EPS)", v. 9.7.0, 2011.

[7] 3GPP Technical Specifications 23.272 "Circuit Switched (CS) fallback in Evolved Packet System (EPS)", v. 11.9.0, 2013.

[8] 3GPP Technical Specifications 24.278 "Service requirements for the Evolved Packet System (EPS)", v. 8.9.0, 2011.

[9] 3GPP Technical Specifications 23.216 "Single Radio Voice Call Continuity (SRVCC)", v. 11.9.0, 2013.

16

HSPA Evolution Outlook

Antti Toskala and Karri Ranta-aho

16.1 Introduction

This chapter presents the outlook of HSPA evolution to Release 12 and beyond. First the topics discussed in 3GPP regarding the HSPA-LTE interworking and regarding the 3GPP radio level interworking with WLAN are covered. Then, on-going studies for dealing with bandwidths smaller than 5 MHz are presented, followed by work on improving the Release 99 based dedicated channel operation. This chapter is concluded with a presentation of the work being done on uplink enhancements as well as addressing briefly the heterogeneous network aspects also being progressed in another ongoing Release 12 study item. It's worth noting that several items in this chapter represent work in progress and many of the issues mentioned may not be necessarily finalized on time, or included in the actual work items targeting finalization by the end of 2014. Thus, some of the items may end up being postponed for Release 13, with finalization scheduled to take place in March 2016. Typically 18 to 21 months is needed for a Release based on experience with recent 3GPP Releases.

16.2 HSPA-LTE and WLAN Interworking

The basic support for interworking between HSPA and LTE was introduced together with the first LTE Release, Release 8. This included the support for PS handover as well as necessary reselection for idle mode operation. Support for Single Radio Voice Call Continuity (SRVCC) was also included, which enabled Voice over LTE (VoLTE) [1] with ongoing PS voice calls to be moved to WCDMA/HSPA (or also to GSM if needed). Various optimization proposals were added later to reduce the additional delay introduced by the use of the CS fallback solution. Later in Release 11, support for reverse SRVCC (rSRVCC) was added too, which enables moving from a CS call in WCDMA/HSPA back to a VoLTE call in LTE.

HSPA+ Evolution to Release 12: Performance and Optimization, First Edition.
Edited by Harri Holma, Antti Toskala, and Pablo Tapia.
© 2014 John Wiley & Sons, Ltd. Published 2014 by John Wiley & Sons, Ltd.

As part of the Release 12 studies, proposals addressing the following areas have been made, as covered in [1]:

- Improvements for the load balancing, aiming to address limitations in the current signaling where only one eNodeB load information can be provided at a time for a given RNC.
- Improvements in mobility signaling in the case of a multistandard BTS, to investigate whether some of the steps could be optimized when it is known that both HSPA and LTE are provided by the same base station.

Besides interworking with HSPA, 3GPP has also been addressing the interworking between 3GPP radio and WLAN networks. In 3GPP radio specifications, there has been no work done so far with the radio level interworking between 3GPP radios (LTE and WCDMA) and Wireless LAN (WLAN) networks. The key issue to address is how to ensure quality of service is retained when moving from a 3GPP network to WLAN network. Currently, if a UE moves from a 3GPP network to WLAN, the resulting quality may be much worse than that provided by a HSPA or LTE network due, for example, to limitations in the backhaul with WLAN or for general load reasons. While at the moment there are no defined measures in RAN for how to fix the situation, such methods are not being specified in 3GPP.

The three different approaches originally considered were, as given in [2]:

- An Automatic Network Domain Selection Function (ANDSF) based solution, which would aim to extend the core network-based ANDSF solution. It has been suggested that a pure ANDSF solution for providing the policies for the UE could not address the dynamic situation on the radio side as ANDSF signaling comes from the core network side. ANDSF was addressed in more detail in Chapter 7.
- The second approach considered is to provide in the RAN side (RRC) signaling offloading rules including the WLAN Traffic Steering Indicator (WTSI). WTSI would indicate whether to steer traffic to WLAN or whether to stay with 3GPP RAN instead. RAN side signaling could provide the list of radio bearers to be offloaded (unless provided via ANDSF). The general principle is a UE receiving guidance from the RAN side on which access points (WLAN SSID) to consider for offloading. Additionally, further information could be provided to help with the selection, such as what the signal strength in the 3GPP RAN network side should be, with the following quality measures to be considered: HSPA CPICH Ec/No and CPICH Received Signal Code Power (RSCP) or on the LTE side the Reference Signal Received Power (RSRP) or Reference Signal Received Quality (RSRQ) as presented in [3]. This approach was chosen as the basis for Release 12 RAN solution.
- The offloading from 3GPP RAN to WLAN could be done based on the dedicated traffic steering commands from RAN. The commands could be derived UE measurements based on WLAN network quality and load. Such an approach with UE measurements and reporting is rather similar to radio handover type operations inside 3GPP RAN. Such an approach has the greatest complexity impact on the network, especially if the RNC area has a large number of WLAN access points which would need to be controlled by RNC.

3GPP has already decided that there will not be, for example, an interface specified between WLAN AP and HSPA NodeB or RNC. Especially for an RNC it would not be desirable to

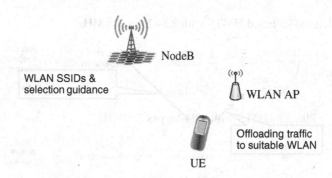

Figure 16.1   Radio level interworking between HSPA NodeB and WLAN AP

interface with a large number of access points, when considering a large HSPA network with hundreds of macro cells belonging to a single RNC area.

16.3   Scalable Bandwidth UMTS

As part of the Release 12 studies, addressing smaller bandwidths than 5 MHz was investigated. As discussed in [4] already, somewhat smaller bandwidths have been enabled in the field with the GSM refarming case particularly, where tighter filtering was used to go down to the 3.8 to 4.2 MHz range when deploying WCDMA in the middle of the GSM carriers of the same operator, as addressed also in Chapter 11. Now the desire is to go down to half of the nominal bandwidth, to 2.5 MHz or even lower in some cases.

During the study, as reported in [5], two alternatives for Scalable Bandwidth UMTS (S-UMTS) emerged, as illustrated in Figure 16.2, for how to address bandwidths clearly below 5 MHz:

- Smaller chip rate, also called Time-Dilated UMTS, where a divider N would be used so that with N = 2 or N = 4 the resulting chip rates would be 1.92 Mcps or even down to 0.96 Mcps.
- Filtered UMTS, where the actual chip rate would remain at 3.84 Mcps, but the transmit (and receive) filter would be narrower than 3.84 Mhz, for example a 2.5 MHz filter instead of the 5 MHz filter could be used to fit the 3.84 Mcps S-UMTS carrier to a 2.5 MHz chunk of spectrum. The transmitted digital waveform could also be further optimized to reduce the impact of tight filtering.

The use of S-UMTS results naturally in smaller data rates than regular 5 MHz UMTS. Also, if the solution is to implement the S-UMTS with a Time-Dilated UMTS approach with longer chip duration and respectively longer TTI, this would result in longer latency both for the user plane and control plane data. Using a 2.5 MHz filter on an unmodified 3.84 Mcps signal with filtered UMTS was shown to have some spectral efficiency degradation on low-to-moderate data rates, and a significant reduction in spectral efficiency for higher data rates. Another approach developed during the study to address the resulting performance issues with filtered UMTS (especially for downlink with high geometries) was using chip-zeroing to reduce the impact of filtering, as shown in Figure 16.3.

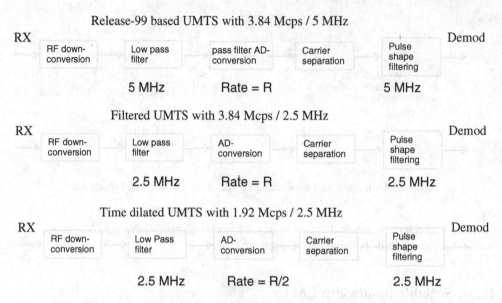

Figure 16.2 Impacts on the sampling rates and filter bandwidths with different S-UMTS approaches

From the performance point of view, the use of smaller bandwidth causes some performance loss as, for example, some of the multipath diversity benefits with 3.84 Mcps are lost, resulting in more sensitivity to frequency selective fading. Figure 16.4 shows the example performance with scalable bandwidth UMTS, with the smaller bandwidth created by a filtering or a filtering and chip-zeroing approach, showing that the use of chip-zeroing reduces the performance degradation compared to the pure filtering solution. As shown in Figure 16.4, the performance with smaller data rates is retained rather well while more degradation can be seen with higher data rates. The results in Figure 16.4 are based on approaches which retain, for example, the TTI length unchanged and do not create additional delay to the data or control signaling. If the approach is based on the time-dilated solution, the extra delay will cause further system-level issues, for example the mobility events are slower due to the longer round trip-time for signaling as well as impacts on the time needed to obtain and transmit the necessary measurements.

3GPP concluded the scalable bandwidth UMTS study at the end of 2013 and agreed to perform further studies on such an option that downlink is based on filtered UMTS with chip-zeroing, with every second chip being then zeroed. In the uplink, sufficient performance for

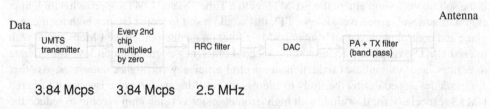

Figure 16.3 Chip-zeroing with filtered UMTS

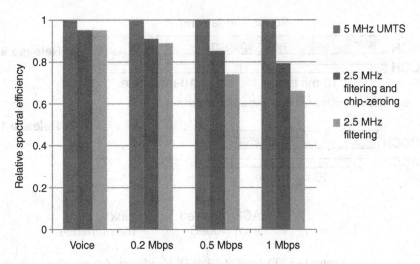

Figure 16.4   Performance of scalable bandwidth UMTS of 2.5 MHz bandwidth compared to Release 11 UMTS with 5 MHz bandwidth

widest use could be reached with a pure filtering based approach as well. Using the chip-zeroing in the downlink is important as otherwise with high geometry values (i.e., close to the base station) there would be greater degradation, especially with data rates in the order of 500 kbps or more. 3GPP ended up not agreeing to include scalable bandwidth UMTS in Release 12.

16.4   DCH Enhancements

Regardless of the fact that HSPA-based radio supports efficient delivery of both packet- and circuit-switched voice, voice traffic is still typically operated using a circuit-switched core network and a Dedicated Channel (DCH) on the UMTS radio network. DCH channels were designed as circuit-switched radio connections capable of simultaneous voice and data services in the very first release of the UMTS radio, Release 99, and were never further optimized when future development became focused on the more efficient packet-access radio. Because of this, there was interest in looking at optimizing the DCH radio specifically for voice delivery and exploiting similar techniques defined for HSPA radio, including DTX/DRX operation, in order to reduce UE power consumption and save on control channel overhead – as well as a variant of HARQ exploiting the varying link quality. With HSDPA, the use of CPC, as introduced in Chapter 4, allows for any service basically to implement DTX/DRX operation saving power and improving system capacity.

The DCH enhancements that were considered in 3GPP included the following: [6]

- Uplink frame early termination and ACK for early termination.
- UL DPCCH slot format optimization.
- Downlink frame early termination and ACK for early termination and shorter TTI.
- DL DPCCH slot format optimization.
- DPCH Time Domain Multiplexing (TDM), which was eventually not included in the specifications.

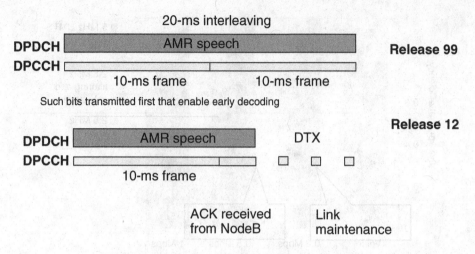

Figure 16.5 Example of uplink DCH early termination

A typical Release 99 operating point for voice service is 1% FER, meaning that 99% of the voice frames need to decode correctly after all the bits in the 20-ms TTI have been delivered. Statistically this means that many packets would be correctly decodable even before all the bits in the 20 ms were transmitted, and the remainder of the bits are transmitted for nothing.

An example of the enhancement in the uplink operation compared to Release 99 is shown in Figure 16.5. The NodeB receiver tries to decode the packet before the frame (or the 20-ms interleaving period) has ended. The receiver may try decoding at different instances to get correct decoding (with a CRC check giving the correct result) and then sending ACK message as early as possible to allow the UE transmitter to stop transmitting the rest of the frame (or interleaving period in general). The earlier the NodeB manages to inform the UE of the successful uplink frame decoding, the bigger the benefit and the more DTX can be applied. In comparison, if one would use HSUPA with a 2-ms frame size to carry the voice data, then 2-ms transmission time would be sufficient if the first transmission was successful.

In order to drive for a higher probability of the voice frame being decodable earlier, all the bits could be encoded to a 10-ms radio frame, and repeated in the second 10-ms radio frame up to the point when the decoding is successful. Further, the outer loop power control could target a more aggressive operating point, for example, 10–20% FER after 10 ms rather than 1% FER after 20 ms. This would give the OLPC more frame erasures for faster adaptation to link quality changes and allow for on average much earlier frame early termination.

The market interest in further enhancing DCH voice capability remains to be seen. While voice currently continues to be provided as CS voice with the WCDMA/HSPA network over the Release 99 based DCH, the additional deployments of VoIP (with Voice over LTE, VoLTE) are pushing the introduction of VoIP for HSPA as well, once the IMS investment for VoLTE is already made due to LTE. When all IP services are run over HSPA, it is foreseen that VoIP will also run on HSPA. In that case all the elements are there to operate CS voice on HSPA too (CSoHSPA) (see Chapter 7). If the introduction speed of VoIP remains slow enough in the marketplace, there may be interest to still optimize the CS voice service separately based on Release 12 to reduce especially the UE power consumption.

If the key drivers to DCH enhancements are considered to be:

- increased voice capacity and
- interference reduction.

then these are both already achieved with HSPA using earlier 3GPP Releases with CSoHSPA (when no further investment to the core network is made) or Voice Over HSPA (VoHSPA) when IMS is available for PS voice support.

In the domain of code resource utilization, the use of DCH enhancement – based on the new physical layer structure planned for Release 12 at the end of 2014 – does not free any additional code resources. Some of the downlink power could be used for HSDPA instead, if the transmission is stopped early enough, but services mapped on HSDPA are of course sharing the code and power resource in a more dynamic and coordinated way as resources are only booked for 2 ms at a time with a fixed booking period.

16.5 HSUPA Enhancements

Study on further Enhanced Uplink (EUL) enhancements aims to improve the uplink performance after many rounds of improvements have focused on the downlink improvements. The uplink changes prior to Release 12 are

- Release 6: Introduction of HSUPA
- Release 7:
  - higher order modulation for HSUPA (16QAM)
  - HSUPA DTX/DRX (CPC)
  - CQI rate reduction (CPC)
  - E-DPDCH gain factor interpolation for more accurate power to data rate adaptation
  - E-DPCCH boosting for better channel estimation
  - CS voice over HSDPA and HSUPA.
- Release 8: Enhanced Cell_FACH uplink (random access over HSUPA)
- Release 9: DC-HSUPA
- Release 11:
  - uplink beamforming
  - higher order modulation for HSUPA (64QAM)
  - uplink 2×2 MIMO for HSUPA
  - further Cell_FACH enhancements for uplink.

The following solutions have been considered in the study [7]:

- Handling more efficiently the situation with multiple uplink carriers, for example enabling faster switching between carriers for better load balancing when UEs are supporting multi-carrier operation in the downlink direction only. Another approach is to aim to have higher Rise over Thermal (RoT) that is, increased uplink interference level operation (such as with the use of 16QAM) on one of the carriers while using another carrier for "regular"

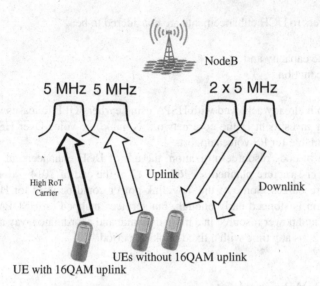

Figure 16.6   Use of specific carrier for higher RoT with 16QAM

RoT operation to ensure the uplink range of cell edge users is not reduced, as shown in Figure 16.6.

- Improvements in the control signaling for better grant handling, facilitating efficient time domain multiplexing to more efficient dealing with bursty data, as well as enabling more efficient high RoT operation with the new carrier type of operation in uplink, and/or with decoupling the uplink transmit power from the used data rate. Also, better access control has been proposed to deal more efficiently with the overload situation.
- Another dimension for enhancing the uplink is considering data compression in the radio layer. For example, encoded video does not compress as it has been already compressed at the original encoder. However, a regular image file or document could be compressed, but attempting to compress a stream that does not get any smaller causes additional processing and memory requirements. In the worst case, trying to compress a stream subject to compression earlier could lead to actually increasing the amount of data needed to be transmitted over the air. On the signaling side, the idea has been to reduce signaling overheads with uplink control signaling, specifically looking at the overhead generated by E-DPCCH and HS-DPCCH.

3GPP is foreseen to start in early 2014 the specification work for the method selected based on the study conclusions in [6].

16.6   Heterogenous Networks

The use of HSPA with heterogeneous networks was studied already in Chapter 7. Release 12 work has looked at further potential enhancements for different scenarios, including both co-channel and dedicated frequency cases, as illustrated in Figure 16.7. Clearly, the existing

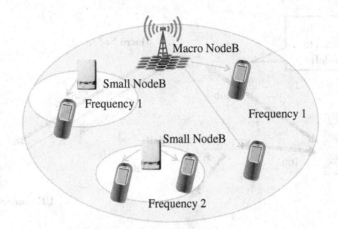

Figure 16.7 3GPP co-channel and dedicated frequency UMTS HetNet scenarios

HSPA deployments have shown that there are no fundamental problems preventing the use of heterogeneous networks with HSPA, but some further optimization could be considered. The study conducted in 3GPP [8] in the area of HSPA heterogeneous networks includes several smaller items, including the following:

- In some cases a mismatch between uplink and downlink can be observed. In order to deal with such a case more efficiently, one could consider solutions such as enhanced quality of pilots in the uplink or boosting the existing channels to compensate for the imbalance, or consider dynamic desensitization of the small cell uplink to compensate for the uplink/downlink imbalance due to smaller downlink transmit power of the small cell, in addition to more traditionally used static desensitization.
- Decoupling the E-DCH scheduling from HSDPA scheduling has been considered to address the mismatch between uplink and downlink as well. With E-DCH decoupling the uplink scheduling grants could be controlled by the small cell with the dominant uplink path to the UE, while the HSDPA data would be transmitted by the macro cell with the dominant downlink path to the UE. In this case the grants could come directly from the small NodeB or via RNC and macro cell.
- With the Network Assisted Interference Cancellation (NAIC), the NAIC-capable UE would get signaling of the parameters used in the interfering cell as well, as discussed in Chapter 8 also and shown in Figure 16.8. The signaling could be explicit, giving details such as modulation, block size, and channelization codes being used; or implicit by providing only the H-RNTI of the UE being scheduled and the UE would decode the HS-SCCH from the interfering cell as well enabling decoding of the interfering signal (post-decoding cancellation), or alternatively a subset of the signaling, for example, used modulation and/or codes could be signaled for more efficient interference suppression (pre-decoding cancellation). As the scheduling is NodeB functionality, there is no need to obtain the relevant information from RNC but rather there should be relatively fast connectivity between the NodeBs involved in NAIC operation, or the information related to the interference is signaled directly over-the-air by the interfering cell to the UE.

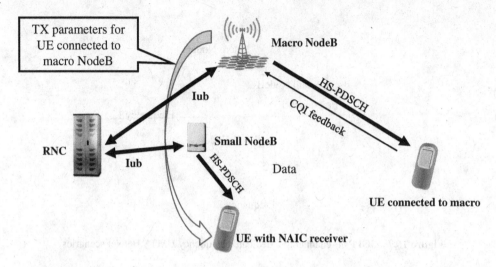

Figure 16.8   NAIC principle for explicit signaling of transmission parameters for another UE

- One further approach is the combined cell with multiple spatially separate but synchronized transmitters seen as a single cell by the UE. New antenna-specific pilots to enable capacity that would be comparable to deploying independent cells would be needed, and unfortunately this also requires new UEs as legacy UEs could not benefit from the situation, but the additional pilot overhead would reduce capacity and generate interference to the legacy UE base.

16.7   Other Areas of Improvement for Release 12 and Beyond

For Release 12, there are also some additional issues that have been raised, such as whether the Broadcast Channel (BCH) structure is sufficient and whether the new features (in Release 12 and beyond) would be needed due to BCH capacity limitations. The proposed approach would rely on the new information elements in the System Information Blocks (SIBs) being transmitted on a parallel BCH to the legacy one, possibly using HSDPA instead of the existing BCH. Alternatively a second Release 99 style BCH could be added. This allows higher capacity than the current BCH, but is only usable with Release 12 and newer devices, thus the amount of information that could be transmitted via such BCH is limited. The use of HSPA for BCH delivery is similar to the LTE broadcast principle, where only the Master Information Block is on a separate broadcast channel but the SIBs are on the data channel (equivalent to HSDPA), as covered in [9]. Further details of the enhanced broadcast of system information are covered in [10]. Other smaller areas include investigation for possible further enhancements for home BTS (femto BTS) related mobility, though the practical interest seems limited for solutions not valid for legacy UEs in Release 11 or earlier Releases.

16.8   Conclusions

In this chapter we have looked at the evolution of HSPA for Release 12 and beyond. Release 12, expected to be finalized by the end of 2014, is foreseen to contain several enhancements for

HSPA, some of them like radio level interworking with WLAN being common with LTE. Most of the solutions to improve interworking or HSPA-specific improvements to HSUPA operation can be introduced in a backwards compatible way supporting both legacy and newer devices, while some of the solutions being discussed relate to the heterogeneous networks or new bandwidth with scalable UMTS. The latter requires new UEs in the market before the solution can be used, as legacy UEs cannot access such a carrier. It remains to be seen if there is enough market interest to roll out solutions which would not allow the 1.5 billion UEs already in the field to use the carrier. However, if some specific spectrum allocation does not have room for two (or any) full carriers, then scalable bandwidth UMTS is an alternative to be considered if one does not want to consider the smaller bandwidth options enabled by LTE, as presented in [9]. Work on UE receiver improvements is expected to continue following the studies on NAIC, which would enable better UE receiver performance regardless of whether all UEs support such a feature or not.

Release 13 work is to be initiated at the end of 2014, with the target milestone set by 3GPP to be March 2016. However, one may assume finalization around mid 2016 based on the roughly 18 months Release duration, as with earlier 3GPP Releases.

3GPP is addressing further areas where performance could be improved with HSPA evolution work, but clearly increasing the peak data rate is no longer the focus in HSPA work. Rather it is addressing easy to deploy improvements for capacity issues, enabling HSPA technology to be the workhorse for providing mobile broadband connectivity for roughly 3 billion (estimated) users globally by 2016.

References

[1] 3GPP Technical Report, TR 37.852, "RAN Enhancements for UMTS/HSPA and LTE Interworking", Version 1.0.0, December 2013.
[2] 3GPP Technical Report, TR 37.834, "Study on WLAN/3GPP Radio Interworking", version 1.0.0, September 2013.
[3] 3GPP Technical Document, R2-133192, "details of WLAN IW Solution 2", NSN, Nokia Corporation, Deutsche Telekom, October 2013.
[4] Holma, H. and Toskala, A. (2010) *WCDMA for UMTS*, 5th edn, Wiley.
[5] 3GPP Technical Report, TR 25.701, "Study of Scalable UMTS FDD Bandwidth", version 12.0.0, September 2013.
[6] 3GPP Technical Report, TR 25.702, "DCH Enhancements for UMTS", version 12.0.0, September 2013.
[7] 3GPP technical Report, TR 25.700, "Study on Further EUL Enhancements", version 1.0.0, December 2013.
[8] 3GPP technical report, TR 25.800 "UMTS Heterogeneous Networks", version 12.0.0, September 2013.
[9] Holma, H., and Toskala, A. (2011) *LTE for UMTS*, 2nd edn, Wiley.
[10] 3GPP Technical Report, TR 25.704, "Study on Enhanced Broadcast of System Information", version 12.0.0, December 2013.

Index

HSPA+ Evolution to Release 12: Performance and Optimization, First Edition.
Edited by Harri Holma, Antti Toskala, and Pablo Tapia.
© 2014 John Wiley & Sons, Ltd. Published 2014 by John Wiley & Sons, Ltd.